FOOD AND URBANISM

FOOD AND URBANISM

THE CONVIVIAL CITY AND A SUSTAINABLE FUTURE

Susan Parham

Bloomsbury Academic
An imprint of Bloomsbury Publishing Plc

B L O O M S B U R Y
LONDON • NEW DELHI • NEW YORK • SYDNEY

Bloomsbury Academic
An imprint of Bloomsbury Publishing Inc

50 Bedford Square	1385 Broadway
London	New York
WC1B 3DP	NY 10018
UK	USA

www.bloomsbury.com

BLOOMSBURY and the Diana logo are trademarks of Bloomsbury Publishing Plc

First published 2015

© Susan Parham, 2015

Susan Parham has asserted her right under the Copyright, Designs and Patents Act, 1988, to be identified as Author of this work.

All rights reserved. No part of this publication may be reproduced or transmitted in any form or by any means, electronic or mechanical, including photocopying, recording, or any information storage or retrieval system, without prior permission in writing from the publishers.

No responsibility for loss caused to any individual or organization acting on or refraining from action as a result of the material in this publication can be accepted by Bloomsbury or the author.

British Library Cataloguing-in-Publication Data
A catalogue record for this book is available from the British Library.

ISBN: HB: 978-0-8578-5452-0
PB: 978-0-8578-5453-7
ePDF: 978-1-4725-2096-8
ePub: 978-0-8578-5474-2

Library of Congress Cataloging-in-Publication Data
ISBN 978-0-85785-452-0 (hardback) – ISBN 978-0-85785-453-7 (paperback) – ISBN (invalid) 978-1-4725-2096-8 (epdf) 1. Sustainable urban development. 2. Food–Social aspects. 3. Sociology, Urban. 4. City planning. I. Title.
HT241.P357 2015
307.1'16–dc23
2014037317

Typeset by Deanta Global Publishing Services, Chennai, India
Printed and bound in Great Britain

For Matthew

CONTENTS

LIST OF ILLUSTRATIONS	viii
ACKNOWLEDGEMENTS	x
PREFACE	xi
Introduction	1

PART ONE FOOD, DOMESTICITY AND DESIGN

1	Starting with the Table	19
2	The Garden and Gastronomy	47

PART TWO GASTRONOMY AND THE URBANISM OF PUBLIC SPACE

3	Food's Outdoor Room	71
4	The Gastronomic Townscape	97
5	Ambivalent Suburbia	130
6	Convivial Green Space	157

PART THREE FOOD SPACE AND URBANISM ON THE EDGE

7	The Productive Periphery	189
8	The Megalopolitan Food Realm	215
9	The Critical Food Region	239
	Conclusion: Food and Urbanism in Review	268
	BIBLIOGRAPHY	279
	INDEX	353

LIST OF ILLUSTRATIONS

CHAPTER 1
1.1 The Frankfurt Kitchen. 31

CHAPTER 2
2.1 Walled kitchen garden, Culross Palace, Fife. 59

CHAPTER 3
3.1 Market-related 'outdoor room', Spain. 72
3.2 Market hall entrance, Porto. 77
3.3 Fruit stall and cart, Alexandria. 84

CHAPTER 4
4.1 Traditional food shop frontage, Paris. 101
4.2 Street food, Whitecross Street, London. 103
4.3 Food truck, Leigh Street, Adelaide. 106
4.4 Street café, Cluj, Transylvania. 112
4.5 Negotiating 'third space' use in a public building, London. 115

CHAPTER 5
5.1 Garden City diagram. 133
5.2 1960s shopping 'street' with supermarket, London. 140
5.3 Suburban shopping centre, Australia. 143

CHAPTER 6
6.1 Traditional productive urban space, Transylvania. 159
6.2 Urban agriculture project in Glasgow. 162
6.3 Second World War Victory Gardens poster, USA. 165

6.4	Community Garden, Adelaide.	169
6.5	Contemporary allotment garden, London.	171
6.6	Skip garden, King's Cross, London.	173
6.7	Community Garden Scheme, North London.	175

CHAPTER 7

7.1	Peripheral food-growing space, Transylvanian village.	190
7.2	Edge-of-town cabbage field, Lincolnshire.	198
7.3	Farmers' market, Norway.	206
7.4	Productive landscape design, southern France.	211

CHAPTER 8

8.1	Super-regional mall 'café' foodspace, London.	223
8.2	'Food desert' located shop, LA, USA.	234
8.3	Chain store 'urban' food shop.	237

CHAPTER 9

9.1	Regional cheese display, France.	244
9.2	Local food products, Porto.	248
9.3	Regional meat products stall, South Australia.	258
9.4	Traditional agricultural landscapes, Transylvania.	263

ACKNOWLEDGEMENTS

I would like to acknowledge the input of a number of people who have helped bring this book to fruition. At Bloomsbury, a wholehearted thank you to Louise Butler for commissioning the book, as an engaged and insightful editor, and more recently to Jennifer Schmidt who has so ably stepped into that role and seen it through to publication. To Sophie Hodgson, Molly Beck and Abbie Sharman, gratitude is also due for your kind and thoughtful editorial support which has been an enormous help in the research, writing and completion phases.

To the staff at the British Library, I would like to express appreciation for all the help in the research process. I am always mindful that my Reader's Pass allows me to enter this Aladdin's cave of books, and each time I visit I am reminded anew of what a privilege it is to be able to work in such a civilized public building.

In relation to the book's cover I want to acknowledge the work of Eric Ravilious and thank his Estate. The cover includes part of a lithograph from J. M. Richards and Eric Ravilious's *High Street* (1938), a book which was made memorable through Ravilious's series of lithographs of food shops, restaurants and other more arcane shop frontages. A detail from *High Street: Baker and Confectioner* (1938) helps illustrate one of the key themes of *Food and Urbanism* – of small-scale food shops as an essential element of the 'gastronomic townscape'. To Tim Mainstone, of Mainstone Publications, and Melanie Leggett of Egmont, UK, thank you very much indeed for your helpful advice on tracking down a copy of this lithograph, and to Bridgeman Fine Art for supplying it to me. To Matthew Hardy and the designers at Bloomsbury, thank you for bringing my cover conception to reality.

To my father, Anthony, and my sisters, Felicity, Jennifer and Christabel, I am immensely grateful for all your love and support over an extensive writing period. Finally, heartfelt gratitude goes to Matthew Hardy, for his boundless love, support, good humour and brilliant critical insights.

PREFACE

This book has had a long gestation. I first started theorizing about design and planning interconnections between food and cities in the early 1990s when I wrote a series of papers on aspects of this interplay, spoke about these ideas at conferences, and did some more journalistic work, including on 'The Comfort Zone', an ABC Radio programme hosted by the late, much-lamented Alan Saunders, which was shaped around the food and cities concerns I had identified at a series of gastronomic symposia in Australia.

By the late 1990s, I had conceived of a structure for a proposed book to explore the ways that food and cities interconnect in design and place-shaping terms that was to be based on a series of expanding spatial scales. And I had made copious notes – when I came to look again at this in 2011, I found that I had already written around 100,000 words – but I realized that I wanted to take a fresh look at food and urbanism, grounded in my academic areas of design, planning, urban sociology and political economy and also to draw on insights and evidence from the many other disciplines that are integral to this extremely broad-ranging subject. The book has kept aspects of the structure with which I started in the 1990s but most of the research on which it is based is more recent: coming both from my own fieldwork on a range of foodscapes and from a huge variety of academic, public policy and community practice sources about food and place that have emerged and grown so rapidly in recent years.

What has not changed to any great degree are the fundamental principles on which the book is based, or the concerns for conviviality and sustainability around which it is framed. With food moving from the margins (described by one reviewer as an explosion of interest in food and cities over the last decade), even more evidence has now accrued that supports a proposition I first made in the early 1990s – that how food is grown, transported, bought, cooked, eaten, cleaned up and disposed of has significant effects on creating a sustainable, resilient and convivial future – and that design and urbanism have a central and sometimes overlooked role in this process.

While critical food studies have increasingly engaged conceptually with this interplay, the design and urbanism aspects need much more attention. We are expecting the future of human settlement to be largely urban, even if what 'urban' means has itself changed. Underpinning the book's analysis of food and urbanism is the view that decisions about how we can live well in these transforming spaces – more convivially and more sustainably – should recognize the fundamental role of food in shaping life's possibilities. In a modest way I hope that this book will make a positive contribution to that process.

Introduction

FOOD, SCALE AND URBANISM

This book explores the complex ways that food and cities interconnect through urbanism: the study of the art of building cities. Topped and tailed by this Introduction, which sets out the framing perspectives, and a Conclusion, which reviews the main themes, the book's three parts and nine chapters move (largely) outwards in spatial terms and forward in time to explore the interplay of food and space. Thus, in Part 1, 'Food, domesticity and design', the book considers urbanism's connections to food in the private domain (acknowledging that this spatial category has been blurred and disrupted), by working outwards from the table to the kitchen, dining room and domestic garden. In Part 2, 'Gastronomy and the urbanism of public space', four chapters deal with what could be described as the 'traditional' urbanism of the outdoor rooms of food streets and shops, food markets, gastronomic townscapes and urban greens, demonstrating how some of the rules for place shaping that have informed these spaces began to break down in the development of 'ambivalent suburbia'. In Part 3, 'Food space and urbanism on the edge', the book moves outward to focus in three chapters on the productive periphery, the food implications of the development of conurbations and the nature of food space in city regions and the countryside. Although there is no 'global'-scale chapter, the work touches on some of the urbanism implications of grossly unequal food relationships between the global north and south while acknowledging that it cannot do justice to this very complex socio-economic backdrop here. At each of these scales the very particular nature of the food spaces that contribute to urbanism's character is identified and explored in a spatialized way, along with overarching themes and topical interconnections, namely, conviviality, sustainability and the changing nature and contested assumptions about private and public space. Insights come from a range of disciplines and primary research, predominantly in Italy, France, Spain, Australia and the United Kingdom, but also in some other locations which offer relevant examples to inform the analysis.

As the *Journal of Urbanism* (Talen et al. 2008) sets out in its inaugural issue, urbanism 'focuses on human settlement and its relation to the idea of sustainability, social justice and cultural understanding'. A helpful working definition of urbanism is 'human settlement that is guided by principles of diversity, connectivity, mix, equity, and the importance of public space' (Talen 2005: 37) and can be counterpointed with features of 'anti-urbanism' which encompass the 'tendency toward separation, segmentation, planning by monolithic elements like express highways, and the neglect of equity, the public realm, historical structure and the human scale of urban form' (Talen 2005: 37). As Murrain (2002: 131) notes, this anti-urbanism appears as a kind of 'death of common sense' in a place-shaping or placemaking sense (terms which are used most often in the applied policy and design literature, although designers, urbanists and geographers have reflected critically on them). Urbanism can be broadly marked

> by qualities, not quantities; by diversity, not size; by intensity, not density; by connectivity, not just location. Urbanism is always made from places that are mixed in use, walkable, human scaled, and diverse in population; that balance cars with transit; that reinforce local history; that are adaptable; and that support rich public life. Urbanism can come in many forms, scales, locations, and densities. Many of our traditional villages, streetcar suburbs, country towns, and historic cities are 'urban' by this definition. Urbanism often resides beyond our downtowns. While urbanism will vary by geography, culture and economy, traditional urbanism always manifests the vitality, complexity and intimacy that defined our finest cities and towns for centuries. (Calthorpe 2011: xiii)

Food, in turn, is central to urbanism, because it is so critical to creating and maintaining this vitality, complexity and intimacy, because it can help make and support walkable, mixed, human-scaled and diverse places and because it can increase the focus of urban space on the public realm. Similarly, in thinking about food and cities in urbanist terms, scale matters. Scale has been widely recognized as a central concept for understanding space in a number of disciplines and thematic areas with a bearing on urbanism. It is of course a central concept in geography (although some geographers have suggested abandoning its use as a theoretical category). In any case, it is not this book's main purpose to review in depth the complex conceptualizing and debates about scale that can be traced within that discipline, including in relation to the political ecology of scale (Swyngedouw and Heynen 2003). Where geographically based insights into scale are of most pertinence is in relation to the geography of food and in reminding us that scale is socially and economically constructed and contested (Valentine 1998; Mandelblatt 2012). In arguing for scale, Jonas (2006: 404) reflects wryly on his unsuccessful 'attempts to educate non-geographers about emerging debates around scale and spatiality' at a conference of urbanists. Yet, this experience may also reflect the need for geographers to grapple more reflexively with design in their own theoretical consideration of scale.

In this book, scale is explored in a socio-spatial sense largely from a design perspective because, unlike geography, design theory does not seem to reduce scale to simply a frame, setting or context for actions or processes that are spatially grounded. It is not surprising to see considerations of scale's implications within the design literature where the interplay between physical form and socio-spatial practices is at the core of conceptualizing space (Cullen 1961: 144; Jabareen 2006; Parham 2012). Scale also acts as a construct for bringing together ideas about city design, planning and sustainability (Jenks and Dempsey 2005). For urbanism to respond to the challenge of unsustainable cities in terms of food (a central concern in this book), working at just one scale is completely inadequate. Rather, 'each scale depends on the others and ... only a whole systems approach, with each scale nesting into the other, can deliver the kind of transformation we now need to confront climate change' (Calthorpe 2011: 3). Scale acts as a useful construct for the analysis of food's interplay with urbanism as it allows detailed examination not only of food elements but also of themes that link different scales, or cut across them, to be teased out. That is an exploration undertaken through this introduction, the nine chapters and the conclusion.

This is, perforce, a look back into urbanist and food history, a survey of contemporary food and design practices and a more tentative foray into possible futures in relation to food and urbanism. Given the book's breadth, it cannot hope for completeness in any of these areas, all of which are subject to substantial theoretical debates in a number of disciplines, nor does it claim to do so. This is not a geography, sociology, political economy or history text, nor solely a design or planning one, but instead seeks to identify some critical themes and examples from across discipline boundaries which have a bearing on the ways food and urbanism intersect. The book seeks to touch on relevant theoretical and applied perspectives, and it raises questions about the interplay of food and space into which further interdisciplinary research might well prove fruitful.

FOOD, EMBODIMENT AND VISCERAL GEOGRAPHIES

As noted in a previous book, Market Place (Parham 2012) the study of food and eating has become recognized not just as a valid area of enquiry, but as central to the way boundaries between nature and culture are being rethought (Atkins and Bowler 2001: ix). Food studies have benefitted from the overall cultural turn in sociology and sister disciplines in the 1990s (Ashley et al. 2004) and from renewed interest in everyday life (Zukin 1992, 1995, 2004; Stevenson 2003); from the construction of human identity (Fischler 1988; Warde 1997); and from the body (Featherstone et al. 1990; Lupton 1996; Probyn 2000). In Parham (2012) nutritional and socio-biological approaches to food and the body in relation to spatiality were reviewed in detail. It was argued that explanations of food practices that rely on the supposed 'wisdom of the body' based on metabolic processes and nutritional efficiency have difficulty in coming to grips with differences in demonstrated preferences, tastes and habits between cultures, ethnicities, genders, ages, classes and *places* in terms of food. Biological needs and food availability play a part, but individual food practices are 'far more complex than a simple nutritional or biological perspective would allow' (Lupton 1996: 7).

Geographers have become increasingly interested in the visceral nature of their discipline (Hayes-Conroy and Hayes-Conroy 2010), and in sociology, some of this interest in the body and in food has taken a spatial turn (Sennett 1994). 'In a world in which self-identity and place-identity are woven through webs of consumption, what we eat (and where, and why) signals ... who we are' (Bell and Valentine 1997: 3). Highly influential work by Bourdieu (1984) on distinction, especially the notion of the habitus, also provides a useful focus on the body, linking taste in food to social stratification. The habitus is understood as a form of embodied capital rather than simply a repetitive habit and as the word implies, is that which one has acquired, but which has become durably incorporated in the body in the form of permanent dispositions ... something like a property, a capital. And indeed, the habitus is a capital, but one which, because it is embodied, appears as innate. [The habitus is] powerfully generative ... a product of conditionings which tends to reproduce the objective logic of those conditionings while transforming it. It is a kind of transforming machine that leads us to reproduce the social conditions of our production, but in a relatively unpredictable way. (Bourdieu 1993: 86)

Ideas about embodiment have become central to theorizing food. Grounded in the notion that we know and experience the world through our bodies, a concern for embodiment has underpinned food-related work in areas including tastes and preferences, civic agriculture (DeLind 2002) and community gardens (Turner 2011), among many others. In this book, a concern for embodiment is tied to ideas about conviviality and thinking about sustainable urbanism in relation to food.

FOOD, SPATIALITY AND POLITICAL ECONOMY/ECOLOGY

Self-evidently, spatialized food relationships operate within global political structures and are subject to broad economic forces (Lang and Heasman 2004; Patel 2007) related to the food system (Tansey and Worsley 1995; Beardsworth and Keil 1997; Atkins and Bowler 2001). These to varying extents impinge not only on food policy (Lang 1997, 2010; Lang et al. 2009) and lived experience but also on space shaping in relation to food. While a book focusing on food and urbanism can only skim the surface of the rich fields of food-focused political economy and ecology, it is worth noting that the global aspects of the food system are not presented as a monolithic force but dependent on 'intricate interweavings of *situated* people', things and connections (Whatmore and Thorne 1997). It is argued that these spatially expressed relationships within the modern food system are highly unequal (Freidberg 2004), predicated on a conventional, industrialized 'agro-food complex' (Maye et al. 2007: 1) in which effects on consumption are yet to be fully understood (Goodman 2002).

For example, the food demands that produce a permanent dietary summer for some in the global north (Morgan et al. 2006: 10) distort the balance between cash cropping and growing for indigenous needs in poorer supplier countries, thus negatively affecting the latter's food space, resilience, health and sustainability. The burden of unsustainable food production practices falls most heavily on the poorest people and the poorest countries (Millstone and Lang 2003), and for both equity and sustainability reasons various strategies to remake the food system have been proposed (Hinrichs and Lyson 2007). So, while there may be no direct interplay apparent between food's spatiality in the first and developing worlds, the different forms of urbanism thus produced are, in fact, closely intertwined. For food, as more generally, 'one of the key challenges of urbanism ... is to find ways to forge a coherent relationship between globalized economic structure and the principles of diversity, mix, connectivity and equity' (Talen 2005: 37).

Of course, neither urbanism nor food trade is a new global element. While archaeologists may have struggled to capture theoretically an agreed definition, settlement forms since the earliest appearance of state-level societies have constituted urbanism (Patch 1991). In the traditional food system, trade in food commodities has existed since at least ancient times, and famous trading cities and routes have clearly shaped global spatiality in profound ways (Hollister 1974: 150). Yet, the massive inequality associated with the modern food system is a result of particular forms of economic globalization arising in our era, and the changes this has wrought on urban, peri-urban and rural food space have been matched by equally profound, and connected, shifts in the nature of urbanism itself. A theme of the book is the way that forms of urbanism associated with modernist ideology have been most closely tied to problems for food-centred space (Parham 1992, 1995, 1996), while the decline and more recent renewal of interest in urbanism that (broadly) supports food space can be seen in twentieth- and early-twenty-first-century debates about what constitutes an appropriate spatiality in a fast urbanizing world.

In contemporary theory and practice, since Louis Wirth (1938: 1) wrote his famous essay 'Urbanism as a Way of Life', in which he explained how urbanization was changing 'virtually every phase of social life', sociological consideration of aspects of making cities has blossomed in many directions including food, just as urban space itself has expanded rapidly in new ways. Wirth argued then that the 'distinctive feature of the mode of living of man in the modern age is his concentration into gigantic aggregations around which cluster lesser centers and from which radiate the ideas and practices that we call civilization' (Wirth 1938: 2). For Fischer, writing in the 1970s, a Wirthian view allowed due recognition to be given to the ecological and social effects of urbanism, something that the influential work by Gans (1962: 625) and others had previously dismissed as unconnected to place, but rather reflecting class and life cycle. More recently, scholars have noted how difficult it is to make connections between the physical design of cities and social goals such as sense of community and equity, but at the same time 'city designers would be remiss if they failed to recognize the ways in which physical design affects social phenomena' (Talen 2002: 165). Given these causal linkages, how possible it may be to institute more sustainable urbanism has concerned a number of theorists of the city (Beatley 1997, 2000, 2004, 2010; Talen 2005; Jepson et al. 2010) in the context of extremely rapid and uneven urban development and the problems caused by climate change (Calthorpe 2010). Conversely, some see proposed urbanism approaches to respond to these concerns as constituting no more than a global strategy of gentrification (Smith 2002), but these latter readings do not enlighten us about the complexity of food's relationship to place.

FOOD AS A SPATIAL DESIGN MATTER

This book sits within the tradition of connecting the physical and social aspects of cities (Soja 1989, 2000), putting spatial design at the centre of the analysis. This interplay has been increasingly recognized. For some geographers and sociologists, however, it remains normative to suggest that design arrangements giving rise to certain spatial forms might influence behaviour and expressions of conviviality. Yet, this can equally be read as a disciplinary incapacity to engage effectively with the nature of design and urbanism, and this book seeks to explore this contentious territory. It focuses on food's influential roles in shaping the interplay between urban form and practice and argues that food's impact on conviviality and sustainability is a spatial concern that is critically important to a predominantly urban future (Parham 1990, 2012: 8).

In a previous book 'Market Place', which explored the design-led renewal of food market areas into food quarters, food was considered as a central aspect of everyday life that reflected a close interplay with forms of urbanism, yet one that had often been treated aspatially (Parham 2012) despite its 'stake in place' (Gieryn 2000: 463). It was acknowledged that there is a rich tradition of food dealt with sociologically (Beardsworth and Keil 1997; Germov and Williams 2008) and suggested that certain geographers and designers had begun to grapple with the spatiality of food (Bell and Valentine 1997; Esperdy 2002; Franck 2005; Bell and Binnie 2005; Viljoen and Wiskerke 2012; Pink 2012). However, it was contended that food remained under-researched at the intersection of social science, urban design and gastronomy (Parham 2012). Viljoen et al's. Continuous Productive Urban Landscapes book of 2005 made 'the case for the design (architectural and urban) of food related spaces in cities, specifically the coherent introduction of productive urban landscapes, but within a wider urban food systems approach' (pers.com). Following the work of Lefebvre (1991), de Certeau et al. (1984, 1998), and reflecting on fieldwork results in London, it was argued that ordinary places in cities, towns and regions should receive more research attention to explore in spatial design terms how food and urban space interconnect.

As in the work by Maitland (2007) on ordinary spaces and developing food quarters in London (Parham 2012), the everyday urbanism advocated by Chase et al. (1999), which explores and values the quotidian, the vernacular and the ordinary, situates food as among its central concerns. Explicitly referencing Wirth, Chase et al. (1999: 6) suggest that rather than reflecting urban design, planning or other specialized areas, 'urbanism identifies a broad discursive arena that combines all of these disciplines as well as others into a multidimensional consideration of the city'. Similarly, acknowledging the influential writings of Henri Lefebvre (1991) on the overlooked nature of the everyday, the interventions of the situationist Guy Debord and de Certeau et al's. (1998) work on everyday social practices, they argue that the everyday is about the ordinary in human experience, and food is central to this picture: 'commuting, working, relaxing, moving through city streets and sidewalks, shopping, buying and eating food, running errands ... the concept of everyday space delineates the physical domain of everyday public activity' (de Certeau et al. 1998).

FOOD, TRADITIONAL DESIGN AND MODERNISM

Since the early 2000s, interest and concern for the multitude of ways that food (especially its production) constitutes a 'place' issue has grown enormously in planning, architecture, design and landscape (Potteiger 2013), even though in sociological work there is still a tendency to treat places as no more than food consumption sites (Burnett and Ray 2012: 147). Despite this continuing problem of underplaying or overly narrowing the breadth of analysis of the spatial, there is a greater acknowledgement from certain theorists that food relationships do not 'just happen' in physical space which is no more than a backdrop to them, but are affected by spatial design and in turn affect spatiality in a multitude of ways. For example, Bosio's (2013) work on 'gastronomic cartographies' 'investigates the everyday life practices (de Certeau 1984) relating to food, and connected to the archetypical architectural food spaces of the domestic kitchen, the café, the market, and the street, through a comparison between the Australian and the Italian contexts'. As Moravansky (2007: 72) points out, 'the architecture of eating is the spatial and temporal organisation of raw materials, it implies spaces and rituals and extends from planning to execution, from shopping at the market to serving the dessert'.

Evidence-based research findings rather than normative assumptions suggest that a number of design elements may contribute to such 'gastronomic' possibilities, that is, possibilities that richly support a food-centred cultural life (Parham 1990, 1992, 1993, 2005, 2008, 2009, 2012; Esperdy 2002; Capon and Thompson 2011). Thus, despite the suggestion that gastronomy (the art and science of good eating) is somehow a pretentious area for study, this book asserts the need to engage with gastronomic urbanism, not as an elitist project but as a work that focuses on conviviality, sustainability and inclusion. In terms of gastronomy's spatiality, urban designers have demonstrated why public space and street design are so central to urban life (Madanipour 1999; Southworth and Ben-Joseph 2003; Moughtin and Shirley 2005) and how to address the decline of the civic realm (Rowe 1997). Such consideration covers a range of experiential qualities including variety, accessibility, vitality, legibility, robustness, identity and richness (noted in Parham 2012: 68), though vitality and civility are thought to be particularly important for 'good urbanism' (Hayward and McGlynn 2002: 127). Mouzon's (2008) proposals for city design argue explicitly for places that are feedable, serviceable, accessible and frugal (noted in Parham 2012: 68). Even where food is not foregrounded, what links these design approaches is the centrality recognized in relation to public space enclosure through which built spaces contribute to an appropriate solid-to-void relationship between positive and negative space (Lozano 1990).

To suggest certain urban forms could have impacts on convivial spatial expressions is, for some, to support an outmoded or imaginary past. This is a criticism tackled head on in Parham (2009), referencing arguments made by Hardy (2009) which reflect on architectural preference for object buildings and particular stylistic devices and materials though to signify modernity. As Hardy (2009) notes in relation to city development, clauses of the widely influential Venice Charter of 1964 'have been used to justify and to require modernist interventions in traditional buildings and places'. Rather than fitting in with their context, the contrasting of the new has become a hallmark of design being 'contemporary' or 'of its time', so that object buildings and spaces appear to pay scant heed to their locations' urban design, architectural or historical associations. In social science research into cities, it is possible to contend that arguments for 'traditional' city form are seen as inherently normative and insufficiently 'contemporary', implying a loss of authenticity in the approach to urbanism in the same way that urban areas that contain traditional building typologies, block patterns and street layouts tend to be seen as remnants or leftovers from the past (Parham 2009). In fact, arguments about nostalgia, pastiche and romanticized places are fundamentally about the perceived problem of authenticity in city design in practice and theory. Of course, authenticity (like tradition) is not a fixed, objective concept but socially constructed, so debates about the appropriate expression of authenticity in a spatial sense are a contested area. When it comes to food, similarly, discussion of authenticity ranges from the technical (are products what they say they are?) to much more complicated relationships to location and tradition. The issue of authenticity in relation to food and place is further considered in a number of chapters but particularly in Part 3 given the rich set of work on food tourism that relates to urbanism.

Focusing in on spatial design in practice – especially through architecture and urban design for city renewal – there is a clear split between the supporters of what can loosely be defined as the European City Model (Clos 2005) and the approach advocated broadly within the terms of the Venice Charter of modernist-inspired interventions. European City Model characteristics are reminiscent of the specific elements that compact cities' theorists propose: fine-grained, mixed-use, walkable, medium-to-high density, public space-oriented and human-scaled cities (Urban Taskforce 1999; Barton 2000; CABE 2000; Davies 2000). This approach reflects how in traditionally shaped cities such design arrangements formed (and continue to form) human-scaled outdoor rooms, which have, in turn, helped create, over the very long term, the physical conditions for social life in the public realm (Trancik 1986).

As Hobsbawm and Ranger (1983: 1) explain, tradition is not a fixed category but subject to transformation and reinvention depending on particular circumstances, and this flexibility is acknowledged in this research. Thus, somewhat ironically, current urban circumstances suggest that city design inspired by modernist perspectives has in fact become the normative, default tradition in place-shaping terms (Parham 2006), replacing the underlying spatial logic of the enclosed urbanism which was abandoned in certain places in the twentieth century. To paraphrase Trancik (1986), by treating the built form as an independent element in a non-spatial vacuum, the collective sense of meaning was lost, as were the rules for connecting parts of the city through outdoor space. This has had widely reported and largely negative effects on public space primacy, being associated in a complex way with a wide-scale retreat to the private domain. Today, there is contestation over whether public space still offers significant opportunities for social engagement (Goheen 1998) in food or other terms. Thus, in exploring food and urbanism, a range of implications of these transformations in spatial arrangements are considered, with reference to concepts of (and issues raised by) tradition and authenticity in food and design and the notion of what is public or private space.

FOOD, URBANISM, TASTE AND JUSTICE

As Morgan (2010: 1852) notes in exploring the idea of the ethical foodscape, food is 'one of the most important prisms for exploring world poverty and sustainable development'. Yet, certain writers have argued that a societal focus on food (especially within Western societies) is no more than an elite obsession, a luxury for the over indulged and over resourced, offering a 'comfortably domesticized high' (Poole 2012: 3). By giving too much value to self-appointed food authorities, and too central a place to food in human life, a cultural obsession with food is made even more repellant when contextualized by a supposedly wilful rejection of industrialized farming methods and the unacknowledged food poverty of others (Poole 2012: 119). Equally, in more academic discourse, while there are many proposals for more local, sustainable food systems which could, in theory, move beyond a niche market (Feenstra 1997; Little et al. 2010), these are often themselves critiqued as exclusionary. This critical scrutiny of the increasingly popular alternative food movement comes from a variety of directions, some of which are sketched in here but explored in more detail in Part 3.

For some, local food has been valorized in ways that insufficiently problematize a developing orthodoxy (Fonte 2008). Others discuss what actually constitutes 'local' (Hinrichs 2003; Giovannucci et al. 2010) and how complicated local food buying practices are when considered in relation to inequalities mediated through place, ethnicity and class (Blake et al. 2010). Local food movements are seen to be culturally and geographically situated and differentiated (Holt and Amilien 2007). Concerns about food inequality are reflected in notions such as the right to food, and food justice and sovereignty theorists have drawn attention to paradoxical elements in the celebration of radical responses to the mainstream food system (Alkon 2008; Alkon and Norgaard 2009; Greene 2009; Levkoe 2011; Clendenning 2011; Sbicca 2012; Alkon and Mares 2012). Thus, it is argued that *locavores* (local food eaters) of fresh, organic food products require significant economic and social capital to take part in such alternative practices, but, within a wider context of neoliberal-inspired economic insecurity, the alternative food narrative framing their actions is 'largely created by, and resonates most deeply, with white and middle-class individuals' (Alkon and Agyeman 2011: 2–3). As DeLind (2011: 273) argues, it may be that rather than focusing in on a regenerative food system, local food discourse has 'three problematic emphases': the individual practice focus of the locavore; mystification connected to the misleading use of the term 'local' by major food retailers; and the focus on particular food 'heroes', all of which are 'shifting local food (as a concept and a social movement) … away from the deeper concerns of equity, citizenship, place-building, and sustainability'. For others, assuming that eating local food is more ethical or ecologically sound may constitute a 'local trap', as, it is argued, local food systems may be no more sustainable or just than those operating at a larger scale (Born 2006). Thus, the so-called citizen-consumer hybrid, which has emerged as a way of defining the practices of ethical consumption, is argued to be more about consumption than about citizenship (Johnston 2008). With a corporatized organic food sector, at least in the North American context, these perspectives may be thought to offer a challenge to food democracy (Johnston et al. 2009).

Similarly, some theorists have critically interrogated the complex nature of food consumption in everyday life (Warde 1997), its increasingly close connections to notions of personal identity (Knox 1992), consumption's role in determining the nature of urban living (Zukin 1995; Miles and Miles 2004) and the landscapes of the post-industrial city (Gospodini 2006). It is suggested that consumption space can become a commodified product (Lefebvre 1991), with foodscapes harking back to an imaginary past (Watson and Wells 2005) while modelling new forms of consumption (Thrift and Glennie 1993). It is certainly true that food consumption can be a way of expressing distinction, as the sociologist Pierre Bourdieu (1984) has influentially explored in detail in his work on taste and the habitus. Naccarato and LeBasco (2012: 2) go as far as to say that (borrowing from Bourdieu) food is now used to form a kind of 'culinary capital' as a marker of social status that also delineates social exclusion.

Yet, theoretical perspectives that problematize food relationships as solely about inequality appear significantly under-powered in exploring the design implications of food's role in the construction of place. Although fieldwork on the renewal of areas in London as food quarters suggests that these are indeed sites for playing out good taste, acquiring cultural capital and spatially expressing aspects of gentrification in which food has an essential role, this is not the whole story (Parham 2005, 2012: 53). Even in these fast-developing food-centred spaces, considerable evidence of authentic interactions and convivial everyday socio-spatial practices based on a more localized food system have been identified in relation to food (Parham 2005, 2012: 147, 237). As later chapters of this book demonstrate, there is a substantial range of evidence from many disciplines about how influential food is in place shaping globally, and this demonstrates a mixed and sometimes encouraging picture about connections to urban sustainability and the chance to lead more satisfying, equitable and convivial lives. In understanding place, food has moved from the margins for good reasons (Super 2002), and this book seeks to explore some part of this area of enquiry from a spatial perspective.

CONVIVIALITY AS A FRAMING ELEMENT

It is argued here that the notion of conviviality provides a powerful framework for exploring food's spatiality in and around cities and regional areas. As noted previously, work on the convivial city offers a critical context for studying food's spatiality as a contributory factor in urban sustainability, of which more details are given in the course of the chapter (Parham 1990, 1992, 1993, 2005, 2012: 10; Miles 1998; Peattie 1998). Urban conviviality is about evanescent, sociable pleasure, reflected in the daily physical and social recreation of the self – including through sharing food – and thus has been situated as nourishing civil society itself (Peattie 1998). One of the most interesting ways through which the concept has been explored is in the context of post-colonial interaction, offering fascinating possibilities for 'cosmopolitan conviviality' (Gilroy 2004: 8) or 'vernacular conviviality' (Gilroy 2004: 3) that 'moves beyond a market-driven pastiche of multiculture' (Gilroy 2004: 163). Gilroy employs conviviality in a very specific and nuanced way that foregrounds openness between people (rather than closed identities), but as a term often invoked, there is elsewhere a lack of precision in its use in relation to food and spatiality. Thus, Ivan Illich's work (1973) is useful in underlining that conviviality has particular forms and settings, encompasses feasting, drinking and good company and is also 'the opposite of industrial productivity ... to mean autonomous and creative intercourse among persons, and the intercourse of persons with their environment, and this in contrast with the conditioned response of persons to the demands made upon them by others, and by a man-made environment'. Conviviality and commensality are also closely linked, with the latter defined as 'a gathering aimed to accomplish in a collective way some material tasks and symbolic obligations linked to the satisfaction of biological, individual need' (Grignon 2001: 24). For Grignon (2001) conviviality arises out of such commensality, and many cultures have institutionalized commensality in spatial ways, such as eating from the common pan (Tierney and Ohnuki-Tierney 2012: 121). Sociable activities focused on food 'make special occasions out of mundane materials of life' (Peattie 1998: 247), encompassing 'events as simple as buying food at the market, sharing a coffee or enjoying a meal together' (Parham 2012: 11). Duruz's (2002: 2) work on secret geographies, the 'everyday rituals and practices attached to food and place' in suburban Clovelly in Sydney seem to fit in with this approach.

Conviviality as an everyday, quotidian matter operates largely beyond instrumental, economic exchange and is intimately connected with the design of the public realm (Shaftoe 2012). In geography and sociology to suggest that certain physical features might support conviviality while others do not can be seen as implicitly normative; such has been the theoretical disconnect between place design and sociability. In such readings, design becomes immaterial or marginal to theorizing place. Here, however, place design is conceived as central to expressions of conviviality although there is no simplistic physical determinism at play. Rather, conviviality and the physical structures of place are understood as closely interconnected, with physical form and socio-spatial practices intertwined and relationally produced in complex ways, as various chapters in this book explore.

It is well understood, for example, that conviviality has a complex relationship with the commercialization and marketing of place, sometimes being used to service these needs in third spaces and 'invented streets and spaces', and thus at risk of becoming 'episodic and vestigial' in the process (Banerjee 2001: 15). Similarly, conviviality may be used to place brand 'hospitable spaces' for food and drink consumption, that is, public, social sites for 'the production and reproduction of ways of living in and visiting cities' (Bell 2007: 7). At the same time, through the development of spaces that demonstrate more 'hybrid hospitality', some urban forms may become more authentically convivial (Bell 2007, in Parham 2012: 11). Places can emerge which represent so-called convivial ecologies (Bell and Binnie 2005), in which sociability revolves around ordinary day-to-day food activities, in particular, public spaces (Maitland 2007).

Equally, as is explored in detail in Chapter 9 in relation to alternative food networks, such expressions of conviviality may reflect like-minded perspectives, including food activism grounded in an 'ethics of conviviality' (Bell 2007: 12). In this context, ideas about conviviality's spatiality are developed through critical consideration of the linked Slow Food and Slow Cities movements, whereby conviviality in relation to food is promoted through local convivia. These are intended to counter 'the loss of local distinctiveness as it relates to food, conviviality, sense of place, and hospitality'

(Mayer and Knox 2006: 322), while explicit connections are made to conviviality's support for achieving sustainability. Underpinning the book's various discussions of the concept is the argument that 'opportunities for conviviality in the city rely upon an extended set of gastronomic possibilities. And these possibilities can be widely conceived in city planning and design. They relate as much to kitchen layout as to market gardening, to the psychology of the café as to policy for metropolitan growth' (Parham 1992: 1).

SUSTAINABILITY, THE MODERN FOOD SYSTEM AND URBANISM

Urban sustainability offers another important framing element to the book, in the context of increasingly unsustainable urban development outcomes, exacerbated by climate change. Within a number of disciplines, there has been increasing interest in the Anthropocene, that is, the way human activities have affected global ecosystems over the long term. A considerable amount of work has been undertaken in areas including agroecology, exploring the ecology of sustainable agricultural systems (Gliessman 1998), and in urban political ecology, which has offered a Marxist theoretical framework for environmental justice research. The latter 'provides an integrated and relational approach that helps untangle the interconnected economic, political, social and ecological processes that together go to form highly uneven and deeply unjust urban landscapes' (Swyngedouw and Heynen 2003: 898). Nature is understood as an urban condition, and the highly unequal and often exploitative interplay between human and non-human actors occurs in complicated scalar ways (Braun 2005). As Kloppenburg et al. (2000: 178) note, the notion of sustainability has been both contested and canonized 'as a kind of cultural shorthand for the "green and good"' by those wishing 'to access the word's discursive potency but whose goals and interests are not necessarily compatible' (Kloppenburg et al. 2000: 178).

Notwithstanding the term's capture in certain cases by those with questionable motives, principles of sustainability can be teased out (Dresner 2008). Attempts have been made to quantify requirements for urban sustainability including in relation to food (Walsh et al. 2006). Equally, it is clear that sustainability issues are apparent all along the food chain in and around cities (Stren et al. 1992; Hough 1990; Lang 1997; Haughton and Hunter 2004). It may be obvious that sustainability in relation to food's spatiality is not only about environmental issues, critical as these are, but also about food's spatiality in social and economic terms, which has often been divorced from exploration of sustainability concerns (Allen 1993). Sustainability in food terms has also sometimes been set up as an 'either/or' proposition which suggests poorer individuals and communities cannot afford to be sustainable, and to argue for sustainability in these circumstances is punitive, ignorant or elitist. Such a view may ignore structural issues in relation to connections between poverty and the food system which are discussed later.

In terms of broader spatial effects, with the overconsumption of resources and the overproduction of waste in Western post-industrial cities, food's dominant spatiality strongly contributes to a negative feedback loop which presents a significant problem for sustainability (Rudlin and Falk 2001). In the context of massive urbanization (Hough 1984, 1990; Patel 2007) cities not only swallow up valuable habitats for food production, but also make intensive demands on environmental resources in the areas around them (Haughton and Hunter 2004: 2), as discussion of ecological footprints and foodsheds in Chapter 7 explores. The shocking amount of waste in the food system in rich countries has been highlighted (as in Bloom 2013) and is seen as an area ripe for consideration within the social sciences (Evans et al. 2012). Thus, despite the argued benefits for food of more compact development, to slow down and ameliorate these effects, as Chapters 5, 6 and 8 show, the narrative of urban expansion has taken a different and greedier path in relation to food and urbanism. This has had a range of problematic effects, which the book explores, including in relation to health, where, to give one example, effects on health and obesity are being experienced at a variety of scales (Jackson 2003). Both a retreat from public space and sprawling spatiality are central to these embodied aspects of food and urbanism, and the idea of obesegenicity encapsulates the understanding that 'food environments can be a powerful and independent determinant of food behavior' (Mikkelsen 2011: 210). As discussions in later chapters demonstrate, the book takes the view that arguments about food and health can be overly focused on individual behaviour when this interplay is much more about structural space-shaping issues. Thus the economics of 'cheap' food, which is implicated in ill health and early deaths, is again predominantly a structural sustainability issue with spatial outcomes the book explores.

Concerns about the spatiality of food are being sharpened by climate change effects all along the food chain: from industrialized, monocultural, technologically and chemically dependent agriculture with high environmental impacts, through just-in-time distribution arrangements that may be weak in resilience terms, to emissions-intensive retailing and consumption patterns and sustainability problems caused by waste. The system is highly capital and energy intensive because of its 'abundant use of external inputs, large machinery, and long-distance transport and communications infrastructures' (Norberg-Hodge et al. 2002). In the United Kingdom, for example, food chains account for about a fifth of emissions associated with climate change effects of household food consumption (Sustainable Development Commission 2008). The main contributors are meat and dairy products, glasshouse vegetables, airfreighted produce, heavily processed foods and refrigeration (SDC 2008: 42). In the United States, it has been estimated that rearing animals for meat adds a fifth to the methane emissions of what is already the largest per capita greenhouse gas producer in the world (Houy 2009: 21). These sustainability concerns are reflected in the economic workings of the modern food system where food production, distribution and retailing are increasingly integrated, consolidated and concentrated into cartels and monopolized structures. Worldwide, as few as six food retailers are thought to dominate global food sales (Lang and Heasman 2004: 160); by 2010, the seven biggest retailers in the United States were expected to control 70 per cent of food retailing (Lang and Heasman 2004: 160). Supermarket practices, including charging manufacturers for shelf space, displays, pay-to-stay and failure fees, were estimated in the late 1990s to generate £9 billion for retailers in the United States. There and elsewhere suppliers' margins are often squeezed to the extent that production becomes unsustainable in financial terms.

At the production end of the system, one strand in the analysis of the food system's weaknesses has emerged in relation to the notion of the end of food, in which food crises caused by unsustainable practices threaten the ongoing capacity to feed the predicted world population of 10 billion by 2050 (Roberts 2008). Especially given climate change, a 'radical rethink' is required (Tudge 2002: 4). The irony is that currently there is sufficient food produced to feed everyone on earth, but because within the modern food system food need is disconnected from food supply, food surpluses and food shortages are mediated by money, not the necessity to feed global populations. Morgan and Sonnino (2010) argue that 2 billion people worldwide are food insecure, while Millstone and Lang (2003: 18) suggest that more than 850 million people 'do not get enough to food to lead active and healthy lives. They are consuming too little protein and energy to maintain a healthy weight, and suffer from deficiencies in the composition of their diet that leave them vulnerable to disease'. Nestlé (2007: 7) puts it very simply: 'even today, insufficient food is a daily torment for nearly a billion people on earth, half of them young children. This lack is especially disturbing because the world produces more than enough food for everyone: it is just not distributed equitably'.

Within the modern food system, the same supply chains and integration that allow season and region to be ignored by some have not only opened the way for the wide spread of dangerous pathogens, but also failed to resolve food insecurity, created alarming externalities and are underpinned by a denial of the finite nature of resources including arable soils, energy and water upon which this production and consumption model is predicated (Roberts 2008). 'At every level, in fact, the modern food sector has become a miniature version of the industrial economy it once inspired' (Roberts 2008: xiv). The hallmarks of this system, and the basis for its success, continuous technological advances, ever-increasing scales of production and endless product innovation present a fundamental paradox in that food is not well suited to this economic paradigm because 'the underlying product – the thing we eat – has never quite conformed to the rigors of the modern industrial world' (Roberts 2008: xiv). Thus, to maintain the benefits of low-cost, high-volume production, producers are locked into a cycle of continually increasing production levels, which has been described as the 'technology treadmill' (Roberts 2008: xiv), while the climate change context now adds increased complexity and urgency to dealing with this problematic approach. It is predicted, for example, that climate change will mean agricultural output falls in both South Asia and sub-Saharan Africa, a problem technological change alone will not solve. Thus 'while technological change can raise agricultural productivity, if the technologies are too expensive for poor farmers they will make the well-off richer and the poor even poorer' (Millstone and Lang 2003: 20).

Gastronomically inclined food space explored in this book can increasingly be seen as restricted by the larger-scale operations of production, distribution and consumption sections of the food chain, where attempts to 'outflank nature'

(Morgan et al. 2006: 10), deny location and struggle against seasonality are combined to introduce a 'permanent dietary summer' (Morgan et al. 2006: 10) for those who can buy in. A helpful way of conceptualizing these dynamics comes from the adaptation of Michael Storper's (1997b) notion of productive worlds, whereby Morgan et al. (2006: 8) speak of different worlds of food. These distinguish between the industrial world signified by standardized industrial process and generic branding; the world of intellectual resources supporting the industrial world (where, for instance, research into genetic food modification takes place); the food market world where standardized production methods are used to adapt mass market food products to niche markets; and the interpersonal world which is marked by specialized, localized food production, as advocated, for example, by Slow Food (Morgan et al. 2006: 8). The food crisis has thus been defined as constituting a form of food war in which health is identified both as a source of sustainability problems and as a focus for solutions (Lang and Heasman 2004). While the notion of ecological public health is further elaborated by Lang and Rayner (2012), it is suggested that we must develop a new vision in relation to food which is in balance with ecological health. 'Food policy needs to provide solutions to the worldwide burden of disease, ill health and food related environmental damage' (Lang and Heasman 2004).

This has profound spatial dimensions. Food resilience is no longer a problem just for the cities of the global south that can be safely ignored by cities of the global north (Morgan and Sonnino 2010). Globally, we have moved from a position in which food became less and less visible in rich cities, because it was assumed that the problem of feeding people had been solved by an industrialized food system, to one in which food price surges and sharp increases in food insecurity demonstrated that these issues affect wealthy as well as poor places (Morgan and Sonnino 2010). Thus, underpinning the discussion of the sustainability effects on food and urbanism at various scales is the view that at this critical stage in the growth of cities, when urban space is expanding enormously worldwide, decisions about this largely urban future should be based on a wider design sense than in the recent past. Cities are inevitably being reshaped at an increasing rate and scale. The argument running through this book is that the settlement forms thus created should be informed by a fuller basis of understanding of how food influences spatiality and urban processes, giving due attention to the gastronomic possibilities for convivial urban life.

PART ONE

Food, Domesticity and Design

CHAPTER ONE

Starting with the Table

INTRODUCTION

This chapter is the first of the two that deal with scales of urbanism predominantly tied to domestic spaces, focusing in on tablescapes, kitchens and dining rooms in this chapter, and home gardens as food spaces in the next. In so doing it situates these spaces as legitimate areas for urbanist analysis of food. It suggests that domestic as well as public spaces are central to developing convivial urbanism, and thus the appropriate place for the book to start in scale terms. As noted in the Introduction, the categories of private and public can no longer be conceived as entirely different and separate from each other but have become blurred, and work in a range of disciplines has explored the implications of these disruptions to traditional dualist notions about city form. What happens at the small domestic scale, nominally private, is often closely intertwined with the events and transformations at wider scales that are understood as public. Thus the inclusion of chapters that focus on (largely) private space is predicated on and explores this connectivity in its approach to food and urbanism.

This chapter begins by examining the way that the table (or its symbolic equivalent in spatial design terms) has been and continues to act as perhaps the fundamental site for expressing conviviality through sharing food. It argues that this sharing is both a material and a symbolic re-creation of ourselves in daily socio-spatial practice that is at the heart of sustainable urbanism. Tracing the historic evolution of the table (and table-like) spaces, the chapter considers both these sites' convivial and exclusionary capacities, particularly in gender terms, with special attention to the contemporary role of design from the micro scale of the platescape outwards. The evolving nature of kitchen design and its complicated relationship with eating space is explored, with insights from across a range of disciplines and places that are particularly tied to the rise of modernism in food design terms. It could be argued that the chapter's examples and the trajectory of the discussion overall are somewhat Eurocentric in nature, but the book does not make claims to universal coverage – nor could it encompass this vast field in one text although efforts are made to demonstrate the spatial diversity of examples that reflect aspects of the discussion. Rather, the aim is to explore what seem important aspects of these scales of urbanism in food terms while acknowledging the need for further research to build a more complete picture in theory and practice.

'WHERE IS URBANISM AND WHY DO WE CARE ABOUT TABLES?'

As set out in the Introduction, urbanism does not start from the front door of the dwelling but is fundamentally tied to spatial expressions within domestic spaces of various kinds. The domestic is a central aspect of human settlement, and to return to the definition of urbanism offered in the Introduction, the way food is represented at home has significant implications for urban sustainability, social justice and cultural understanding. Moreover, in food terms conviviality through food sharing seems to be universal to different human cultures, and it is argued that a critical site for such expressions is the table or its spatial equivalent. In exploring instances from historical practice before moving on to the contemporary, the intention is to demonstrate longitudinally some of the recurrent themes of conviviality and sustainability tied to this scale of food and design from a wide range of times and places, reinforcing that this is not simply reducible to a historically or locationally specific phenomenon.

Exploring sources for table-based conviviality provides a diverse range of examples. Spatialized expressions of conviviality through feasting together are evidenced from early Greek and Roman sources (Wilkins and Hill 2006: 63), as well as from Mesopotamian representations of table-bound banqueters found on dynastic seals (Bray 2003: 22) to the patio-based Mayan elite feasting of the 650–1000 AD period (Hendon 2003: 210). As is noted in relation to the Roman banquet, 'convivial eating and drinking formed one of the most significant social rituals in the Roman world, inextricably interwoven into the fabric of public and domestic life' (Dunbabin 2003: 2).

The design arrangements and location of the table or table-like space may have shifted culturally over time, with eating together moving from the hall or evolving living space where the cooking fire was located, to the separation of the cooking function into the kitchen and the dining room specifically constructed for food consumption, but the convivial notion of the shared meal in a particular physical space has remained fairly constant. This shared meal around a material or metaphorical table expresses conviviality by reaffirmation and solidarity, marking daily life, the seasons and stages in agricultural production, especially the harvest. It is symbolic of family, kinship and other social ties. Mealtime is the site for socializing children into an understanding of commensality (Ochs and Shohet 2006). The shared meal provides a process for social reinforcement and cohesion that is fundamental because it occurs over and through food. Social isolation, by contrast, can be signified by complete retreat from the conviviality of the shared meal, as in the extreme example of more than two million *hikikomori* of Japan – young people who never come out of their bedrooms at all and subsist on food left for them on trays by their mothers (Ronald and Hirayama 2009).

The table or table-like space as the site for conviviality is also of great religious significance (Sered 1988: 135; Feeley-Harnik 1995; Fernández-Armesto 2002), as by 'establishing who eats what with whom, commensality is one of the most powerful ways of defining and differentiating social groups' (Feeley-Harnik 1994: 11). Just as fasting practices reflect religious belonging (Bynum 1988), every major religion has a table-based sacramental meal representing new life or rebirth (Visser 1987). In Christianity the presence at the table is crucial to the religious symbolism of the Last Supper; in Judaism the Passover ritual has its specially set Seder table, while the Sofreh cloth is central to religious observance across religions in Iran (Shirazi 2005). More broadly, for many, the ancestors are present at the table which operates as a space for collective memory. In contrast to hierarchical dining norms, Christian dining and worship in Roman houses was unusual in its time because the diners were perceived as equal before God and table seating status related only to the level of conversion into Christianity (Sennett 1994). In design terms, too, the Roman dining room became the basis for church design in the West: in the meal for materialists the sacramental table represents life itself. The table is also the site for acting out dietary habits and food proscriptions that by creating absences seek to reaffirm religious belonging (Anderson 2005: 156). Even though the historical antecedents of halal and kosher foods have been argued to have originated in specific agricultural, regional and hygiene needs (Harris 1985), they may now be followed as an article of faith, not through necessity.

Yet dining at the table or in a space that represents this in design terms can also present a fundamental paradox: on the one side conferring conviviality and solidarity through familial, social or religious belonging, on the other capable of representing status differences, subordination and control. The lack of a table-like spatial structure therefore is not inherently unconvivial, as the flexible dining arrangements of some Japanese dining spaces make clear through their dynamic response to seasonal conditions (Ashkenazi and Jacob 2000; Kazuko and Yasuko 2001; Ronald and Hirayama 2009). However, it is in those places where the physical structure of the table is strongly connected to sharing food that its absence, or absence from it, can undermine conviviality. Those given an inferior place or excluded altogether express through their placement the unequal power relations that dining together reaffirms. The table thus provides the space for power plays over status and control; ritual expressions of domination and subordination are demonstrated, and unequal relations of race, class and gender are magnified by the rituals of dinner (Visser 1993). As Malcolm X wrote, 'I'm not going to sit at your table and watch you eat, with nothing on my plate, and call myself a diner. Sitting at the table doesn't make you a diner, unless you eat some of what's on that plate. Being here in America doesn't make you an American' (in Feeley-Harnik 1994: 13).

Obviously, this is not new. The Spartan *sussita* (messes) offered sometimes-minimal rations of barley-cake, wine and 'a small piece of meat in the infamous black soup', but only male citizens could eat here (Wilkins and Hill 2006: 64). In Greek cities large public dining spaces in civic buildings, the *prytaneia*, were sites for honorific feasting among men (Wilkins and Hill 2006: 64), and, while it came to mean a large public feast, the Roman *convivium* was originally a meal for friends which was 'intended to express a relationship between equals as well as an opportunity for patrons to entertain clients and probably show off their wealth' (Alcock 2006: 197). This was an occasion in which women sometimes took part (Alcock 2006: 197), yet even where a table per se was not a part of the dining configuration, as in the Roman *triclinium*, where guests reclined on couches, placement was considered to be of exceptional importance (Amery and Curran 2002). The less important or hangers on of the host were not allowed to recline but humiliated by being made to perch on stools (Visser 1993) while the most important were placed at the right-hand end of the couch (Bowes 2010: 57). Roman dining rooms were, in fact, spaces that could be used in a variety of ways, not just dining, and space syntax analysis of Roman domestic spaces suggests that there was 'an enormous diversity of spatial patterning' (Bowes 2010: 41). Some scholars take the view that the emergence of particular dining room configurations owed most to the development of the curved dining couch or *stibadium* (Bowes 2010: 28) which reflected increased rigidity in dining behaviour whereby portable couches were replaced by permanent semicircular couches (Bowes 2010: 55). In later Roman houses of the elite, where patronage was both symbolized and acted out through dining 'even interactions between peers took on more hierarchical arrangements. The apse of the dining room held guests in a vice like grip, as dinners became ritualized theatre, while the apse of the reception room framed the dominus as mini-emperor' (Bowes 2010: 55).

Status gradations in the medieval hall were similarly expressed through diner placement. The bulky hall table comprised trestles and boards which could be set up and dismantled at need, and not only the quality of the linen that marked out those spaces reserved for the master and his guests, but also a strict hierarchy according to status governed the seating placement (Henisch 1986: 147, 151). As the elite tended to distance themselves from others they sat at high table, and eventually physically withdrew into a separate room. Diners in the hall needed to be able to see the important people (and be aware of their symbolic power), and seating arrangements down one side of the trestle allowed this. It mattered who was at high table and who placed beneath the salt. As Visser (1993) has noted, arrangements for modern banquets tend to hark back to this layout, but here there may be denial that status is an important matter so that there is great difficulty in achieving seating arrangements at the table to everybody's satisfaction, whereas in contemporary practice in Japan, such spatial hierarchy is explicitly and strongly formalized in the way that Sumo wrestlers eat *chanko* (stew): the top wrestlers or important guests eat first (Tierney and Ohnuki-Tierney 2012).

The table is also the site for playing out food denial, inequality and food disorders. It provides a compelling dramatic space for disturbed or negative practices and relationships in which food may be used to express confusion, rage, pain and exclusion. Anorexic and bulimic behaviours, for example, are now more common than generally recognized (Hoek and Van Hoeken 2003), and while they resist clear explanations, family dynamics at the table are generally thought to be involved (Ackard and Neumark-Sztainer 2001; Haworth-Hoeppner 2004). Work on dining in settings like old peoples' homes and hospital wards for those with dementia shows conversely that dining together helps those thus institutionalized to want to eat (Day et al. 2000; Wright et al. 2006). For some the table is the pre-eminent symbol of decadent or exclusionary consumption. Writing *Down and Out in Paris and London* in the early 1930s, George Orwell viewed the tables of diners seen from the gruesome kitchens of the Ritz as quintessentially expressing a decadent, corrupt urban life, contrasted with a more morally correct austerity. Historians have likewise drawn attention to the trajectory of the London club from its coffeehouse antecedents (Cowan 2004, 2005; Burnett 2004; Ellis 2011). Women were largely excluded (except as servers or prostitutes) from this intellectually influential masculine space of debate and curiosity (Clery 1991; Cowan 2001). This exclusion had significant knock-on effects given the coffeehouse's argued basis for the development of the public sphere (Habermas 1989). The related growth of clubland dining in the nineteenth century (McDouall 1974) reflected a muscular, great house and public-school-inflected gastronomy, very much a male domain that continued into the twentieth century (and was amusingly parodied in Wodehouse's Drones and Junior Ganymede clubs for master and man). Similarly, while dons famously dined well, Virginia Woolf (1984) described bad food in women's colleges as signifying contempt by those serving them for the temerity of women seeking 'unfeminine' education (a theme receiving extreme expression in Dorothy Sayers' *Gaudy Night* of 1935).

Historically, for women a place at the table has been hedged about and conditional. Expected at once to play out a role ascribed as natural in ensuring conviviality is maintained, and in socializing children to be convivial by instilling table manners, women have themselves often been given an inferior position, eaten less or less favoured foods, or been barred from the table altogether. In ancient Greek examples dinners were only for peers (although subtle gradations of status and power were evident); the lower orders and women were not invited. In elite Roman dining, while women were present and shared couches with husbands, they only did so because their husbands provided legitimacy for their presence there (Sennett 1994). Much later, in nineteenth-century elite house plans, the rituals of women's post-dinner withdrawal from the table were clearly expressed, with architectural arrangements becoming more complex further up the social scale. The conditions of the servant hall table similarly mirrored these mores above stairs, with a strict hierarchy determining table presence and position.

Like women, children similarly have had their lower status reinforced by exclusions from the table, including being required to dine at different times from adults or by being given a special and lesser part of the table at which they are sometimes expected to stand or perch rather than sit. This artificiality may be used to give adults separate space and time away from children and serves as the basis for deciding some foods, and later dining hours, are unsuitable for them (Visser 1993). That such arrangements are culturally constructed, not natural, is evidenced by substantial differences between, say, French, Italian and Anglo traditions (Visser 1993). In some places children eat the same foods and experience the same hours of dining at the same table: in Italian families it was traditionally considered proper for children to experience the full rituals of dinner.

What and how people eat at the table is also of planning and design interest. From the food service industry, although untutored in architectural design, those looking at food space have recognized that the eating atmosphere matters (Edwards and Gustafsson 2008). Designers interested in these issues have similarly proposed physical arrangements for the table that would support a pleasant eating atmosphere and thus, it is argued, a high level of conviviality (Alexander et al. 1977: 844).

While some table spaces invite such leisurely conviviality, 'others force diners to eat as quickly as possible, so they can go somewhere else to relax' (Alexander et al. 1977: 844). It is argued that not only is a heavy table and comfortable chairs required, but the quality of the light must hold the group together (Alexander et al. 1977: 844), with the light creating a pool over the people and providing them with a sense of protected enclosure (Alexander et al. 1977: 844). Inside the circle of light over the table the diners bask in the convivial glow.

From early archaeological evidence of food-related ceramics (some mass produced) as in Mesopotamia (Bray 2003), by way of the central platter and individual trencher of medieval times (Albala 2003: 103; Brears et al. 1993), through the introduction of individual servings and table settings in the society of the Ancient Regime (Finkelstein 1989: 33), to the present-day profusion of eating bowls, plates and drinking vessels, the 'microgeographies' (Sobal and Wansink 2007: 1) of the table and its accoutrements are highly influential on consumption behaviour. They are thought to influence choices and intake and obesity and health in ways that are 'subtle, pervasive, and often unconscious', and it is therefore argued that 're-engineering built environments' could allow the chance to reshape intake in more positive ways (Sobal and Wansink 2007: 1). Thus the notion of the tablescape has been developed to explore how the spatiality of the table reflects and shapes the interplay between physical space and the social relations of dining. In the development of the habitus, for example, tablescape design denotes distinction (Bourdieu 1984). The design of table furniture and dining implements is capable of displaying both beauty and symbolism (Bendiner 2004; Pegler 1991). These can express cultural norms about consumption and provide physical territories for that consumption (Becker and Mayo 1971). Table settings meanwhile define the boundaries of the individual's food space. It is argued that tablescapes are part of micro-scale-built environments which include the table's surroundings, such as 'rooms, furniture, objects, and foods themselves' (Lawrence and Low 1990).

Food visibility on the table matters in a design sense because it brings attention to the availability of food for consumption and provides cues for people to eat (Lawrence and Low 1990). As Visser (1993) notes, egalitarian sharing from the same pot is symbolically represented by the common serving dish from which each diner helps herself, whereas the notion of plating food is related to dining in which there is a separation between the server and the served. Conviviality would imply no such distance, and in thinking about what explains the pleasure of the central big dish or bowl from which all partake, the following tablescape parameters need to be taken into account: 'The abundance (number of sources), ampleness (size of sources), and fullness (emptiness or saturation of sources) of the tablescape and vessels on the tablescape are parameters that can influence food intake of individuals who are served or have access to these communal containers on the tablescape' (Sobal and Wansink 2007: 6).

Equally, where there is a diversity and complexity in the tablescape, food intake can be influenced by the way that both food and non-food items are spatially configured on the table. For example, at the buffet table a causal link has been shown between higher food intake and the presence of a diversity of disorganized serving dishes as compared with food displayed in clearly defined patterns and fewer bowls (Sobal and Wansink 2007: 7). At the micro-spatial scale, the so-called platescape, which 'represent[s] the sum of the visible attributes of a particular plate or similar food container' (Sobal and Wansink 2007: 7), has pervasive effects on food intake and the social relations around it. It is thought that around 70 per cent of all calories consumed are after food has been transferred to the platescape of cups, bowls, plates and other containers. The physical design of these containers unsurprisingly influences intake: with greater consumption from larger or wider rather than taller and smaller containers and the increasing size of platescapes associated with the rising levels of obesity (Sobal and Wansink 2007: 7). However, while bowl size affects intake, diners are not generally aware they self-select dramatically more food and eat more when they have a larger bowl to eat from (Sobal and Wansink 2007). In socio-spatial practice terms, where there is a central bowl to take food from, it is argued that people do not converse, whereas when the food is plated, conversation will occur as people are not worried about getting enough to eat (Visser 1993). Design also matters at the level of the configuration of serving portions. Various public health and sustainable gastronomy models for healthy plate configurations have been developed, including the 'Eatwell' plate from the United Kingdom's Food Standards Agency.

Of course, table, food container and implement design cannot be reduced to simply the technical matter of supporting or controlling food intake. Table-related design reflects not only the physical needs of diners but also the social relations that govern table-bound eating. Far from simply being a utilitarian matter of implements that aid eating, for the 'socially competent bourgeois individual of modern times' table manners 'police the emotional timbre of exchange' (Finkelstein 1989: 130). Thus, for example, as the decorated cloth- and leather-covered trestles and benches favoured by medieval aristocracy were replaced by tables, dressers and chairs, their different sizes and styles came to denote status and underscore the roles of host and guests at the table. Likewise, dishes, bowls, cups, plates, glasses, knives, forks and spoons have all been freighted with rich cultural meanings and associations, as well as attracting guidance in relation to their spatial configuration on the table for symbolic social purposes (Rebora 2001). Elaborate table manners may have developed to deal with the issue of potentially dangerous implements in a confined space where conflict and even violence was possible (Visser 1993). Implements with murderous capacity are arranged within a set of conventions and usages intended to contain threatening situations (Visser 1993).

THE PARADOXICAL KITCHEN

It is difficult to consider the nature of table design in urbanism terms without exploring the kitchen as its predominant location, and the use of fire as central to its varied spatial expressions (Davey 2008: 100; Montanari 2006: 30). Many terms for the kitchen come from the Latin verb, *coquere*, to cook, and its related nouns, and 'the kitchen is always where cooking is done' (Davidson 1991: 10). Historical accounts include those of the Pompeiian *culina* in which meals and banquets were prepared: these were 'surprisingly small, with usually only enough room for one slave' (Amery and Curran 2002: 112). They contained a raised cooking hearth and often sinks with running water, but no ventilating chimney: in fact, kitchens were among the only rooms in Roman houses with fixed installations, while other rooms served many purposes (Bowes 2010: 40). The kitchen's traditional separation from the rest of domestic living space historically removed the hazards of smoke, fire and smells from the house, but gave rise to spaces that have very often been badly designed, ill lit, cramped and poorly oriented (Upton and Vlach 1986), their low status within the dwelling congruent with 'spaces dealing with food, dirt, women and servants' (Davison 1978, in Johnson 2006: 125).

For obvious reasons of heat dispersion, kitchens were often built as detached structures in hot climates, while in colder places, cooking space might move with seasonal conditions from a more open summer kitchen to a more protected winter one. In Mexico, for example, some households still maintain a traditional separation between the *cocina de diario* (everyday kitchen) and the outdoor *cocina de humo* (kitchen of smoke) for preparing feasting meals (Tierney and Ohnuki-Tierney 2012). According to some accounts, with the advent of masonry houses, fire migrated from the centre of kitchens to stoves at the edge, while kitchen chimneys may have evolved in Northern Italy in the thirteenth century (Davey 2008: 101) but did not become common until the mid-sixteenth century (Albala 2003: 91). Serlio's sixteenth-century architectural treatise discussed kitchen placement and design, from the separated cantina for the peasant, through the 'habitations within cities, of all ranks of men' where plans for attached houses showed kitchens and vegetable garden spatially linked on ground floors (Serlio 2001: 92) to the vertical division between kitchen beneath and living spaces above in his influential house plans for the well-to-do (Davey 2008: 102). Palladio, meanwhile, banished kitchens to the lowest part of the house as one of those spaces that was at once necessary for daily life but foul in character (Davey 2008: 102). In Italy, the kitchens of Venice, with no opportunity for basement space, were often located in attics to reduce fire risk. Elsewhere, kitchens were placed on the ground floor in outbuildings, while in the increasingly popular terrace house typology only sunken courts gave any light and air to basement kitchens (Davey 2008: 102).

The fundamentals of kitchen design do not appear to have altered significantly until the early modern period, and despite descriptions such as Bartolomeo Scappi's of the range of kitchen utensils a good kitchen would boast, most kitchens remained quite rudimentary in their *batterie de cuisine*. From the later part of the eighteenth century though, new kinds of cooking equipment began to appear, including iron kitchen ranges designed by Thomas Robinson and brick range structures developed by the improbably named American, Count Rumford: by the mid-nineteenth century, these were available 'to all but the poorest households' (Davey 2008: 102). Isabella Beeton's (1861: 25) principles for good kitchen design included instructions for the room's configuration, dimensions, ventilation, accessibility, desirable level of remoteness from living areas, fuel and water sources and connections to food storage areas. In the houses of the rich, elaborate separate circulation systems were designed in order to move food from remote kitchens to dining spaces without being seen, as shown in plans illustrating Robert Kerr's 1864 *The English Gentleman's House* (Davey 2008: 104). By the turn of the twentieth century, gas-fuelled cooking was becoming common in upper- and middle-class houses and reinforced the separation of cooking from living spaces as it allowed the heat source for cooking to be divorced from heat needed to keep householders warm (Davey 2008: 104).

In a survey of a large range of kitchens from different cultural traditions, Davidson (1991: 11) found that kitchens may be used for diverse functions besides cooking: as meeting places, living rooms, eating areas and linking spaces to kitchen gardens and orchards. The layout and equipment of kitchens and pantries has been influenced not just by food storage requirements, but also, unsurprisingly, the nature of local resources. Traditional food and cooking practices and design arrangements reflected these resource constraints and opportunities, as described by McFeely (2000: 12) in relation to Mary Abney's kitchen in rural Missouri, where vegetable growing, preserving and butter and jam making were all part of the food preparation year. The oak icebox was important in this example, while elsewhere equipment ranged from the wood of the Haute Savoyard kitchen to the cloam oven for making clotted cream in Devon, the butchering table in India (Davidson 1991: 11) and the Katsuo-bushi shaver of Japan for grating dried fish for dashi flakes (Kazuko and Fukuokoa 2001). While there is no absolute relationship between sophisticated infrastructure and good cooking, equipment innovations, such as the introduction of the 'black sow' cooking stove in Sweden and the rice steamer in Japan, are argued to have profoundly altered cooking practices. However, kitchen traditions generally tend to be slow moving: the resilient food preferences of diasporas demonstrate how particular cultural and religious mores shape both space and practice in particular ways, as in the immigrant Muslim kitchens of the United States (Metcalf 1996).

Despite their importance as sites for cultural continuity in urbanism, across the twentieth century, kitchens have been models of technological change, metaphors for modernism and microcosms of new consumer regimes (Oldenziel and Zachman 2009: 10; Houy 2009). They have also been places where 'resistant users' have withstood technological changes (Hessler 2009: 164). With antecedents in the kitchen science movements of the nineteenth century (McFeely 2000; Shapiro 2009), kitchens in the twentieth century have been expressions of proselytizing in which the general public were 'won over' by designers to the merits of transformations including the fitted kitchen (Freeman 2003: 50). While a too deterministic reading of kitchen space underplays the fact that kitchens are as capable of diverse readings as any other room (Floyd 2004: 62), it seems valid, as Mock (2011) has argued in relation to the United States' experience, to view the home kitchen as 'the most technologically saturated room in the early-twentieth-century American home *and* a cultural space imbued with strong values, customs, and nostalgia'. In the United States and elsewhere, kitchens have enjoyed a somewhat paradoxical position as part of a shifting ensemble of space thought to reflect multi-tasking roles for middle-class women, as the domestic servant class disappeared (Lees-Maffei 2007) and more women took up paid work outside the home, putting in a double shift.

Read by geographers and sociologists as spaces strongly reinforcing traditional female gender roles concerning domestic consumption (Domosh 1998; Llewellyn

2004; Blunt and Dowling 2006; Casey and Martens 2007), kitchens have also been sites for expressions of conviviality. Kitchens are where women have created a sense of home and felt comfortable and in control (Moran 2003; Counihan 2004; Supski 2006) through cooking and passing on home recipes (Ballantyne and Benny 2009). They have been both spaces for the indirect power of the kitchen cabinet and where women might express their oppression through the use of food as a form of control, manipulation or abuse of themselves or others (McIntosh and Zey 1989; Counihan 1999; Counihan and Kaplan 2004). The kitchen has perhaps above all been understood as the prime location for signifying women's paradoxical domestic role. Food preparation and cooking are both a release and a trap, and the kitchen a place at once of empowerment and containment in which the woman acts as both the servant and the manager of her family, of which more below (Murcott 1982; Johnson 2006: 124; Burridge and Barker 2009).

At the same time, in food terms, kitchens have become increasingly central to more artful narratives of convivial social life, with a return to dining at the kitchen table as a cosy mark of respect and intimacy (Visser 1987). This might be achieved through the nostalgic tropes of eating peasant food in the country-style kitchen (Lyon and Colquhoun 1999; Duruz 2001) or within a more streamlined modernist design approach, with molecular food to match, in either case demonstrating an individual's stylish domestic habitus (Myhrvold 2011). Kitchen-based dining is encumbered with class and cultural meanings, from David Cameron's kitchen suppers (Craig 2013), through the conspicuous competence of the middle-class cook's appropriation of a sophisticated cultural range of food dishes and serving styles (Visser 1987), to the aspirational dinner party thrower reliant on entertainment guides and ready meal subterfuge ('cheat' to impress your friends). As each demonstrates varying degrees of skill and tastefulness in Bourdieu's (1984) terms, it is not hard to see cultural commodification, the expression of status and the acquisition of symbolic capital occurring through food and entertaining in the kitchen. Work on changing food tastes suggests not only that the 'occupational class of the individual remains a powerful predictor of food choice' (Tomlinson 1998: 295), but also that the design of kitchen space is clearly implicated in these performative aspects of domestic urbanism.

Cookbooks, the cooking pages of newspapers and, increasingly, online food sites may thus be read as 'manuals of taste' that trace changing mores and food fashions (Gallegos 2005). These texts and images are reinforced by architectural journalism which also displays currently fashionable design tropes. Together, these forms of 'aestheticized consumption' (Brownlie et al. 2005) allow home cooks to keep up with and model gastronomic trends in domestic cooking, kitchenware and spatial design deemed most suitable to meeting these cultural aspirations. The reach of these influences seems to have grown through a current media preoccupation with cooking. Research from Australia situates television cooking programmes as 'worrying away at the boundary of the mundane and the distinguished', in this case allowing their viewers to negotiate tensions in relation to multicultural approaches to Australian foodways (Newman and Gibson 2005: 83). Research from Greece somewhat similarly suggests that such cooking programmes offer a new space for a large audience to explore questions of authenticity and societal change through their relationship to old and new cooking practices (Sutton and Vournelis 2009). An often repeated assertion is that cookbooks, cooking guides, websites and cooking programmes are viewed rather than used as a basis for actually preparing such meals (Naccarato and LeBasco 2012: 41), as, arguably, food skills decline (Giard 1998; Lang et al. 1999; Lyon et al. 2003; Engler-Stringer 2010; Rousseau 2012). Empirically based work does support the view that a culinary transition to food prepared commercially outside the home is in process (Lang and Caraher 2001: 2), requiring various externally prepared 'meal solutions' (Pritchard 2000: 208). Certainly, research by economic psychologists suggests that resorting to 'convenience' foods is a strategy responding to an actual lack of time and income (as well as perceived resource constraints) and a connected preference for time- and energy-saving approaches to meal production (Scholderer and Grunert 2005).

My review of house plans from the 1920s to the 2010s demonstrates both areas of continuity and significant shifts in kitchen design arrangements that reflect these issues and transformations. Early-twentieth-century plans (and writing about the role of kitchens) showed a trajectory in which kitchen space was intentionally shrunk to more minimal dimensions than it had traditionally enjoyed. In this minimalization process, kitchens were expressly designed to be too small to allow space for the worryingly working-class tendency of dining, receiving visitors or even sleeping in the kitchen, as part of an overall programme linking house design to social improvement (Hessler 2009: 166). In the United Kingdom, kitchen design in the early part of the twentieth century could be understood as reworking Victorian notions of the domestic through the application of scientific methods that called for kitchen rationalization. Raymond Unwin's housing designs, for example, saw the kitchen conceived as solely a workspace (with associated pantry, scullery and laundry rooms) as part of a utilitarian ensemble of service rooms (Jackson 1985: 131). Unwin also explored the idea of the 'house place' where the living room contained a cooking stove, dressers and cupboards but not washing up and food storage, which were still separated out as dirty rather than clean activities (Jackson 1985: 131). Not only were ideas about hygiene and social improvement embedded in this design schema, but this separated configuration of 'work' and 'social' space reflected a desire to return to the supposed innocence of the farmhouse kitchen with its tiled floor, oak settle, shelves and dressers (Jackson 1985; Davidson 1991; Duruz 2001), with the untidy and dirty aspects of food preparation and clean-up work that was undertaken out of sight. The advent of technologies including electricity again altered kitchen design, underpinning manufacturing of electrically powered kitchen appliances, with these argued as 'replacements' for servants (women were still expected to do the work) and tied to modernist architectural design discourses about the appropriate shape of domestic space (Worden 1989: 140), of which more below.

The separation of work-oriented kitchen from leisure-focused dining space was also clear in American house plans in the period between 1900 and 1940s. While, as Cohen (1986: 269) reports, working-class households 'rejected small kitchens and kitchenettes in favour of ones large enough for dining', increasingly, this food service area comprised a set of workspaces that might include a tightly confined kitchen, a small scullery, a pantry, laundry, icebox and cellar, with direct access to the backyard by way of the kitchen door. In some larger house designs, a breakfast nook or alcove was attached to the kitchen space, being clearly labelled as such to emphasize it was an appropriate location only for this least important meal of the day. The kitchen's working status and lower level of amenity was accentuated, for example, in the Sundale plan of 1930, where the kitchen is 'cut off from the rest of the house so cooking odours need not concern you' (Antique Home Style, undated); a comment clearly directed to the presumed male buyer.

From the early to middle years of the twentieth century, kitchen design in industrialized countries was subject to various apparently scientifically grounded, supposedly technical solutions to the problem of healthfulness and housework. As early as the turn of the twentieth century, increasing 'interest in scientific values and rational planning for every sphere of life' was bringing the kitchen into prominence as the centre of the modern house, a shift reflected in the attention given to model kitchens in architectural pattern books, domestic science textbooks and women's magazines (Wright 1980: 239). Increasingly there was both a levelling up as working-class women were drawn into industrial labour and out of the servant class, and a levelling down in which middle-class women could no longer afford domestic help (Roberts 1991). By the 1930s, despite ongoing discussion of the 'servant problem', in reality, domestic servants had already moved out of the financial reach of many middle-class households, reinforcing the tension between women's roles as mistress of the house and also their role as worker within it (Lupton and Miller 1992).

Just as production at this stage of industrial capitalism was being broken down into its constituent parts in the development of the assembly line through Taylorist and Fordist techniques which owed a conceptual debt to the work of Charles Babbage (Braverman 1974: 89), so too was the idea of breaking down domestic labour in the kitchen into a series of separate, discrete tasks. These were based on time and motion studies connected to labour-saving devices, and were influential in shaping design responses to kitchen space (Frederick 1918; Bose et al. 1984; Banta 1993; Rutherford 2003). Design for the kitchen as a laboratory for the professional housewife, in turn, influenced utilitarian, gendered assumptions about the kitchen's place and the cook's role in the design of home. In this context, small kitchens designed without eating tables at their centre were conceived as workspaces for women in the 'biotechnic' urban age (Mumford 1938). Text accompanying the Hillcrest house design of 1930, for example, explains that 'entering the kitchen through a swinging door from the dining room, you find a most convenient arrangement – compact and splendidly planned. The housewife can prepare a meal with very few steps, as sink, case and range are just a step away, when one stands in the center of the floor' (Antique Home Style, undated).

The advent of the kitchen as a laboratory-style work centre reflects how thinking about scientific rationalism and modernism influenced designers of real kitchen spaces (Banta 1993: 233). The American material feminists' time and motion experiments of the late nineteenth and early twentieth century made explicit how poorly planned kitchens were in rational terms and 'centred on the potential change from an unplanned, illogical kitchen to a rational, efficient workspace'

(Llewellyn 2004: 45). Reformers including Lillian Gilbreth and Christine Fredericks applied Taylorist principles to the kitchen, breaking down tasks into the smallest constituent parts (Lupton and Miller 1992: 13; Williams Rutherford 2003: 43; Freeman 2003: 31). Similarly, kitchen rationalization was part of a wider programme of the New Dwelling (*Neues Bauen*) movement of 1920s Germany, sweeping away the 'irrelevant clutter' of outdated lifestyles, domestic design and equipment to 'free' the housewife for a more creative domestic role in family nurturing and education (Bullock 1988: 177), while similar tropes were evident in Czechoslovakia (Guzik 2004). A government-sponsored committee investigating domestic design standards in Germany of the late 1920s demonstrates this thinking quite plainly: 'The starting-point for their approach was again the view that the kitchen should only be used for cooking and washing-up. The kitchen was therefore to be small – indeed this was a necessary virtue to meet the committee's fears that a larger kitchen might simply become a general family space, or worse still, might be used for sleeping' (Guzik 2004: 198).

In Weimar, Germany, kitchen design gave space for a range of ideologies to be imposed (Jerram 2006), but the mass application of Margarete Schütte-Lihotzky's prototypical 'Frankfurt Kitchen' (see Figure 1.1) in which the architect combined

FIGURE 1.1: The Frankfurt Kitchen.

Photo: Christos Vittoratos. Image available at http://commons.wikimedia.org/wiki/File:Frankfurter-kueche-vienna.JPG under a Creative Commons Attribution-Share Alike 3.0 Unported license.

Taylorism's rational techniques with a functionalist architectural approach was widely influential (Walser 2005). Its advent was followed by the Congrés Internationaux d'Architecture Moderne (CIAM)'s 1929 discussion of efficient living in small spaces (Llewellyn 2004: 45). Different social actors, including a coalition of modernizers comprising architects and housewives associations, participated in making the kitchen; and Schütte-Lihotzky's design work heralded an approach that was built into much public and private housing in pre-war northern Europe (Hessler 2009: 164; Melching 2005; MOMA 2011) and was an unacknowledged influence on American, Australian and other kitchens too (Betts 2004; Hessler 2009). The technical innovations of the rational kitchen were not just about step reduction, but the integrated configuration of internal fittings and appliances on the basis of scientifically justified efficiency and hygiene arguments.

In the United Kingdom, designers including Wells Coates at his Isokon (Isometric Unit Construction) flats began to explore the functionalist kitchen as part of the minimum flat designed for efficient living in small spaces (Cantacuzino 1978), while housing reformers such as Elizabeth Denby were active in promoting European ideas about rational, scientific kitchen design in dwellings conceived as machines for living in a Corbusian mode (Llewellyn 2004: 46). The British architect Jane Drew's (c. 1944) book *Kitchen Planning* offered detailed design advice (and wonderfully evocative architectural drawings) along 'rational' lines. In Australia, meanwhile, the minimal kitchen's work triangle, again based on time and motion studies, was codified into kitchen designs to minimize unnecessary movements between food storage, sink and cooking facilities (Johnson 2006: 125). Six basic kitchen layouts – the one-wall single-line kitchen, the parallel or galley kitchen, the U-, L- and F-shaped kitchen and the island kitchen – emerged (Johnson 2006: 125, from Levene 1978: 7–16). In Japan, meanwhile, a tiny kitchen space became the norm (Ashkenazi and Jacob 2003) although this may also owe something to dense urban conditions generally, while in Finland, kitchen design became a functionalist matter in which 'cleanliness in the home converged on the kitchen. The kitchen was turned into a clearly demarcated laboratory-like workstation for one person, the new kind of active and practical housewife. Its space was reserved strictly for cooking and washing dishes, with no place for eating meals or interacting socially' (Saarikangas 2006: 163).

As prefigured by the Frankfurt Kitchen, the streamlining of kitchen furniture and objects also became ubiquitous by the middle of the twentieth century as the design details that made up the kitchen space and advances in kitchen-based appliances were subject to this kind of design thinking (Cromley 2010: 169; Lupton and Miller 1992). The modular cabinets that became prevalent in twentieth-century kitchens had their antecedents in nineteenth-century philosophies of storage and systems in which systemization and interchangeability derived from the mass production techniques and commercial applications of modernism (Beecher 2001: 27). By the 1930s, the idea of the continuous kitchen had taken hold in the United States, with architecture and furniture seamlessly merged into the 'new form of the fixture' (Lupton and Miller 1992: 41).

Such approaches suggested claims to scientific objectivity and acted to underplay or ignore the fact that kitchen design was also about shaping and defining gendered social and economic roles and spaces in relation to food. The kitchen was to be solely for food work, and the planning of kitchens as tight workspaces was intended to disallow any connection with social activities at home. Moreover, as 'labour-saving' kitchen devices proliferated, so did the requirements of the rising standards of hygiene, wiping out any supposed gains for women in free time away from domestic tasks (Cowan 1987). New domestic food technologies could also bind their users into wider consumption regimes, as in the case of the domestic freezer, associated with 'out-of-town supermarkets, frozen food producers, global transport systems and agricultural practices' (Shove 2003: 178). In tracing these changes in the Netherlands from 1920 to 1970, van Otterloo (2000) suggests (to paraphrase), there was a kind of increasing interdependency between households and manufacturers, through the production of food and cooking equipment. In this way the urbanism of the private domain of the dwelling became ever more entwined with the public, economic spaces beyond it.

While changing design perspectives about food preparation and consumption were transforming kitchen space in the nineteenth to twenty-first centuries, the notion of kitchenless flats and houses was also, at certain points and places, influencing the theory and practice of architectural design for domestic food space and labour and reflecting wider influences on domestic urbanism. Public kitchens, along the lines of Berlin's *Volkskuchen* (people's kitchens), were advocated in the United States by Mary Hinman Abel and her associate Edward Atkinson who aimed to reform the wasteful and nutritionally suspect eating habits of the poor (Levenstein 1988: 48). The nineteenth-century material feminists' proposals for better city form meanwhile were intended to allow middle-class women into the public domain by socializing domestic production and thus relieving them of the burden of domestic work (Hayden 1981: 229; Williams Rutherford 2003: 42). Child care, laundry, kitchens and domestic tasks would be socialized and shared and co-operative eating houses established (Hayden 1981: 229; Williams Rutherford 2003: 42), while housing would be built to foster communal dining and well-paid domestic labour would be understood as a work specialization like any other (Hayden 1981).

These schemes had a strong focus on rethinking the continued existence of the domestic kitchen, through design and management proposals for kitchenless cottages, co-operative kitchens, residential hotels for women and families, model villages, tenement 'refits', community dining clubs and wider urban forms to encompass these radical respatializations (Hayden 1981: 229). Although not described as such, in urbanism term, I would suggest this meant reorganizing domestic space along more convivial lines. The overarching theme was to challenge (some) women's entrapment by privatized domestic labour and consumption in many forms and at a variety of spatial scales, and rethinking the kitchen was key to this approach. Within the Garden City movement in the United Kingdom, too, interest in collective housekeeping saw kitchenless houses built as co-operative quadrangles at Letchworth, Welwyn Garden City and Hampstead Garden Suburb (Ravetz 1989: 192). Barry Parker's plans for schemes based partly on the design of university quadrangles, which in turn reflected monastic design sources, would have common rooms in which cooking and serving meals would replace the 'thirty or forty little scrap dinners of individual housewives and do so considerably better and more cheaply' (Davey 2008: 106). Raymond Unwin similarly proposed communal laundry, cooking and dining rooms in the design of some dwelling units, rather more prosaically to reduce expenditure (Jackson 1985).

The increasingly common design form of the service flat from the early part of the twentieth century took up this idea of communal provision of kitchen services for urban dwellers at an elite level, while in more radical experiments in conviviality such as Moscow's Dom Narkomfin, designed by a group led by Moisei Ginsberg and Ignaty Milnis in the early 1930s, workers in the People's Commissariat for Finances shared an apartment block which boasted a canteen and communal kitchen run by professional cooks (Davey 2008: 106). Although this experiment fell victim to changing political circumstances, other communal kitchen designs included those of the Kollektivhus in Stockholm, which incorporated centralized cooking and food transferred to virtually kitchenless apartments by dumb waiters (Davey 2008: 107). In the post-war period, Cromley (2010) has shown how downtown studio apartments with mobile or hidden kitchen units and pantry spaces were designed and marketed in the United States to middle-class single people. From these minimal kitchen spaces, their occupiers were (it was suggested) able to produce gourmet meals that could happily rely in large part on high-quality tinned goods and convenience foods (Cromley 2010). Similar tropes emerged in the 1990s Australia (documented by Duruz 2001: 22) where the end of the kitchen was again heralded in popular discourse. Here, with the development of houses and apartments appealing to Generation X lifestyles, it was suggested that breakfast was the only meal that would actually be prepared at home. Once again, in the early twenty-first century, designs for kitchenless (or minimally provided) houses or flats have emerged in various guises. The way, for instance, that kitchen space now flows into open plan living and eating areas of the house, and the advent of kitchen technologies that support family members to dispense with cooking and simply heat their own food, could allow these dwellings to be viewed as essentially kitchenless (Floyd 2004: 62).

More formally, in contemporary design practice, kitchenless dwellings and housing with minimal kitchen provision exist at both ends of the housing spectrum. At one extreme, these are an aspect of the inadequacy of the worst-quality dwelling spaces, and, at the other, they represent design forms intended to attract inner-city aspirational singles who will dine out or only 'cook' ready meals at home. Not only does the annual IKEA catalogue generally feature furniture and appliances for such nominal kitchen provision (although still modish for those with more symbolic than financial capital), at the more elite consumption level represented by the downtown kitchenless apartment, the focus is on demonstrations of stylishness where home cooking is no longer a valued or assumed life skill, or a required element of domestic life. While recent plans for apartments where there is no kitchen at all match these alterations in lifestyles for some modern urban dwellers, ironically, these same trends may be driving the development of kitchen-'styled' spaces outside the domestic domain, as is further considered in Chapter 6.

Of course, developments in kitchen design, and meanings attached to them, are geographically and culturally located, as global surveys of the design of 'the cook's room' demonstrate (Davidson 1981, 1991). Italian immigrant women in post-war United States and Canada, for instance, often organized two kitchens for themselves in their new homes: one upstairs for the display of success in their adopted lives; the other larger and more traditionally configured in the basement for the real business of cooking and feeling completely at home (Pascali 2006; DeSalvo and Giunta 2003). In post-Second World War Western Australia, too, for immigrant women the 'kitchen as home manifested in particular practices; it was a place where myriad social relations were enacted; it was a place of work; and a place of comfort' (Supski 2006: 133). Kitchen space was shaped according to these women's varying notions of domesticity and efficiency, including prepared food at a central table, in direct contravention of the 'work triangle' design norm (Supski 2006: 133). Although the kitchen's role in speaking of home might remain similarly important, spatially, the situation could be quite different in non-Western dwellings, especially those in hot, humid climates, where the space for cooking may be much more amorphous in terms of fixed wall structures. A study of kitchen spaces in Hanoi demonstrates that while a cooking stove is a fixture in the area for food preparation, there may not necessarily be a walled room around it (Nystrom 1994). Moreover, issues including poverty also tend to intersect with design. Physical dangers to health that have been reduced or eliminated in Western kitchens may still be present, with women and children disproportionately killed by smoke in badly ventilated kitchens while chronic obstructive lung diseases is a common hazard exacerbated by poor-quality cooking stoves (Bruce et al. 1998; Warwick and Doig 2004; Bailis et al. 2009).

Across the twentieth century, a recurrent theme in the intersection of the design, architecture and engineering of kitchens has been the conflict between designers' technocratic, modernist assumptions and the unruly tastes and cultural preferences of kitchen users themselves. The development of a 'consumption junction', that is, 'the place and time at which the consumer makes choices between competing technologies' has reflected kitchen users' attempts to directly influence kitchen design (Cowan 1987). It has equally demonstrated the strong resistance to these challenges by architects and engineers who have sought to exclude users' perspectives and irritating behavioural norms from the design process. Users could, of course, undercut architectural intentions as in Frankfurt where residents

> subverted and tinkered with the modern kitchen layouts they encountered when they moved into their new apartments. To the horror of modernist designers, users tried to squeeze in the dining tables and beds that modernist ideology had banished, to erase the functionalist inscription of the separation between living and eating by razing kitchen walls, and to fill their lean-and-clean and efficiently-inscribed workspace with knickknacks. (Oldenziel and Zachman 2009: 14, based on Hessler 2009: 177)

This resistance to modernist urbanism was occurring elsewhere too, as users attempted to return to the multi-purpose nature of kitchen space that had preceded the rational kitchen (Wright 1981: 124). For example, Finnish kitchens from the 1930s to 1950s changed in response to the resistance of Finns to the technocratic imposition of the kitchen as small workspace. Finnish cultural traditions of eating in the kitchen were so deeply embedded that design for more space for dining in the kitchen was judged a necessary response to these persistent social preferences (Saarikangasa 2006). In the United Kingdom, too, kitchen users refused to behave in the 'rational' way that proponents required of them in their new, efficiently planned kitchens. In the Kensal House flats (for which Elizabeth Denby helped design kitchens), it was reported that the women residents were 'often unhappy with the small size of the kitchens they came to inhabit; and in spite of their unsuitability for the purpose, the kitchen remained the place where the family would sit to eat meals even if this meant perching children on top of work surfaces and the cooker' (Llewellyn 2004: 48). In Frankfurt, too, 'despite the resistance and complaints of inhabitants, the architects failed to admit that something should or could be changed in the kitchen. Instead of changing the design, they wanted to change women's routines, and they were convinced that what they did was best for the tenants' (Hessler 2009: 177).

Class as well as gender issues were clearly evident in this contested transformation of kitchen space. In Belgium's 1930s experience of the 'rational kitchen', it was within architectural discourse and among bourgeois and middle-class women that these ideas were received most positively (Van Caudenberg and Heynen 2004: 24). Belgium's working-class and countrywomen, meanwhile, proved less receptive, continuing to prefer a living-kitchen space in which family eating, socializing and relaxing together could all take place in one large room. Intriguingly, socialist women's groups took a similar design perspective because, in keeping with an egalitarian worldview, a larger space allowed each family member to get on with their own interests without disturbing one another. The Belgian modernist architect Angeline Japsenne became persuaded of the importance of respecting these cultural preferences and began designing such a family kitchen in her 1939 plans for a working-class house (Van Caudenberg and Heynen 2004: 24). However, the mainstream architectural view was in support of the superiority of the rational kitchen, if women would only realize it (Van Caudenberg and Heynen 2004: 33). The architect De Konnick wrote at the time: 'The modern architects stress unanimously that, without an appropriate education [for the housewives], their efforts to realize a better use of the rooms and materials and finally to obtain a maximum of comfort in the house could remain sterile' (Van Caudenberg and Heynen 2004: 33).

House designs in a number of countries by the late 1930s and early post-war period began to show kitchens that flowed more openly into living and dining spaces. The breakfast bar typology derived from the architecture of the American diner began to be designed-in during the 1940s (Davey 2008: 108) and while kitchen workspace might still reflect assumptions about the 'work triangle', this formerly closed-off space now started again to be physically reconnected to living areas, as in Allmon Fordyce's Living-Kitchen, shown at New York's 1939 World's Fair (discussed below in relation to the evolution of design for dining space). Going somewhat further, Nelson and Wright (1945: 71–2) in *Tomorrow's House* challenged some of the arguments of the 'efficiency boys' and argued rather that

> [t]he kitchen cannot be a small room. It must be a big room – possibly the biggest in the whole house. It should contain all the cooking facilities, all the laundry

equipment, probably the heater, and certainly necessary space and facilities for family meals, and even meals when guests are present ... the housewife spends a disproportionate amount of her time working around the kitchen, and there is every reason why this room should be designed to be a completely livable, as well as workable, interior.

In the United Kingdom, the early post-war *Dudley Report into Design of Dwellings* concluded that there must either be space provided to eat in kitchens, or kitchens must be closely connected to dining space (Ravetz 1989: 199). Thus while *Planning Our New Homes* (1945) from the Scottish Housing Advisory Committee continued to make a clear demarcation between the 'living unit' and the kitchen 'work unit' considered proper for the post-war houses to come, greater connectivity between these units was beginning to be discussed. In a reference aimed at homemakers in the private market, *Modern Homes Illustrated* (Yerbury 1947: 24) shows plans from the United Kingdom in which combined living room kitchens, but more predominantly, modern open plan living rooms, are presented as aspirational spatial models. In a plan by Ernö Goldfinger for a 'modern small house ... here again is an open plan with a generous living room, a dining alcove adjacent to the compact working kitchen'. Referencing Scandinavian, Czechoslovakian and German house plans of the mid-century period, *Modern Homes Illustrated* similarly shows at once some loosening up of the design categories of living and dining room, reflecting a desire for a more relaxed and open architectural arrangement, while tiny, cramped kitchens remain to one side (Yerbury 1947: 24). Whitehand and Carr (2001: 135), reporting on morphological work in the United Kingdom, document the playing out of these changes in the 1945–93 period. In their sample of some forty houses studied, most kitchens no longer had their butler's sink, while dressers and larders were also disappearing.

Considered in urbanism terms, the shift to a more open kitchen design arrangement was a profound alteration from the closed, mechanistic kitchen that had gone before, yet also reflected changing notions of modernity in housing design (Attfield 1999). In North America and elsewhere, an alteration in both women's role and the domestic economy is argued to explain this dramatic spatial shift in both architect-designed and other dwellings (Cromley 1996). In Australia too, similar design changes to kitchen space and shifts in conceptualizing women's domestic labour were underway and were strongly linked. While women entered paid work in increasing numbers during and after the Second World War 'the answer offered by *The Australian Home Beautiful* [magazine] to the double work shift was not to redistribute domestic labour, but to redesign the space in which it occurred' (Johnson 2006: 128).

Like cities, domestic interiors were being planned using zoning principles, with an 'activity zone' emerging (Wright 1981: 254) in which the 'food axis' of the house was transformed (Cromley 1996: 8). Yet, paradoxically, the kitchen as a work zone started to break down and merge into a larger kitchen-dining-living space (Cromley 1996: 8). A number of wider trends, including the advent of products to deal effectively with cleaning and kitchen odours, technological advances in kitchen appliances and the increasing availability of ready prepared foods, meant that working in the kitchen, 'once dirty and smelly, was reinterpreted as a pleasurable and public activity, linked to the pleasures of dining and sociability. Women's housekeeping labour was brought out into the open as part of the general activity of the social zone in a house' (Cromley 1996: 19). Householders began to employ a variety of strategies to reconfigure separated kitchen and dining spaces and activities, including physically merging food service and social areas by taking down partitions and walls (Cromley 1996: 18). As noted in the British *Modern Homes Illustrated* (1947) this made female multi-tasking easier, with the mother's gaze employed rather in the style of Foucault's panopticon:

> The line of demarcation between the working areas and the living areas of the house is not so rigid. Cheap and efficient machinery for air extraction means that a solid barrier between the remainder of the house is no longer essential; and the housewife preparing her meals and doing her household chores can keep a maternal eye on the children who may be playing in the dining alcove or living-room.

Yet this integration was not completely satisfactory to all households. In post-war New Town housing in the United Kingdom, residents found that the integrated kitchen-dining room designed in to their dwellings created its own shared-use difficulties, and one of the first modifications householders made was to build in walls to separate off open plan kitchens from living spaces, despite the very small spaces that resulted (Attfield 1989: 218). The kitchen portion of these spaces did not necessarily get any bigger through these design changes. Illustrations and text in contemporaneous home magazines and architectural journals show that where space was allowed for eating in the kitchen, a range of configurations of kitchen counters, benches, stools and fixed tabletops spatially marked these areas as informal eating space for family and close friends, and these zones were often counterpointed by a more spacious, formal dining area, of which more later in this chapter. Perhaps influenced by Marcel Breuer's 1949 house design commissioned by the Museum of Modern Art, design for the work-triangle-based 'control centre kitchen' of the 1950s opened up the space to enable the housewife to both undertake domestic tasks and maintain visual oversight of children at the same time (Cromley 2010: 187). With reference to contemporaneous Australian house plans, Johnson (2006) argues that this open plan space allowed women to direct their gaze not only to living and dining space, but also towards the bedrooms, hall, front door and carport, to manage familial relationships and be watched – and judged – in the performance of this very traditional food-related role.

In older dwellings, meanwhile, it was often basement space formerly given over to the humblest domestic activities that received the most kitchen-related retrofitting attention. For example, in her study of immigrants' food and design practices in Toronto, Montreal and New York, Pascali (2006: 686) explains that for Italian families buying old, unrenovated houses, 'the basement kitchen developed from the need to accommodate habitual, extended family gatherings as basements provided more space for big groups'. Similarly, Cromley (2010: 211) reports on research into the kitchen practices of Indian Muslim families in Buffalo who removed walls between kitchens and dining rooms to fit in larger dining groups. In both cases these changes to increase opportunities for kitchen-centred conviviality contrasted with wider norms and practices in kitchen design.

ASK THE EXPERTS

These design shifts act to contextualize designers' advice to kitchen users about the spatial arrangements that might best support kitchens as appropriate food spaces. These can be considered a kind of extension of the proposals for convivial tablescapes discussed earlier, reflecting a largely negative critique of the social effects of the rational kitchen, yet often maintaining assumptions about the nature of women's domestic roles in which the kitchen should be isolated from the rest of the house, and the women a kind of servant in their own family (Alexander et al. 1977: 661). In keeping with their 'kitchen-living' view of this space, one that seems rather counter-culture Seventies inflected, Alexander et al. (1977: 663) argue that what is needed (to paraphrase) is to make the kitchen big, bright and comfortable enough to include family space; locate it near the centre of the house; include a large table in the middle (surrounded by both hard and soft chairs); and provide a stove, counters and a sink round the edge of the space. One source for this approach is the design of the large country kitchen, with its central table and collection of chairs surrounded by preparation spaces edging the walls. This is a design trope that suggests a longing for what might be understood as a more innocent time, and to kitchen shaping and use within a strong anti-urban tradition (Wilson 1991; Duruz 2001). As noted earlier in the chapter, nostalgia for the country kitchen is seen to reflect cultural remembrance of 'home, nurturing and food' (Duruz 2001: 26) while Lyon and Colquhoun (1999) point to even more recent and ironic sources in visual culture for nostalgia in kitchen design. Yet at the level of everyday practice, rational kitchen design assumptions are still embedded in normative advice to those remaking their kitchen spaces in the early twenty-first century, as well as being ubiquitous in the design of actual, contemporary house plans. 'Much is spoken of the "golden triangle" principle of kitchen design and there are many variations on the theme, but the over-riding common sense principle is that the most used equipment and work areas should be grouped together to allow the user to operate effectively and efficiently while cooking' (Homes, Design, Kitchens undated, early 2010s).

Just as the logic of mass-produced industrialization took over more of everyday urbanism (and kitchen objects proliferated) in contemporary kitchen design, another significant development has been the advent of more automated systems, appliances and infrastructure, through the development of domestic interactive technologies (Hughes et al. 2000). Alongside these shifts, advice continues to be offered by knowledgeable cooks about the best design solutions for kitchen space (Whitaker et al. 2001). There remains a fascination with the creativity of cooking as expressed through these experts' skills (Lyon et al. 2003: 174). In this context, the notion of cooking from scratch has become a contested area in sociological considerations of kitchen space, as well as at the level of popular discourse in relation to actual practice (Caraher et al. 1999). It is argued that the 'lack of definition perpetuates taken-for-granted assumptions about the nature and scope of the skills in question' (Lyon et al. 2003: 168) and there have thus been calls for more grounded sociological work to understand the relationship between technological change and 'doing cooking' (Short 2007: 557).

In fact, the position seems somewhat complicated and even paradoxical in its spatial effects. A range of research reported by Bisogni et al. (2006: 218) certainly supports the view that more food is made and consumed outside the home, and there is a lower frequency of meals shared as a family, although why the supposed decline in family meals is a focus for research concern has been posited as a kind of panic about the nature of societal change in foodways (Murcott 1997: 33). Reports are emerging of widespread malnutrition among elderly people (BBC 2013), a result, it is argued, of more individuals living alone, and the pleasure of sharing meals being less available to those who wish do so. The undoubted popularity of gastronomic advice from celebrity chefs (Mitchell 2010) coexists with both a tranche of home cooks who display conspicuous competence (Visser 1987) and the development of the previously mentioned 'culinary transition' (Lang and Caraher 2001: 2) in which the way food is prepared, served and eaten is altering at a societal level. This is being driven by a number of factors including 'changing work roles, family organization, household structures, lifestyles, and food systems' (Bisogni et al. 2006, based on Cullen 1961) which have been explored by a number of researchers (including Riley 1994; Poulain 2002).

Many households now rely on food acquisition and assembly in which pre-prepared components are brought together at home to construct a meal, implying a loss of cooking skills in the process (Lang and Caraher 2001: 2) and a reliance on industrialized foods of very poor quality (Blythman 2006). While some recent research from British Columbia discerns an ongoing interest in home cooking (Simmons and Chapman 2012), other findings from Scotland provide cold comfort by suggesting that although cooking skills have atrophied, especially among younger population cohorts, we may have also substantially overestimated past cooking capacities in both extent and diversity terms (Lyon et al. 2003: 167). For them, the 'real tragedy would be loss of interest in cooking, and there is evidence to suggest that is not the case' (Lyon et al. 2003: 167). In somewhat sideways fashion it is argued that the decline of home cooking may actually allow families more time for other forms of socializing, again making this not necessarily a bad thing for family connectedness (Short 2003, 2007). It also seems that while cooking skills within the kitchen are declining among certain groups, ironically, these shifts may also rely on a greater capacity to undertake complex food planning and grocery buying practices outside it (Engler-Stringer 2010: 4).

In the context of an increasingly commercialized kitchen culture subject to this culinary transition, kitchen users are at risk of being alienated from the creativity of skilled cooking as one of the core practices of everyday life, and instead reduced to the position of spectators watching machines function in their place (Giard 1998: 212), or, if not machines, then celebrity chefs (Rousseau 2012). At the same time, industrial designers of kitchen objects and systems have sought to insert themselves into the kitchen as the 'sustainable link between products and people'

(Sherwin et al. 1998: 53), arguing 'they can have influence over how people treat objects and artifacts' (Sherwin et al. 1998: 53). Design ideas for smart sinks and data walls forming 'the brain of the kitchen' (Sherwin et al. 1998: 53), for instance, not only seem to suggest a view that such technologies replace the need for kitchen skills, but also reflect a continued certainty by designers about the wisdom of imposing their own design solutions on others' kitchen spaces and practices.

A review of latter-twentieth and early-twenty-first-century house plans, meanwhile, demonstrates several persistent trends in kitchen design and placement in the dwelling space, as well as some newer shifts in this scale of urbanism. From around the mid-1970s, in mass house design (rather than the elite architectural level), kitchens were required to be both efficient workspaces and capable of living up to their marketing as the heart of the home, while the separate kitchen came to be seen as a problem requiring renovation (Cromley 2010: 208). By the 1980s and 1990s, larger kitchen spaces were evident, but these still tended to have an elaborate workbench rather than a table at the centre, with this bench used to semi-separate the 'working' part of the kitchen from informal table-based eating space seen as 'social'. Such kitchens were routinely being retrofitted into existing older houses as these had a stock of more generously proportioned rooms and exterior spaces (than did new housing), allowing them to be so reconfigured. The substantial kitchen-dining-living spaces (the 'extension') created by these interventions increasingly replaced outdated, cramped combinations of kitchen, scullery and pantry, as well as sometimes superseding tightly configured kitchens of rational design. Open plan space, in which the kitchen, dining and living space became one large ensemble (in the United States of the 1990s sometimes known as 'the great room') emerged as a standard design element in aspirational home building and renovation, often with a consciously modernist styling that signalled modernity and thus cultural capital. Clearly, at least some of these transformed kitchens became designed sites for displays of conspicuous consumption that would perhaps be rejected by those with a more assured habitus. For those aspiring to reproduce expert templates for kitchen design, books of 'great kitchens' offer 'design ideas from America's top chefs' on how to shape kitchen space and model food consumption within it (Whitaker et al. 2001).

FOOD AT HOME

As Chevalier (1998: 48) notes, 'inside the house, two spaces are related; the kitchen where "nature" is transformed, and the dining room where it is consumed'. Thus the discussion of table and kitchen spaces acts to contextualize the evolution of dining rooms as transformative food spaces in the domestic domain. As we have seen in the earlier discussion about the table (or its spatial equivalent), dining spaces have over the long term been important domestic foodscapes. Equally, a central theme in the development of the dwelling as an aspect of urbanism has been designers' various attempts to remove eating space from kitchens into separate dining rooms and users' equally determined efforts to reintegrate cooking and eating areas. Ironically, once dining space became physically reconnected to kitchen and living spaces, shifts in food preparation automation and commercialization again started to remove cooking from this combined space.

As can be seen from the analysis of pre-modern dining rooms, at the elite level these often showed a layering of exclusion from the table, broadly but not solely along lines of gender, race and class, with design arrangements for dining space reflecting this separation of servers and diners (Grover 1987; Clark 1987). Just as kitchens could be seen as feminine, with delicate and even playful decorative features to match, dining rooms could be understood as masculine spaces, with heavy, sombre furniture reflecting a patriarchal position in the house as the site of male relaxation and female service (Ponsonby 2003: 201). In late-nineteenth- and early-twentieth-century house designs, as the role of server and diner coalesced (Lupton and Miller 1992), designers sought to develop or maintain a functional separation between women's domestic work and the family's social areas, with dining space placed firmly within the latter category.

As noted above, there was strong resistance from designers of this period to accepting the kitchen as the heart of the home, and they actively sought to design out 'the worryingly working class tendency' of dining in the kitchen (Jackson 1985: 25). For Raymond Unwin, concerned with the design of small houses for the middle classes with few servants, and influenced by the thinking of William Morris and the Arts and Crafts movement, 'spectacular living areas' were advocated instead (Jackson 1985: 26). In these house designs, a traditional hall and inglenook opened out to occupy a large proportion of the house, 'firmly emphasizing the crucial importance of the social living area with all its opportunities for family conversation and recreation' (Jackson 1985: 26). As Cromley (2010: 164) similarly reports from the US experience of this time, poorer households were more likely to prefer to eat in their kitchens while the middle class attached their dining to living-room space, as 'historically clean reception or entertaining space'.

Yet, by the 1920s, in the United States, the dining room started to be identified as wasted space that was used for only a few hours a day, and thus at the budget end of the housing market constituting an unnecessary expense (Cromley 2010: 162). The disappearance of butlers' pantries, the building in of 'efficient' space-saving devices such as the dining or breakfast nook in the kitchen, hidden china storage cabinets and kitchenettes in apartments and hotels were all design results that developed alongside changing social attitudes to the place of dining at this period (Cromley 2010: 163). With the loss of the butler's pantry that had been used to control movement back and forth between kitchen and dining room, as well as limiting smells and noise from the kitchen (Cromley 2010: 166), early-twentieth-century house plans in the United States often showed the dining room separated from the kitchen workspace by swing doors or a service hatch. This facilitated movement between dining room and kitchen, but emphasized too the physical separation of functions between food preparation and cooking, on the one hand, and eating, on the other. Gender roles similarly remained undisturbed. The service opening between kitchen and dining room 'puts the percolator and toaster within reach of the table, giving the master of the house something to do while he is waiting for his eggs' (Nelson and Wright 1945: 87).

At the same time, in the early post-war period, profound social changes began to be reflected in the evolution of dining space placement and design. The question 'where shall we eat?' began to occupy the minds of designers who pointed out that the traditional arrangement of kitchen and dining space needed to be rethought (Nelson and Wright 1945: 87). Thirty or forty years previously, in middle-class houses there had been 'a large kitchen with a table in it for the hired girl ... for dining, the family had a dining room' (Nelson and Wright 1945: 39). As this configuration of dining space was overtaken by the socially and economically driven design changes described here, families began to eat in the breakfast nook or alcove because it meant 'less work for the housewife' (Nelson and Wright 1945: 40). The dining room became 'a kind of architectural vermiform appendix, which was kept because the operation of removing it had not yet become fashionable' (Nelson and Wright 1945: 40). During and just after the Second World War, architectural interest in living-dining rooms was in part a response to limited incomes, but this makeshift design solution to resource constraints was short lived. Both architects and householders started to see the kitchen as again an appropriate place to dine, but one that needed design attention to make it 'the warmest, most cheerful room in the house' (Nelson and Wright 1945: 41) rather than ruthlessly efficient and soullessly technocratic.

A new kind of kitchen space, designed by Allmon Fordyce for the New York World's Fair of 1939, showed to an admiring crowd a revised version of an essentially traditional configuration: the kitchen-living room (Cromley 1996). Foreshadowing a common design arrangement today, this room contained both a dining table and a kitchen counter (with cupboards and shelves above) dividing the space between cooking and dining. Appliances were not in clinical white, but in attractive colours and wood finishes. While the kitchen-living room brought its own technical issues of appropriate lighting, as well as dealing effectively with cooking odours, noise and mess, design advances were proposed to deal with these. The kitchen-living room began to be seen within upper-middle-class or bohemian home kitchens, prefiguring more universal application, as a site for friends to dine in the kitchen and even help with the washing up. 'People saw in this design not just a good-looking kitchen, but a brand-new new way to live in a house ... the kitchen-living room not only lightened the burden of housework, but it was good looking enough for guests' (Nelson and Wright 1945: 42).

As discussed earlier in the chapter, open plan arrangements of kitchen, dining and living space became increasingly ubiquitous in both architect-designed and the much larger volume of mass market house plans in the post-war period up until the 1970s. Looking specifically at the dining aspect of this change, it was clear that with more relaxed social mores, middle-class households no longer needed to show distinction in Bourdieu's (1984) terms by separating out dining from kitchen space: rather, such distinction was now expressed by dining in the kitchen. Yet despite its design dominance, just as the opened-out kitchen had caused some issues for users, householders sometimes found the 'merged food functions' of integrated dining space problematic, including in relation to privacy, arrangements for party dining and the failure to segregate adults' and children's activities (Cromley 2010: 202).

All the same, integrated dining space continued to be seen as stylish, and in loft living spaces created in downtown New York and elsewhere (Zukin 1982) dining in the style of Allmon Fordyce's living-kitchen was pushed to its logical extreme (Cromley 2010: 202). From the mid-1970s, as social life became more informal, dining and entertaining was mostly located in the kitchen-family room, and this design approach to dining space was appropriated by mass market house builders. It is worth remembering that the preference for a convivial kitchen-dining space had been explicit in the resistance to the rational kitchen at the level of women's input into design discussion and practice from an even earlier period (Van Caudenberg and Heynen 2004), and the living-kitchen refused to entirely disappear as dining space despite the best endeavours of modernist rationality.

In home dining in the early twenty-first century, the trends noted earlier in relation to food's commercialization and automation continue to reflect structural shifts in social and economic life impinging on domestic dining design. One of the most intriguing spatial changes to dining at home is that in the world of snacking and grazing in front of the computer, tablet or television screen, domestic eating sites have dispersed, while the technology of food appliances and preparation space has become most concentrated in a single small area of the house (Cromley 2010: 204) to which prepared food is brought in. Culinary transition examples which blur the economic and cultural boundaries of private food space are diverse. Yasmeen (1996), for instance, refers to the 'plastic bag housewives' of Bangkok who collect prepared meals in plastic containers on the way home from work, to which they only have to add home-cooked rice (a practice Yasmeen connects in this and later research to Thai women's very high incorporation into the paid labour force, and thus double shift). Elsewhere, with the near ubiquity of home dining based on fast food, ready meals and 'cheating' short cuts marketed as benefits to hard-pressed households (Short 2007), food is often eaten separately in a range of home locations, with individuals consuming alone, not dining together in a dining room or at a dining table.

Yet, as Valentine (2001) argues, the contested spaces and practices of 'eating in' are strong markers for the formation and expression of personal and familial identity. Mass market house plans now offer often baroque assemblages of spaces open to the kitchen, with possible dining sites called variously 'meals', 'dining', 'family', 'family/meals', 'activity', 'great room', and 'living/dining', among others (Sunday Mail 2013). With the advent of postmodernity in domestic design (Dovey 1994; Johnson 2006), in the early twenty-first century, the primary sites for dining at home have extended to an enfilade of meal spaces. Dining may occur at an island or peninsula bench for snack meals and at the table next door for those fewer times when people do actually sit down to eat a meal together, in 'family-meals' areas, 'living-meals' spaces, or even 'home theatres', among others. The descriptions of these food-related spaces found in the newspaper real estate lift-outs and home supplements reflect these spatial arrangements and play on similar high-specification kitchen design features and assumptions.

With this spatial dispersal of domestic food space, these design shifts can be seen as at once upsetting the conviviality of the table but offering the arguable benefit of speed in a context of the perennial problem of scheduling food into everyday life, despite ambivalence about such prepared foods (Warde 1999: 518). The dispersed nature in time and space of dining in, though, may present particular issues for women, as they tend to retain primary responsibility for family health and ensuring convivial familial relations. Through the combined effects of food commercialization, domestic deskilling and spatial change, women no longer command the physical or psychological space of family dining, and may be able to exert this authority only over their own bodies.

The decline of the formal dining room also begs the question whether dining rooms as separately configured spaces have become unconvivial in any case. As the kitchen-dining-living ensemble has gained ascendency, dining formally in a dining room, except on special occasions demanding more ceremonial eating arrangements, is now seen as a stuffy, leftover kind of activity, no longer congruent with relaxed modern lifestyles. As an aspect of practising modernity, it is perhaps unsurprising that this perspective was reflected in exhibition houses of the Bauhaus in the early 1920s (Bullock 1988: 183), was explored as a new way of thinking about dining space in design literature and was expressed in approaches to architectural practice in the United States as early as the 1930s and 1940s (Nelson and Wright 1945). Ironically, as so-called McMansions substantially inflated housing floor areas in the latter part of the twentieth century, more formal dining rooms were also returned to the house plan mix, mirroring informal dining spaces and suggesting a continued desire to display a form of food space of a different, less relaxed kind. In Bourdieu's terms these might be thought to speak of a rather uncertain habitus rather than the reverse and to a less rather than more convivial domestic food space configuration.

DOMESTIC FOOD SPACE IN REVIEW

As this chapter has shown, the table or its spatial equivalent has through settlement history offered a design locus for convivial food practices and for modelling inequality in food terms. While tablescapes have been analysed in technical terms as sites for encouraging or regulating consumption in both private and public domains, their design configurations have shown much broader meanings relating to cultural and familial practices. Kitchens, too, have been sites for contested meanings about food, and the subject of competing design prescriptions that reflect contradictory attitudes to gender and class in particular at the scale of domestic urbanism. Similarly, a theme of contestation between expert' and users' perspectives has run through the design history of the twentieth-century kitchen, that has been a de facto argument about the nature of conviviality.

The reintegration of kitchen food space into the 'social' areas of the home has been, ironically, accompanied by an argued decline in both cooking skills and the increasing commodification of consumption. This has occurred in ways that closely tie apparently private spaces into wider scales of activity and spatial design (as in the connections to industrialized food provision) and may have negative effects on convivial design. The location and nature of dining, too, has been reflected in radical changes to this food space as mores change and house designs try to keep up. In addition to the extensive alterations to kitchen-living-dining arrangements, domestic food space has been transformed to include outdoor dining zones that flow from the dwelling interior. In the next chapter, this exterior portion of the domestic realm is explored in urbanism terms through a discussion of the shaping of private gardens for food and its urbanism implications.

CHAPTER TWO

The Garden and Gastronomy

INTRODUCTION

While making no claims for completeness, this chapter sketches out some of the designed roles private gardens have played in urban food productivity in the past – their widespread loss of this status in twentieth- and twenty-first-century cities as gardens became predominantly food consumption sites and aspects of the revival of domestic gardens for informal food production. Conceptually, the chapter reflects urban design arguments that the wider context for growing food at home is a basic human desire for proximity to green things, trees and water which is deeply felt across cultures and times and is part of the magic of cities (Alexander et al. 1977: 798). Conviviality, it follows, is also dependent on the quality of urban green spaces, including domestic gardens, as it is in these spaces that connections are made to the natural order. That relationship is crucial to physical and psychological well-being and can be supported by urban design, planning and architecture (Alexander et al. 1977: 801, 819). While the more collective resurgence of urban agriculture as an element in convivial public green space is considered in Chapter 6, historic and more contemporary examples in this chapter build on the analysis of the design of interior dwelling spaces for food and eating discussed in Chapter 1. The focus is on the increasing centrality of food-related spaces to garden design, exploring different ways that food and gardens interconnect, including immigrant gardening experiences that have helped maintain design for a productive urban landscape. Above all, this chapter explores how private gardens as spaces for food and eating can reflect convivial urban design principles and practices that contribute to sustainable urbanism.

THE PRODUCTIVE GARDEN SITUATED

In exploring the spatiality of food, it is clear that over the very long term the domestic garden has both reflected and shaped many food relationships in design terms. The *hortus conclusus* or contained domestic garden has been an important source of space for growing and preparing food since the earliest indications of agriculture and settlement began in East Asia around 10,000 BC (Aben and de Wit 1999: 21). Garden history encompasses the extraordinary history of paradise gardens in the Islamic world (Ruggles 2008), alongside modest kitchen gardens and walled gardens historically situated in towns from the earliest urban times (Campbell 2005: 18). Egyptian sources show enclosed, walled garden spaces that functioned as outdoor rooms given over to fruit and vegetable growing from around

6,000 BC (Turner 2005: 25). In Persia, meanwhile, walled gardens or *pairidaeze* from the time of Darius used hydraulics to obtain scarce water (Mougeot 1994: 2). Early examples show complex irrigation systems supporting wadi farming and arboriculture (Ruggles 2008: 20–3), while in water-scarce areas like Persia, irrigated channels were used to water productive plants and structure garden space (Turner 2005: 86). Although space for private gardens in ancient Greek towns was highly constrained, archaeological sources and other accounts demonstrate that in the Classical period horticultural spaces behind town houses known as *xystus* were used to grow vegetables and flowers (Turner 2005: 74), and a rich belt of suburban vegetable gardens and fruit trees watered by nearby rivers and streams were clustered around the polis (Carroll-Spillecke 1992).

The private garden was an integral part of Roman domestic space, and frugality, simplicity and productivity were highly valued (Henderson 2004). Cato dwelt on the moral value of growing cabbages, noting that flowers should be planted only for religious rites, and described an ideal productive suburban garden (Carroll-Spillecke 1992). By the third-century BC even modest Roman houses had a small vegetable garden, the *hortus*, behind the house, with the presence of aqueducts made town garden watering feasible (Carroll-Spillecke 1992: 94). Like Cato's *hortus*, Pliny's garden was described as productive space, which was irrigated to grow vegetables and useful plants (Von Stackelberg 2009: 12). Columella's description of the range of vegetables and herbs grown in the Roman garden, meanwhile, included onions, cucumber, cabbage, lettuce, leeks, fennel, mint and dill, while Pliny added saffron crocus, sage, thyme and rosemary (Von Stackelberg 2009: 45). However, with increasing wealth and leisure Romans' garden tastes changed. Evidence from Pompeii (where about 500 town gardens offer a rich source of archaeological material) suggests that many houses had large gardens, some with vineyards, fruit trees and a rear courtyard or vegetable garden behind the house, while medicinal plants were grown in the more central and private peristyle (Amery and Curran 2002; Turner 2005). The elaborate gardens of Pompeii also demonstrate that for some, the garden was no longer 'the poor man's farm' as designated by Pliny, but could be the site of seductive luxury (Von Stackelberg 2009). As in the late twentieth and early twenty-first century, a distinct nostalgia for a lost, productive simplicity seems apparent in Roman accounts of a striving for balance between the needs of the body and the soul to be achieved through the domestic garden.

Islamic Spain, meanwhile, offers some of the best preserved more recent evidence of garden space for productive planting, although sources are clearer in relation to elite garden design than those for domestic production, lying within a paradise garden mode. Between the ninth and the eleventh centuries, various agricultural manuals and treatises identifying plant species were produced, including the Kitah al-filaha of Ibn Bassal which listed a considerable number of edible plants (Ruggles 2008: 33), while kitchen gardeners in the Islamic tradition advanced food production through their scientific innovations (Ruggles 2008: 33). The exotic plant varieties in Granada's Nasrid Generalife palace garden, for example, 'were a reflection of the rich diversity of crops for cultivation—sorghum, hard wheat, rice, banana, sugar cane, eggplant, cotton, artichokes, and varieties of citrus—that people enjoyed as a result of agricultural innovation and exchanges in the early Islamic world' (Ruggles 2008: 44).

Medieval garden evidence demonstrates the existence of both monastic cellerers' gardens for growing vegetables and medicinal plants as well as town gardens, although information about specific plantings within the latter is thin (Turner 2005: 134). By the ninth century, European sources suggest, in relation to domestic food production, raised garden beds for food crops were being employed (Turner 2005: 134). More commonly in towns of the medieval period, productive gardens of irregular shape defined by built form and incorporating raised beds were used intensively for productive purposes, while the *hofje* or small courts of medieval cities in the Netherlands could well house orchards and vegetable-growing space (Aben and de Wit 1999). The northern European *potager*, depicted in medieval woodcuts, illustrated gardens that were enclosed by walls of stone or wood, within which fruit trees were planted and espaliered along sunny, south-facing walls.

While its design origins lay in the humble, domestic garden, where the *potager* consisted of a modest array of vegetables grown round the house for use in the daily stew or hotch-potch (Tilleray 1995: 10), by the mid-seventeenth century vast aristocratic *potagers* at both Versailles and Villandry became the most widely copied gardens in France. A commercial gardening trade with an urban presence had in any case already become clear by the thirteenth century, with guilds for gardeners including *courtilliers* for courtyard gardens and *maraîchers* for vegetable cultivation, while productive town gardens of bourgeois families or burghers were also expanding by the thirteenth and fourteenth centuries (Jones 1997). Domestic gardens in medieval cities were sometimes a significant source of fresh vegetables: in certain locations the economies of growing locally or shipping food tipped towards the latter very early on (Sennett 1994). For example, by the twelfth century, inner Paris was importing many of its vegetables from considerable distances, as urban land became more valuable for building (Sennett 1994). Evidence from medieval France also demonstrates that urban food growing was profoundly tied to agricultural life and practices, and this was reflected in its urban design:

> Of necessity [towns] retained links with agriculture. They were embedded in a pastoral economy, and through their thoroughfares wandered cattle, sheep, poultry and pigs, the latter performing the very useful function of cleaning the streets. These stood in great need of cleaning, given the cramped conditions of urban life, the embryonic character of municipal authority and the absence of paving stones. Within their walls, and [close] outside them, the towns were responsible for vineyards, kitchen gardens and even fields. (Braudel 1986: 420, quoting Pitte 1983: 149)

In design terms the work of garden historians shows very clearly that the quality of spatial enclosure discussed in the book's Introduction has been central to garden spatiality over the long term, with kitchen gardens functioning as outdoor rooms defined by surrounding built form, or walls of trees, espaliers or hedges (Braudel 1986: 420, quoting Pitte 1983: 149). These private outdoor rooms have historically been places shaped not just for utilitarian purposes but also for pleasure (Campbell 1996: 11). In cities where space is at a premium, owning a garden for leisure has always been an expression of power, and the rich history of the domestic garden demonstrates its role in signifying wealth as well as offering productive space. In some places the constraints of location or climate have made urban gardens the exception, but the powerful have often used their privileged domestic gardening capacity to produce vegetables and fruits out of region and season. In Europe these activities were predominantly associated with great estates in the countryside, with their forcing houses for tropical fruits and increasingly vast extent (Campbell 2005; Petherick and Eclare 2006), as the fascinating history of domestic pineapple growing demonstrates (Grigson 1982: 335).

Urban design that integrated food growing also produced profound social benefits in the early modern period. In the Swedish town of Uppsala, for example, the availability of locally grown potatoes and grains was connected to social welfare systems that made life possible for urban widows (Björklund 2010). In the early modern London of the sixteenth century, meanwhile, Tressell's surveys showed, 'suburban' gardens in places like Southwark and Bishopsgate 'could reach considerable lengths and were no doubt a source of income from market gardening' (Schofield 1987: 27), and the map record tends to morphologically support this contention. By the Georgian period, the ordinary town garden at the back of the terraced house was developing as a site for a large variety of edible plants including fruits, vegetables and flowers that may have been mixed in a rather indiscriminate way (Cruickshank 1990: 201). Plantings of apple, pear and mulberry trees, as well as vines, were reported in inner urban areas of Bishopsgate and Ludgate, and nectarine trees in Gower Street (Cruickshank 1990: 201).

Seventeenth- and eighteenth-century accounts in the United Kingdom showed that village and town gardens for food were also the subject of conscious consideration that explored the kinds of productive plants thought suitable for different classes. One writer declared that the village cottager's garden should be large enough to allow the family to eat healthily, although with a somewhat meagre range of vegetables, suitable to their lowly station in life. Similarly, vegetable gardening was an urban domestic concern in the American colonial period. Around the middle of the eighteenth century, 'if the housewife had been following standard practice on these matters, the herbs and vegetables that were added to the stew would have come from a kitchen garden that she had planted and tended herself' (Cowan 1983: 23). Particular urban conditions influenced housing typologies, and thus the kind of food production undertaken. In the side courtyard houses that developed in New Orleans on long, narrow plots configured to deal with the threat of fire, a closely connected kitchen and paved outdoor courtyard space was ubiquitous and sometimes included areas for raising poultry and a minimal vegetable garden (Keister 2005: 22). In the New World of Australia, in the early nineteenth century, fruit trees and vegetables might be displayed in the front and sides of the cottage, a practice that can be seen reflected in the colonial art of the period, such as an 1835 painting by the artist, John Glover, of his extensive domestic garden and vegetable plot at Mills Plains in Van Diemen's Land (present-day Tasmania).

By the mid-nineteenth century, John Claudius Loudon was writing about town gardens and food growing in his *Gardener's Magazine* (Cruickshank 1990: 201). Certain urban settings became centres of super-productivity in vegetable growing, including the Marais in Paris, and more broadly, domestic gardens of the late nineteenth and early twentieth century were extremely productive, for reasons of economic necessity rather than as concerns in relation to nutrition that animate contemporary debates and exhortations about fruit and vegetable consumption. In Scotland in the first part of the twentieth century, for instance, those with access to vegetable gardens or allotments may have made better meals 'almost by accident' in a context of the dietary degradation of the industrial town. This was marked by the passing of staple foodstuffs including porridge, salt herring and potatoes and their replacement by products that were bought in shops and traders' vans, such as bread, tinned foods, tea and sweets (Lyon et al. 2003: 169).

The productive domestic gardens of the settlers of the New World, meanwhile, were dense with cultural meaning, spatially constituting 'a convenient package for a bundle of virtues tied to the social and economic circumstances of the colonies and enacted "in miniature" in food-producing backyards' (Gaynor 2004: 242). Vegetable gardening at home was commonplace in such cities of the interwar period, and its spatial design had particular, shared expressions. In Australia, for example, vegetables were banished to the backyard of the suburban villa, making it a productive area encompassing vegetable garden, a fowl run for 'chooks', vines for grapes, and fruit trees from whose crops dried fruit and preserves were made (Duruz 1994; Head and Muir 2007: 18; Hall 2010). Growing one's own food was equally represented as the domain of the male breadwinner (Johnson 2000b: 101). Research in Perth and Melbourne demonstrates that domestic scale poultry raising, dairying and goat keeping, along with vegetable gardening, were primarily portrayed as masculine spaces and activities, although cookbooks of the late 1930s noted the usefulness to housewives of home-grown foodstuffs normatively produced by their menfolk (Gaynor 2004).

At times of particular hardship, home vegetable growing became even more ubiquitous, although gender stereotypes in relation to that production remained intransigent (Kirby 2012). Driven by propaganda campaigns and the exigencies of war (Gaynor 2004: 251), in the United States, Victory Gardens were promoted as a patriotic necessity (Miller 2003) and, by 1943, there were 20 million of them in which were grown 40 per cent of America's vegetables (Mintz and Kellogg 1988: 160). In the United Kingdom both rationing (Zweiniger-Bargielowska 2000) and digging for victory (Kynaston 2010) were employed as necessary practices of wartime austerity, although the latter is now subject to critical reinterpretation (Bramall 2011; Ginn 2012). In Australia, the Melbourne University Social Survey of 1941–2 showed that home vegetable and fruit growing were widespread, with around 48 per cent of all households sampled growing some food, and as many as 88 per cent doing so in some suburbs (Gaynor 2004: 250).

While the locus of domestic food production was primarily within private gardens, allotments and vegetable plots elsewhere also played a part as a kind of spatial extension of productivity beyond but closely connected to that private space.

Although the design of allotments, summer gardens and community gardens is predominantly explored in detail in Chapter 6 as part of the discussion of green space mostly in the public domain; it is worth noting here the ways in which these food spaces have been designed, understood and promoted as spatial and social extensions of domestic gardens, or responses to the lack of such private productive space. Allotments and similar food spaces, including the working-class *jardins ouvriers* in the *banlieues* of French cities and towns with their *cabanons du dimanche* (Faure 1991: 197; Cabedoce 1991; Cabedoce and Pierson 1996) played an important part in the diet and sociability of many working- and lower-middle-class households during the interwar and wartime periods in the United Kingdom and the rest of Europe, for reasons of economic necessity (Lyons et al. 2003) rather than for nutritional, gastronomic or stylistic habitus reasons, as have become more common today (Clark and Clark 2011).

In the first half of the twentieth century, allotments or summer gardens were commonplace around Europe's towns and villages, often taking up leftover spaces and offering both productive and recreational food-growing opportunities that augmented meagre dwellings and domestic food resources. In Sweden, around 1900, a movement led by Anna Lindhagen founded a number of allotment garden areas which were aimed at working-class and low-income people to improve their living conditions and food, and thus it was reasoned, result in moral improvement (Nolin 2006: 122). In Helsinki, a city influenced by Garden City ideas, the allotment movement is argued to have arrived from Germany in the 1910s, and allotment garden areas were then established, particularly to deal with food shortages (Lento 2006: 198). In the United Kingdom, local authorities were required to provide allotments under the Allotment Act of 1918 (Reeder 2006: 36). Across Europe, and in some other countries, allotments remain an urban form evident to varying degrees in cities and towns today, gardened by both working- and middle-class families, as in Berlin's 80,000 *kleingärten* (Groening 2005: 22). These take a variety of forms:

> They are called kleingärten in Austria, Switzerland, and Germany, allotment gardens in the United Kingdom, ogródek dzialkowy in Poland, rodinná zahrádka in the Czech Republic, kiskertek in Hungary, volkstuin in the Netherlands and in Belgium, jardins ouvriers and jardins familiaux in France and Belgium, kolonihave in Denmark, kolonihage in Norway, kolonitraedgard in Sweden, siirtolapuutarhat in Finland, shimin-noen in Japan, community gardens in the USA (during the Second World War they were often referred to as victory gardens in the United States, Canada, and Australia). (Groening 2005: 22)

CONTEMPORARY PRACTICES IN DOMESTIC FOOD PRODUCTION

The private garden's design relationship to conviviality is rather a complex one. In contemporary practice private gardens as expressions of leisure and homemaking centred on consumption have become dominant: these are places where gender relations are played out, a search for paradise is undertaken and space for food production is constrained (Bhatti and Church 2000; Hewer 2003). Private gardens are also sites for conspicuous consumption and display connected to food, although they may also allow for more unmediated conviviality as people gather to eat in them. Despite the intimate and ancient domestic design interconnection of food and gardens, in more contemporary urban experience, vegetable gardens (and solar clothes drying) have been replaced by swimming pools, electric dryers and 'designer landscapes' (Dovey 1994: 137). While Hall (2010) laments, in an Australian context, their loss as part of the decline of the backyard, this spatial and cultural trend is evident in the representation of suburban housing design across the Western world, where the utilitarian garden given over to food productivity is seen to have lost its dominance in the urban realm. In this reading, domestic gardens broadly, although not uniformly or completely, moved from sites of production to consumption, reflecting a new mass consumption reality in which food goods could be cheaply bought rather than produced or made at home (Hewer 2003). In the light of such structural changes, gardens themselves shrank in size, with the combined effects of smaller blocks and larger houses further squeezing out potential food-growing spaces.

Conversely, a continuity of small-scale domestic food production has also been evident, running counter to this dominant trend (Gaynor 2004; Morrow 2011). Although gardens as sites for dining have come to dominate the use of outdoor domestic space and meaning for many in recent times, in some places, in some cultures and in some households, private gardens continue to be or are again designed to play a productive part in the urban food system. The food-growing activities of so-called urban peasants discussed in the next section offer examples of such productivity in practice, although their (contested) existence is argued to only have occurred in Australian cities in a particular historic era from the nineteenth to the middle of the twentieth centuries (Mullins and Kynaston 2000: 160). Notwithstanding an ongoing role in domestic food production for some, it is certainly the case that for many the lived experience of garden space in the post-Second World War era shifted considerably away from food production for use and a display of 'social rectitude' (Daniels and Kirkpatrick 2007: 314). The front garden was generally the first to become an area of formal display for 'identity marking' that represented the household to visitors and neighbours, while the back garden, by contrast, continued to express a more inward-facing individual and familial identity that could still encompass food growing and the domesticating appropriation of nature (Daniels and Kirkpatrick 2007; Chevalier 1998: 48). Thus,

> [t]he front garden (often very small, perhaps only a yard or so in width) was a celebration of ornamental non-functionality, as befitted its opposition to the routine of the back yard and back entrance ... The front garden was a place for fantasies of rockery in quartz, for flowers and shrubs, and for careful display of propriety and cleanliness. (Chapman and Jarnal 1997: 7)

As Chevalier (1998: 50) notes from her fieldwork on the gardens of Jersey Farm, north of St Albans in the United Kingdom, these areas for display are always 'gardens of delight', not vegetable or fruit gardens. Plants are flowers or shrubs, above all evergreens, reflecting a view since the Elizabethan period that the 'mere' vegetable or herb garden is of a lower, more utilitarian order than delight in nature for its own sake. In fact, attempts to 'beautify' or regulate suburban space have often come into conflict with productive use of private gardens, as food-related-activities such as poultry or goat keeping are deemed unsightly or unhygienic (Gaynor 2004). A functional design typology of private gardens, such as that developed by Seddon (1990) for Australian conditions, saw such space as predominantly arranged for display of different kinds, from 'nostalgic' to 'tidy suburban', 'native', 'prize specimen' and 'nice show of colour' in character. In this identity-marking spatiality, food growing did not feature as a central element of urbanism, except for the '"Italian" garden of a productive or "hose-down" style, leaving most owners plenty of time for leisure, hospitality and social interaction' (Seddon 1990). More recently it has been acknowledged that front gardens (and gardens generally) can play positive roles in planting for aesthetic biodiversity and micro-climate support (Hall 2006: 14; Cooper 2008; Cameron et al. 2012). They have potential to contribute to environmental protection from flooding by offering soft drainage spaces (Parham 2007). They can also provide capacity for urban agriculture, as discussed in Chapter 6. Normatively, though, domestic gardens became prime sites for cultural appropriation, offering urban dwellers ready-made design forms to signal stylishness (Bhatti and Church 2004). Gardens thus provide spaces for performing food-related consumption including dining, of which more later in the chapter.

The nineteenth and twentieth centuries also experienced waves of economically driven migration to new cities, with impacts on the design of domestic gardens and the nature of home production of food. In the United States, Canada and Australia, for example, settlers were often from rural backgrounds, and brought food-centred cultural values to bear that reflected requirements for self-sufficiency. These included strong traditions of vegetable growing and food processing of grapes for wine, flour for bread making and fruit for preserving. The capacity to make do in food-producing terms, and to establish informal bartering in food, were particularly necessary in the face of poverty and inadequate or absent infrastructure and services in new locations like the 'Ukrainian' village of Ardeer on the Western fringes of Melbourne (Morrow 2011). The design of garden space was very much focused on the utilitarian; such settlers made productive use of every part of the outdoor space that was available for this activity, and they also supplied food and gardening services to others. From the Chinese gardeners in nineteenth-century California (Chan 1996) and the Australian goldfields (Jack et al. 1984) to the Mexican garden labourers employed on low wages by middle-class Los Angeles residents to maintain their manicured gardens (Huerta 2011), immigrant gardeners have often formed a kind of invisible substrata in urban life (Peattie, in Huerta 2011: preface).

Work on the urban gardens of Hmong communities who emigrated to California in the aftermath of the Vietnam War demonstrates how Hmong women continue the practices of the tropical home gardens that they previously tended in Laos (Corlett et al. 2003). The house-lot urban gardens of women in central Mexico equally reflect ways of renegotiating traditional roles in domestic garden space (Christie 2004), while research into Greek and Vietnamese gardens in Marrickville, an inner urban working class (but now gentrifying) suburb of Sydney, found that for these gardeners, 'garden produce and the type of environment created by the garden helped to emphasize and maintain cultural relationships, provide a space of nostalgia, and give a sense of ownership and control' (Graham and Connell 2006: 375). Such 'reproduction of place and practice' was also identified in a study of Macedonian, Vietnamese and British gardeners in Wollongong and Sydney (Head and Muir 2007).

High levels of garden productivity were similarly found among post-war Italian and Greek immigrants to the South Australian city of Adelaide, who maintained cultural traditions of food growing that were declining among the host community. Such immigrants were particularly notable for the productiveness of their garden space and thus sustaining a direct connection to at least some of the food they consumed, while their productivity contrasted with broader social trends affecting Australian urban food space (Parham 1993). By the 1970s and 1980s, such productive inner suburban Italian gardens were being replaced through gentrification, giving way to the changing values and priorities of younger generations of immigrant Australians. In the inner Adelaide suburb of Maylands, for example, houses that Italian migrant gardeners had previously complemented with highly productive gardens, comprising neatly tended vegetables, vines, fruit trees and wine-making facilities were bought by gentrifiers intent on imposing a more expansive suburban built-form model on modest inner urban cottages; building ensembles of new rooms and outdoor spaces at the backs and sides of their houses for display and leisure rather than food production.

The broad spatial trend in food-growing terms here appeared to be a loss of cultural and gastronomic diversity nearer the centre of the city and only the chance of its improvement nearer the edges where those displaced from the inner suburbs have retreated. Housing forms changed too, and with smaller block sizes, and the growth of McMansions in new areas, less space was available for gardens per se, even if there had remained the desire to grow vegetables, fruit, herbs and grapes for wine at home. Yet constituting a counter process of identity formation through productive gardening, in the early twenty-first century, young Australians of Italian background are reclaiming and reshaping their cultural heritage by once again growing grapes, making wine, producing fresh vegetables for food and building bread ovens for wood-fired bread at home (pers.comm). This younger cohort may also be undertaking these food practices as a stylish middle-class pursuit, juggling professional jobs with productive garden activities and operating within a much more self-conscious gastronomically inflected mode than the hardship-driven approach employed by their parents, and especially their grandparents. These individuals appear to be expressing a particular food-centred habitus, different from that of earlier generations who engaged with food production in a more utilitarian way. More broadly, it is clear that productivity and creativity is a feature of the way that back gardens are used in today's Australian suburbs. Food is grown; just not at a subsistence level (Head and Muir 2007: 92).

CONNECTING THE HOUSE AND GARDEN THROUGH FOOD

Chapter 1 explored how, in the post-war period, there was an opening up of food space that very directly connected the house interior and its external garden in design terms. This was not new, however; it reflected a design approach that has gone in and out of fashion over the centuries. At the level of elite garden design, in the Roman villa plan, for example, colonnades, loggias, galleries, pavilions and balustrades were all used to link the garden to the villa, with a clear interest in transitional and edge spaces being employed architecturally to extend the plan of the very formal house to the disorder of the natural world. The villa garden was an important dining space, representing both domestic conviviality and the conspicuous consumption of leisure (Graham 2005). Roman town houses excavated at Pompeii show a variety of designed options that connected dining to garden space, including rooms for summer dining, often an *exedra* or partially open space giving on to the peristyle courtyard or a loggia that could be closed off with shutters at need (Amery and Curran 2002: 105–6). Used 'for entertaining and for larger banquets' the generously proportioned garden room 'was decorated with frescoes depicting natural scenes with animals and plants, extending the theme of the garden' (Amery and Curran 2002: 110). At a more modest level, the interplay between convivial home dining outdoors and the productive garden was apparent in both a design and social practice sense:

> Stone seats and tables were used for outdoor meals. They may look harsh without cushions but most of us would find more comfort in a small Roman court than in the typical suburban garden today. They were secure and sheltered. Plants were grown: for the kitchen (e.g. cabbage, parsley, fennel), for making medicines (mustard), for making drinks (grapes, apples, medlars), for feeding bees and keeping them healthy (rosemary), for making perfumes (roses), for making garlands (ivy, myrtle), for decoration (acanthus, periwinkle, laurel, rose) and for shade (pine, cypress). (Turner 2005: 69)

Outdoor dining spaces could be highly elaborate. During the Renaissance, garden dining at the aristocratic level was reflected in complex landscape design. Borromini conceived a design for villa Pamphili in Rome that delicately played with the *giochi d'aqua* (water jokes) of the time, where 'the guests having enjoyed the garden in the morning by strolling there, it would be possible to flood the shady paths while they were dining [so that] rising from table, they could go everywhere in little boats, which would turn out most wonderful' (Agnelli et al. 1987: 20–1). Cato's digs at the vulgarity of pretentious villa gardens of the Roman bourgeoisie have their echoes in criticisms of the nouveaux riche gardens for wealth display found in suburban and megalopolitan locations worldwide (Parham 1993: npr).

As discussed earlier in the chapter, and touched on in Chapter 1, post-war generations were subject to broad structural shifts in domestic consumption, which mostly altered their relationship to the garden and changed its design configuration.

Garden space became 'an expression of individuality and ideology in a consumer society' (Daniels and Kirkpatrick 2006: 314) rather than judged important for productive food capacity. For many, the domestic design relationship between production and consumption was broken. Vegetables now appeared to be included on sufferance, yet as Yerbury notes in *Modern Homes Illustrated* (1947: 304)

> it is no longer considered essential to hide the vegetable garden from the house [as] most of the vegetables likely to be grown will be no more than eighteen inches high, and a shield (if one is necessary) of currant or gooseberry bushes should be enough: this will give a greater feeling of space than the old fashioned trellis and bring the vegetable garden into the whole design.

Whereas in the first half of the twentieth century indoor food spaces were seen as civilized while outdoor areas were still understood as utilitarian and food production related, by the middle of the century dining outside *al fresco* had become an area of increasing design interest to architects and those advising on lifestyle in the United States and elsewhere (Cromley 2010: 172). Garden space was increasingly conceived as a spatial extension of the domestic consumption realm of the house and a site for expressions of conviviality (or at least its appearance). Mid-twentieth century designs demonstrated considerable architectural attention to outdoor dining space, with terrace dining plans integral to bespoke domestic architecture and eventually to mass housing too. In a late 1930s architectural competition sponsored by Macy's department store in the United States, house designs featuring space for dining outdoors were shown, and argued to be possible in certain temperate climates for around seven months of the year (Cromley 2010). At the top end of the housing market, outdoor dining space was starting to be designed in as part of a seamless array of interior and exterior eating spaces. The 1940 'House of Glass' by Landerfield and Hatch, for example, had a built-in breakfast nook with benches and a table linked to the kitchen and a dining room with sliding walls to the living room, which in turn opened on to an outdoor patio which featured an additional table and chairs (Cromley 2010: 171). The top-down nature of the introduction of outdoor dining space into popular culture and mass housing design practice is exemplified by the work of notable American architect Louis Kahn (Marcus 2013). From the 1930s to 1950s Kahn showed examples of outdoor barbecue areas as an integral domestic design feature, creating outdoor rooms that formed essential parts of overall architectural compositions in his Weiss House of 1947–50 and Genel House of 1948–51 (Marcus 2013).

These transformations started to become more widely publicized. Ford's (1942) *Design of Modern Interiors* and Nelson and Wright's (1945) *Tomorrow's House* were both confidently planning houses for a successful post-war world, and included in the latter case some particularly beautiful modernist examples of domestic architecture in which terraces, patios and porches provided sheltered space for outdoor dining in a range of American settings. Similarly, at the mid-market level, Cliff May's modern ranch houses, with their integration of exterior and interior elements, 'took America by storm' (Gregory 2008). As outlined in the 1946 book *Sunset Western*

Ranch Houses, a collaboration between Cliff May and Sunset magazine, this indoor-outdoor design integration and its food focus was illustrated in 'pace-setter' houses such as the Californian Riviera Ranch House:

> Every major space opens onto a patio or garden. The large outdoor fireplace at one end of the loggia, beside the kitchen, is the house's most emblematic feature, combining the most traditional and progressive elements. Its large brick opening includes a built-in barbecue with an automatic spit, grill, and 'gas-heated plate', and turns the loggia into an outdoor cooking, dining and entertaining space – the heart and soul of the home. (Gregory 2008: 78)

Thus outdoor dining became increasingly embedded as a cultural practice that connected to a growing informalization in family relations and domestic spaces. The garden patio offered a new setting for domestic conviviality, but one that needs to be approached with some critical distance, given that its expression was so tied into maintaining certain gender roles. For instance, the birth of patio-based grilling culture in post-war America was supported by various accoutrements offered to ease the outdoor cooking and serving experience (Miller 2010). Through their manly tools and gadgets these emphasized that gender roles remained undisturbed by the otherwise transgressive act of men cooking. In Canada, similarly, outdoor barbecues were marketed to men for whom such cooking was defined by advertisers as a legitimate, masculine pursuit, extending fatherhood's realm into the domestic space of cooking (Dummitt 1998). As Dummitt (1998: 212) notes, barbecuing was subject to a 'veritable orgy of linguistic posturing that linked outdoor cooking to symbols of virile masculinity and manly leisure'. By the late 1950s, cookbooks included sections on outdoor cookery, and men were offered easy, 'age-old' recipes in which meat was a central ingredient that they would cook over a primal fire using a tough, sturdy barbecue located in their 'natural' outdoor domain (Dummitt 1998: 212).

These design and cultural shifts also marked social changes in garden design for dining in other Western countries (Johnson 2006: 129). Thus in Australian cities, the 1950s housewife's gaze might take in her children's outdoor play area rather than a formal outdoor entertaining space, by the 1960s 'as modernity [was] displaced by post-modernity', the Australian house became larger, more segmented and, crucially in this context, had 'stronger linkages to increasingly formal outside areas' (Johnson 2006: 129). From the 1970s onwards, the linking of outdoor patio eating spaces to the indoor cooking and entertaining areas of the house had become a common design trope at a range of house price levels, from the individually architect-designed house to the much larger market of volume house design and building. In the 1990s and 2000s, with food spaces further inflated in size, came ever more ornate designs for outside dining. The overblown real estate language used to describe these often lavishly detailed outdoor dining and entertaining zones sums up aspirational, consumption centred perspectives on domestic garden space. 'Exemplifying contemporary Australian living, with light-filled rooms, high ceilings and informal living areas, the unique floor plans of the Urbis and Domo are designed to enhance and nurture a casual and comfortable easy living indoor-outdoor lifestyle' (Homes. Adelaide Matters 2013: 3). Under the heading 'Monaco Matters' advertorial text extols the charms of new house design, the 'Monaco' (Vlach 2013: 4). The text emphasizes the dining-related integration of interior and exterior food spaces and reflects the need for these to facilitate conspicuous consumption centred on food both within the family and in relation to exhibiting that lifestyle to others.

REANIMATING THE PRODUCTIVE GARDEN

Despite these broad design-related trends towards gardens performing as food consumption rather than production space, in certain modern-day circumstances garden owners and others have sought to explore and reanimate the garden as a productive domestic landscape: sometimes creating food gardens of exceptional beauty and productivity. *Potagers* and walled kitchen gardens are being rediscovered (see Figure 2.1) and brought back into productive use, or developed from scratch (Clarke 1988; Dorian 2007; Geddes-Brown 2007), with design advice offered for 'foodscaping' (Creasy 1982: 5) for both utility and beauty (Bartley 2006) and recipes which reflect their historical associations and the abundance now on offer (Tilleray 1995). In some cases, too, literary examples of the vegetable-growing year have been used as a structure within which to explore wider food and sustainable placemaking concerns through the vegetable garden, as in Kingsolver et al.'s *Animal, vegetable, miracle* of 2007. Kitchen gardens historically have had strong design schemas, including planting beds, paths, walks, quarters and walls, tied to their utility as productive sites but also allowing scope for considerable beauty, as in the ornamental

FIGURE 2.1: Walled kitchen garden, Culross Palace, Fife.

Photo: Elisa Rolle. Image available at http://commons.wikimedia.org/wiki/File:Culross_Palace,_2003_%282%29.JPG under a Creative Commons Attribution-Share Alike 3.0 Unported license.

kitchen gardens of the late sixteenth century in England which were designed to be walked in and enjoyed as well as used for food (Campbell 1996: 83).

Notable examples of such productivity at the grand scale include the previously mentioned *potager du roi* at Versailles, which began by producing vegetables, fruit and flowers for Louis XIV but still offers produce for sale. Influential in more stylistic terms are examples of the renegotiation of kitchen gardening at the elite level such as at Glin Castle in County Limerick, the Chateau de Bosmelet in Normandy (Larkcom 1998; Geddes-Brown 2007), or the redevelopment of the garden of the sixteenth-century chateaux of Villandry in the Touraine, built for Francois 1's finance minister Jean Le Breton. Restored in the early twentieth century, the huge expanse of formally planted vegetables, medicinal herbs, flowers and fruit is described as a kind of *jardin de curé* (Lévêque and Valéry 1995: 28). With a playful mixing of quotidian vegetable planting and a manicured landscape at the monumental scale, the enormous expanse of *potager* garden at Villandry is impressive, but unlike the usual *potager* it is purely ornamental, not grown for use. Villandry thus perhaps expresses something of a gastronomic disappointment, despite its formal design beauty and sheer scale. An example like the chateau of Bosmelet in Normandy, by contrast, offers greater satisfaction as the aesthetics of its design are matched by productive use (Geddes-Brown 2007). McCulloch (2011) surveys a number of such reanimated or newly created kitchen gardens, mostly in Australia or with Australian connections, including that of early twentieth-century artists John and Sunday Reed at Heide outside Melbourne, Petersham House, in South London, a rooftop garden at a Melbourne hotel and Stephanie Alexander's school-located kitchen garden (part of a wider kitchen garden programme) at a school near Bendigo in Victoria. The importance of design of the productive garden as an outdoor room for both utility and beauty in these renegotiations is clear from the account Geddes-Brown (2007: 103) gives of her creation of a large walled kitchen garden at Columbine Hall, in Suffolk, about which she says:

> We wanted a large working vegetable garden, and inspired by visiting that at the Chateau de Bosmelet, we determined that this would be as ornamental as it was productive. ... Like many garden owners before us, we decided that the regimental lines of vegetables, however ornamental, were best kept within walls. The walls would provide a microclimate for the vegetables and a site for lots of climbing roses and clematis, and the area would come as a surprise to visitors and would not clash with the formal seventeenth-century gardens, much of which was designed by George Carter.

More broadly, the state of the domestic garden and its urbanism possibilities in relation to the future of food in the city are tied to structural trends in urban development that are discussed in later chapters, but it is obvious that in contemporary conurbations private gardens are very unevenly distributed in a spatial and economic sense. Access to private garden space is largely determined by economic capacity rather than need; while changing work patterns, driven by the extraction of increased surplus value, are leading to greater time compression for food-related activities (Warde 1999), making gardening viewed as a luxury for many. The culinary transition, discussed in Chapter 1, meanwhile, is contributing to the reshaping of food consumption and production spatially, including in relation to skills and inclination for food growing. Given these shifts, many urban dwellers do not have the capacity to grow food at home even if they want to do so.

However, there are also signs that many householders are interested and concerned to make changes where they can that allow them to reconnect with a more edible urbanism focused on food production. Research from places including the United Kingdom demonstrates that domestic gardens are strongly associated with a sense of well-being, and growing vegetables and fruit is a significant component in that finding (Dunnett and Qasim 2000). In the context of home gardening for food, interesting initiatives include front yard farms, schemes to rent out home-growing space to those without access to private gardens and crop-sharing projects. Some of the urban food-growing initiatives that are underway connect the domestic domain to broader urban spaces, projects and movements, which are discussed in Chapter 6's exploration of convivial green space in cities. It is worth noting here though that a large amount of private vegetable production still occurs in cities, despite changes to domestic space. London Food Link recently reported that around 30,000 people in the capital rent allotments to grow vegetables and fruit, and 14 per cent of households grow vegetables in their gardens (Donovan et al. 2011: 58). Considerably older data from the early 1990s Australian Bureau of Statistics Home Production Survey asked respondents about production over the previous twelve months and home-grown vegetables totalled 153,000 tonnes, compared with 2,725,000 tonnes in the Agricultural census (that is, commercially produced vegetables). Among other findings, more vegetables, fruit and nuts were grown in metropolitan Adelaide than country South Australia in terms of tonnage.

DESIGN REQUIREMENTS FOR VEGETABLE GARDENS

Harking back to the discussion in the Introduction to this book about the lack of sustainability of our current food system, certain designers have thought it possible to set out requirements for more convivial garden spaces that would contribute to sustainable urbanism through food growing. Increasingly, these reflect the perceived need to shift diets from their current, arguably unsustainable focus on meat (Spencer 1996; Goodland 1997; Welch and Graham 1999; Leenaert 2012). Instead, the trajectory is towards a more vegetable, pulse and grain-based gastronomy (Duchin 2005; D'Silva and Webster 2010; Macdiarmid et al. 2012). In the late 1970s, Alexander et al. (1977) proposed some fundamental design requirements for domestic gardens, which were brought together in various spatial design patterns (patterns 168–77), around which their book is structured. These sought to conceptualize utilitarian and convivial aspects of home food production in an integrated way. In their schema, domestic gardens should be (to paraphrase) a little bit wild, rather than overly domesticated, manicured and trimmed, reminding their users of the natural world and its processes (Alexander et al. 1977: 802). The garden would need to be protected and partly enclosed by walls, with trellised walks to give positive shape to outdoor space and contribute to its beauty (Alexander et al. 1977: 802). After making the case for urban vegetable growing, the authors suggest that every 'house cluster' and 'every household which does not have its own private land attached to it should have a portion of a common vegetable garden close at hand' (Alexander et al. 1977: 820). By including fruit trees, terracing slopes, adding greenhouses and developing vegetable plots, 'in a healthy town every family can grow vegetables for itself. The time is past to think of this as a hobby for enthusiasts: it is fundamental part of human life' (Alexander et al. 1977: 819).

Food's centrality to this design vision is evident; greenhouses are used to trap solar energy for growing vegetables and flowers, and the vegetable garden is treated as an essential component of design for urban space, reflecting vegetables' role as the most basic foods for a sustainable diet. The design requirements for a vegetable garden are defined as around one-tenth of an acre for a family of four for an adequate year-round supply of vegetables, and users are reminded to 'make sure the vegetable garden is in a sunny place and central to all the households it serves. Fence it in and build a small storage shed for gardening tools beside it' (Alexander et al. 1977: 821). Some composting capacity would be required, while grey water from sinks and drains would be used for watering (Alexander et al. 1977: 821). It is worth noting that since this design prescription was developed, such grey water reuse has in fact become more mainstream as part of sustainable urban drainage systems.

All this begs the question as to how much of their domestic food needs a household should or could deal with through home food growing. Clearly, there is no one answer to this question. For some urban geographers the idea is in itself part of the so-called 'local trap' in which local production and consumption of food is valorized without real evidence of its utility in meeting urban sustainability or equity requirements. For feminist geographers, railing against the double or triple shift, home food growing may be conceptualized as yet 'more work for mother' (Cowan 1987). Another issue is the implied design preference for a low density of urban development to allow every individual dwelling to have relatively substantial private garden space for such productivity. In fact, 'transect'-based approaches suggest broadly that density (and thus the amount of private garden space) would follow urbanism principles of congruence with urban function from low densities close to the peri-urban edge grading to much higher densities near activity centres and urban cores (Duany 2002; Bohl and Plater-Zyberk 2006). Various schemes noted below and in Chapter 6 also show how private garden space can be adapted and shared to much greater food effect as part of urban retrofitting actions.

IMPLICATIONS FOR THE GASTRONOMIC GARDEN

The fundamental importance of private gardens in urbanism terms has increasingly found a reflection in design guidance as well as in both organizational and individual practice in relation to food production. In Canada, guidance for masterplanning a new mixed-use community at Southlands in British Columbia includes among its productive design elements proposals for private kitchen gardens, vegetable plots, window boxes, balcony and roof gardens (Donovan et al. 2011: 43). In the Australian city of Melbourne, a guide to *Food Sensitive Planning and Urban Design* (Donovan et al. 2011: 23) offers a 'conceptual framework for achieving a sustainable and healthy food system' which proposes to ensure consideration of home food productivity in new subdivision design and suggests giving awards for the best food garden to celebrate local food production. While these strategies are to be applauded in environmental sustainability terms, a more transect-led approach as discussed in the Introduction seems likely to build in a better balance between appropriate density and productivity. Otherwise, the problem of continuing to build unmixed residential areas at low densities could outweigh the benefits of water-sensitive, food-accessible design noted in the *Guide*.

At the level of public policy and professional guidance, city-wide food plans and strategies for food which may include support for backyard gardening have become relatively common in Canada and the United States (Chiang 2008; Neuner et al. 2011), as in the Food Plan for Oaklands (Unger and Wooten 2006). National policy guidance from professional peak bodies, such as the *Policy Guide on Community and Regional Food Planning* (2007) from the American Planning Association and the Planning Institute of Australia (2009), offer advice and guidance on productive urban space. In the United Kingdom, meanwhile, a national-government-led food strategy to the year 2030 was released in 2010, and a policy guide *Good Planning for Good Food* (2011: 12) argued for landscape design principles which allow food growing to be integrated into urban open spaces including private gardens. City planning authorities, often in association with health partners, including in London (2007), Manchester (2007) and Liverpool (2010), have also developed strategies that offer policy proposals across the food chain from production to waste. *Food 2020*, for example, recognizes the importance of urban food growing in home gardens: noting that the 'the popularity of "grow-your-own" has risen significantly over recent years. An estimated 33% of people already grow or intend to grow their own vegetables. Growing food – at home, in a community garden or allotment – can produce a number of other benefits including better mental and physical health, bringing people together and improved skills' (HM Government 2010).

Almost uniformly, though, there is a greater emphasis on food production in community-based settings, as opposed to individual home-based food-growing opportunities. In the United Kingdom, the *Liverpool Food and Health Strategy* (2010: 9), for example, offers proposals for the 'development of 31 growing schemes enabling participants to learn how to develop land mainly in small plots, to grow fruit and vegetables'. In Australia, similarly, state and some city governments have developed food policies within their strategic plans or as stand-alone strategies. However, food growing at home is mentioned in passing, if at all, possibly because such activity in the private domain is seen as beyond their purview although its urbanism implications are profound. It is perhaps to the third sector that we should turn for a more developed interest in the possibilities for home-grown food, with the stress on growing opportunities connected to wider interests in urban food resilience, encompassing food poverty and health concerns. For example, the London Food Link (2012: 2) which is part of the charity Sustain, has rated London Boroughs on their performance in securing 'a healthy and sustainable food future for everyone' including community food growing and found that in many London boroughs, 'tens of thousands of local residents have better access to green space for growing food' although it is unclear what proportion is private-garden based.

The revitalization of interest in home gardening can be situated within the social movement of urban agriculture (explored in detail in Chapter 6). One example is the burgeoning of so-called front yard farms where households decide to dig up the ground in the front of their houses, and turn this into productive food space. This renegotiation of space which is more usually given over to formal planting displays (and in many cities is blighted by hard space for car parking) provides an interesting shift in design terms from the front garden as purely ornamental in character, as discussed earlier in the chapter. The transitional space between the private realm of the house and the public space of the street is an area designers have argued could be used more productively as part of urban biodiversity networks. The notion of an attack on the front yard has been advocated and documented in Haeg's 'Edible Estates' project (2008: 16), which has sought to reconfigure the front garden to more productive food growing ends through 'full frontal gardening' that replaces the resource depleting and polluting lawn. Soler (2011: foreword) similarly asks, 'are you brave enough to step out your front door and rip up your lawn?', proposing an end to the 'toxic monoculture' of the front lawn and its remaking into a 'slightly wild, edible garden' that is 'sustainable, beautifully designed, and edible: a modern day victory garden' with 'curb appeal' (Soler 2011: 8, 13). As part of an urban social movement in which individual households take the lead, front yard farming examples are now increasingly common in the United States, Canada, Australia and elsewhere. Some of their proponents have set up Facebook pages or blog about their gardening activities.

Front yard farming can sometimes come into conflict with local planning regimes. Those deviating from planting norms and the class identity derived from these may also face more informal social sanctions (Blunt and Dowling 2006: 117). There are a number of examples where shifting to productive front gardens or planting crop bearing street trees has been undertaken in the face of prohibitions by city regulations or ordinances (with Portland, Oregon a notable exception): 'These residents are vigilantes, as almost every city in the nation prohibits fruit bearing trees in the public right-of-way. It is ironic that in places like San Francisco and Berkeley (municipalities that in many ways are leading the charge for better access to healthy, locally produced food), fruit- and nut-producing street trees are outlawed' (Nordahl 2009: 54). In a case from Missouri in the United States a would-be front yard farmer was forced to fight his local planning authority as his front yard farm was said to be in violation of city ordinances. A similar case occurred in Quebec, where a street facing *potager urbain* became the subject of dispute. In media reporting of these shifts, there is awareness that front yard farming is, in a sense, a return to the design of the productive cottage garden of the nineteenth century, as this account from the *Charlotte Observer* in the United States explains:

A growing number of Charlotteans whose backyards can't support a garden are turning their front yards into mini-farms. And it's a case of a trend coming full circle: Long ago, small cottage gardens in front of homes contained a riotous mixture of flowers and vegetables, but over time, food became something to hide out back. Now growing your own food isn't just socially acceptable, it's hip. And for many people, there's no reason not to put it on display. (http://www.charlotteobserver.com/2012/08/16/3459793/front-yard.html)

Designers and those interested in domestic food self-sufficiency have given thought to the optimal-sized house block in which sufficient food space can be achieved to help feed a household of four annually. In Australia, for example, in the late 1990s, bigger house blocks were about an eighth of an acre, with about 35 per cent of that space taken up by the house and 65 per cent by the garden. A process of reducing block sizes down as low as 300 square metres saw more maisonettes, terraces and multi-household apartment blocks built to conserve land and avoid too rapid incursion into the rural hinterland in the context of some of the worst global sprawl conditions. While at those block sizes and housing configurations it was not possible to achieve vegetable gardens for each dwelling, as was seen earlier in the chapter, more dense, compact city models have traditionally provided shared garden spaces like allotments instead of individual, private gardens for all. On a larger suburban block from the 1970s, growers like Gavin, a food blogger, show a wide variety of vegetables and fruit trees in the front, side and back gardens. Fruit trees include pear, nectarine, plum, apple, peach, cherry, feijoa, olive, lime, lemon and mandarin while berry and 'veggie plots', a 'chookhouse', a greenhouse, a wicking bed, a cob oven, water tanks and solar panels are tucked into other spaces. He notes in answer to a question from a site reader that, 'My suburban block is 779 m^2 (0.19 acre). It was the average block size back in the 1970s, when my home was built. Not all is arable, as the pool takes up about a fifth of the land and the house another two fifths, leaving me the remaining two fifths to grow food and raise chooks'.

In many cities, garden owners have begun to effectively rent out their garden space to food growers and then share the resultant crops. Sometimes these vegetable gardening and fruit growing arrangements for private garden space have become somewhat more formally organized, as part of community-based not-for-profit schemes, like Landshare in the United Kingdom which connects would-be growers with spare urban land including private gardens. Launched in 2009, and by late 2012 with over 70,000 members, Landshare describes itself as bringing together 'people who have a passion for home-grown food [with] those who have land to share with those who need land for cultivating food' (Landshare). There are also a series of initiatives from communities in particular towns and cities. Transition Cambridge in the United Kingdom, for example, asks

> Do you have space in your Cambridge garden which is under used? Or do you want to grow your own veg but don't have a garden to do it in? If so, the Transition Cambridge Garden Share scheme is for you! So, you're a garden owner, with not enough time to do your garden justice? Or perhaps you're hampered by a lack of knowledge or experience and don't know where to start? You don't need to

allocate much land: even four square metres (about 6 paces) would be useful, and you can also get a share of the produce. (transitioncambridge)

Similarly, in Edinburgh, a garden sharing scheme, *Edinburgh Garden Partners* (undated) explains that it promotes and enables 'the growing and consumption of local and organic food. This has many environmental benefits, including carbon saving. We encourage the sharing of produce between the owner and gardener'. The *Grow Your Neighbour's Own* scheme in the English town of Brighton and Hove, meanwhile

> pairs up gardeners who have nowhere to grow their own food with garden owners or allotment holders who have the space to grow but for whatever reason are not able to. We want to help form lasting (gardening!) relationships between people, preferably people who live near each other – the garden owner/allotment holder and gardener arrange between them what they will grow and how often the gardening will take place, and share the produce as it is harvested! (Grow Your Neighbour's Own)

Those taking part in such schemes are often in the position of not otherwise having access to food gardening space. One participant describes her experience as part of a garden share scheme run by community-led charity Transition Town Totnes in the United Kingdom. In this case the scheme provides the chance to garden locally, and the opportunity to grow affordable fruit and vegetables, while leaving one-third of the resulting produce with the owners of the private garden space:

> I live in a flat with no garden but I've always liked the idea of growing my own fruit and veg. I registered for the scheme as soon as I saw it and got fixed up with a fantastic garden down the road. It's about 25 x 18 ft and I spend a good deal of time there, especially in summer. I grow all sorts, including strawberries, raspberries, redcurrants, and loganberries, and loads of vegetables: potatoes, beans, salad leaves, red peppers, tomatoes and squash ... I live on my pension so money is important and this has saved me a lot of cash. You can have a bit of an initial outlay if you need things such as tools, slug pellets or a water butt – I paid about £25. But since then it's cost virtually nothing. I'm usually given seeds and cuttings for birthdays and Christmas. (*The Guardian*, Friday 2 September 2011)

It is clear from these accounts that there is considerable enthusiasm from a grassroots level to be involved in food growing in private garden space, and that the planning system can either help or frustrate opportunities for maximizing the use of such gardens for food growing. Planning regulations reflect social norms about spatial design, and in some cases, such as front yard farming, the return of front gardens to productive use challenges societal expectations of these as transitional spaces that act as passive settings for domestic dwellings, or have become informally given over to car parking space, as cars dominate the public and private domain. Where there is consciousness at the level of urban government about the need to protect and enhance private garden growing space, this is likely to provide a useful basis for the kinds of schemes reviewed here. As a recent report from the charity Sustain (2012: 6) argues in relation to London, there is a need to:

> Create and protect food growing spaces in and around a locality. Allotments, community growing spaces and a range of other under-utilized public and private space could be used to increase the availability of healthy and sustainable food by expanding the space allocated for food growing. Such spaces can also enhance the quality of productive green spaces in neighbourhoods thereby contributing to biodiversity.

THE CHAPTER IN REVIEW

In this chapter the private space of the garden has been considered as an aspect of sustainable urbanism: both as a productive food space and a designed site for food-centred conviviality with ancient antecedents. The ongoing centrality of this scale of food production and dining space in cities has been explored through a diversity of design and urbanism examples including walled courts, kitchen gardens and *potagers*, demonstrating its near universality in these domestic roles. In current practice it can be seen that for some the garden is still an important place for food production; with resilience strategies stemming from a variety of cultural and economic causes including migration and supporting some intriguing outcomes in relation to conviviality and sustainable design. At the same time, the private garden's predominant meaning, especially in Western cities, as a space for leisure and display, shows how it has altered its focus from food production to consumption with spatial design to match. In physical design terms the garden has become an extension of the domestic interior and for playing out gendered roles in relation to food. The burgeoning of domestic food growing can in part be seen as an aspect of the rise of urban agriculture which has in turn invoked changes in garden design and use. These food-centred alterations in private garden space have begun to connect to broader societal ambitions in relation to the sustainability and resilience of the urban food system. Private gardens can form part of an overall set of convivial green spaces that contribute to sustainable urbanism, of which more in Chapter 6. In the next chapter meanwhile, we move to another spatial scale, turning to the outdoor rooms of food markets and related food spaces in both traditional practice and contemporary cities.

PART TWO

Gastronomy and the Urbanism of Public Space

CHAPTER THREE

Food's Outdoor Room

INTRODUCTION

In this chapter the focus shifts from urbanism's characteristics at the scale of individual gardens to that of the outdoor rooms of food markets and related land uses, as attention to food's interplay with the city moves from the private to the public domain. It is argued in this chapter that while the construction of the city reflects a range of food relationships in public space from joyful to uneasy that profoundly affect the quality of urban life, it is the everyday interplay between people and food in and around food markets and food shops that is perhaps the most important way that food has been central to urban life, to ordinary city space and to urban social practices over time. The chapter thus focuses in on the historic role of the food market as a critical element in urbanism, before exploring the decline of market-based food retailing, and the regeneration of urban food quarters that have emerged in places that have managed to maintain or renew a rich gastronomic quality as successful outdoor rooms, a concept well developed in the urban design literature, discussed in previous chapters, explored in Parham (1992, 2008, 2012) and further interrogated below.

As the chapter demonstrates through a diverse range of examples, many of the market spaces that work well for food are 'traditionally' shaped in urban design and morphological terms (see Figure 3.1). As argued in the Introduction, invoking the notion of tradition in design terms is neither normative nor nostalgic. Instead, it is based on the reality that such food spaces reflect location-responsive spatial qualities which have been developed and expressed over the long term – sometimes hundreds or even thousands of years. These time-tested design qualities are argued to contribute to a food-related sense of place that can be discerned across many times and places. It is acknowledged that the focus is primarily on Western urbanism (forms such as the floating markets of Bangkok are not explored) although recent work on markets in a Southeast Asian context seems to reflect similar themes of conviviality through everyday, public-space-focused practices based on food (Duruz et al. 2011). Reflecting on the market design examples, the latter part of the chapter gives some thought to possible futures for food's outdoor rooms in contributing to urban conviviality and acts as a springboard to Chapter 4's exploration of other elements of the 'gastronomic townscape' (Parham 1992), including the design qualities of food streets, cafés, restaurants and street food.

SITUATING FOOD MARKETS AS OUTDOOR ROOMS

In the contemporary city, a number of theorists in different disciplines have explored aspects of both continuity and change in urban food sites including food markets.

FIGURE 3.1: Market-related 'outdoor room', Spain.
Photo: Susan Parham.

These aspects of spatiality have been variously conceived as the gastronomic townscape (Parham 1992), spaces for gastro-tourism (Parham 1996; Boniface 2003), indulgent streets (Valentine 1998), foodatainment places (Finkelstein 1999) and hospitable spaces (Bell and Binnie 2005). As described in the last chapter, with reference to the micro scale of the table, one of the most useful approaches, predominantly associated with public health literature, is the notion of the foodscape (Yasmeen 2006; Sobal and Wansink 2007), which allows researchers to analyse 'how food, places and people are interconnected and how they interact' (Mikkelsen 2011: 209). A definition of out-of-home foodscapes clearly points to the essential spatiality of this concept by referring to 'the spatial distribution of food across urban spaces and institutional settings' (Johnston et al. 2009: 512). Some of the many food spaces associated with these constructs are discussed in detail in the following chapter, but the next part of this chapter, meanwhile, focuses in on the urbanism of the food market in both its historical and contemporary forms. In so doing, it explores the design qualities and social and economic meanings of food as an element shaping urban space through expressions of conviviality and material culture. In this analysis, food markets are seen to be both the symbolic and material heart of the city. As a dynamic location for conviviality, the food market is a place where we are free to act, to take political action, or not to consume if we so chose. The market is central to public space which is, in turn, defined as

> a destination; a purpose built stage for ritual and interaction. Broadly the reference is to places we are all free to use, as against the privately owned realm of houses and shops. They are themselves often defined, these public places, by the private architecture of house and shop. But the distinction of purpose holds – the fact that in public places we act in ways we cannot, or do not, in the private realm. (Kostof 1992: 123)

The location and configuration of the food market is thus both a design issue and a crucial site in the public focus of city life. As Mumford (1961, in Lilla and Glazer 1987: 68) explained, the marketplace in the medieval town was not a formal square but was derived organically: 'an irregular figure, sometimes triangular, sometimes many-sided or oval, now saw-toothed, now curved, seemingly arbitary in shape because the needs of the surrounding buildings came first and determined the disposition of the open space'. Mumford went on to say that this ample space allowed room for not just market trading but civic gatherings and ceremonies, and in this way 'recaptured, in fact, the function of the earliest forum or agora' (Mumford 1961, in Lilla and Glazer 1987: 68). In design terms, Camillo Sitte's (1945, 1965) nineteenth-century figure-ground studies of various European cities demonstrated this spatial truth and prefigured more recent urban design work on positive and negative space in which the appropriate balance of spatial enclosure and openness helps define city form, and gives pre-eminence to the public realm (Bacon 1982; Trancik 1986; Broadbent 1990; Lozano 1990). In Sitte's (and others' later) work on positive space, the road becomes the floor and the trees and the buildings the walls of the outdoor rooms of the town. Sitte (1965) worked out a mathematical relationship between height-to-width in this urban public space, a ratio that defined not just physical dimensions but contributed to social use and ambience. Within a certain ratio the space was experienced as pleasant, while outside the ratio it felt either too closed or too open (Bacon 1982; Trancik 1986; Broadbent 1990; Lozano 1990). Thus, streets that are over scaled – too big and too wide – do not provide comfortable 'positive' space for their users.

Over the more than one hundred years since Sitte's research was completed, Trancik (1986) was able to contrast the strong solid-to-void ratios of traditional city form, with the weak and incoherent arrangements of modern cities in the late twentieth century, and to find the latter structurally wanting in creating positive outdoor rooms. Through a contemporary adherence to functionalism, in which land use separation is the rule, modernist urban form lost the three-dimensional understanding of the relationship between buildings and space that was critical to traditional cities, yet failed to replace this spatial sensitivity with human-scaled alternatives for space shaping. In other words, 'functionalism laid the ground work for our loss of traditional space, but the ethics of modernism have been inadequate to provide a dominant framework for city design' (Trancik 1986: 11). Although much contemporary urban space does not successfully create outdoor rooms (with deleterious effects on food, as is further discussed in later chapters), the importance of such enclosed spaces has been recognized in recent design guidance. To give one example of the centrality to urbanism, this suggests a need to reflect these place-shaping principles in the development of public space:

> Open space should be designed positively, with clear definition and enclosure. There should be no ambiguity or left over space. This can be done by giving each outdoor space a clear function, character and shape, and clarifying boundaries through the positioning of adjacent buildings, walls, fences, trees and hedges. ... It is the three dimensional mass of each building which defines the public realm. Building elevations and the cross-sections of public spaces should therefore be scaled to foster a sense of urbanism so that streets, squares and parks are defined by appropriately scaled buildings and/or trees fronting onto them.
> (Davies 2000: 87–8)

In design terms, central to a food sensitive approach then are

> streets, squares and parks [that] can be conceived as a linked variety of 'outdoor rooms', whose character varies according to whether they are ... go to places, or destinations for staying, eating, meeting or events; go through or past spaces, such as favoured streets or squares; stop in places, to sit and watch the world go by; or indeed a combination of all these things – providing multi-functional spaces where people live, work and are entertained. (Davies 2000: 88)

Equally central to food's relationship to urban form is that the food space of the outdoor room is associated with creating urban conviviality and vibrancy, the latter defined by Bentley et al. (1985) as perhaps the most important urban design quality necessary to good city form. As Jane Jacobs (1961) argued, traditional cities were characterized not only by the enclosure allowed by a dense urban fabric; they were vibrant because they also had highly complex, mixed land uses. Historically, food space was one of the fundamental components in this complex mix (Morris 1994).

FOOD MARKETS AT THE HEART OF URBAN SPACE

Cities became possible as modes of settlement only where and when there was an adequate supply of food from agriculture and trade to support a non-agricultural population. Markets were often the basis of town formation, and their role as places where food was sold has been one of urban space's most fundamental characteristics since earliest settlement history (Wilkins and Hill 2006). From the beginning of the development of villages, towns and cities, the food market has taken place in the main public areas of the settlement, and the cultural relationships between state, religious and civil society that underpin urban development, have been closely connected in a publicly accessible set of spaces in which the food market is central.

Ancient Greek town plans, for example, show food markets as practically and symbolically important to the life of the town, often abutting the agora as the 'multi-purpose everyday heart' of the city (Morris 1994: 46). Thus, archaeological findings show that in Priene, a Greek city on a spur of Mount Mycale, the food market was located just to the west of the main agora, at the opposite end of the space from the temple (Morris 1994: 28). In Roman Ephesus, similarly, the market building formed part of the armature of the city's central civic spaces, facing a public plaza and close to both commercial and religious structures and spaces (Yegul 1994: 97). That centrality largely continued uninterrupted until the mid-twentieth century, with civic space closely related to the food market in the same central location. In the French city of Marseilles, for instance, 'the site of the ancient Phoenician-Greek agora remained the site of the later Roman forum and even of the medieval market square' (Zucker, in Morris 1994: 46) while in Lucca, in northern Italy, the

Roman forum became the medieval town square, and continues to function as a food market today.

There are mixed indications from other urban traditions. Evidence from Tang Dynasty China (618–906) shows the town plan of Chang'an with a formal grid in which 'fortresslike' East and West Markets are critical elements in that city's spatial structure, at a time in which it is argued the commercial street emerged here as an urban form (Kiang 1994: 46). Such strict spatial and temporal control over market space was to break down, and in cities like Kaifeng of the Song period (960–1127) 'artisans and merchants flocked to the city and shops spilled beyond market confines into the congested streets' (Kiang 1994: 50). A contemporaneous account of a lively food trade in Kaifeng from just prior to its fall in 1127 notes a very food-centred urbanism:

> East of the Accipiter Inn and on the north side of the street is Pan's Tower Wine Shop. Below the wine shop, commerce goes on day and night, with the things traded varying throughout the day. ... At dawn, food items such as sheep's head, tripe and lungs, red and white kidneys, udders, tripe, quails, rabbits, doves, wild game, crabs and clams arrive at the market. (Kiang 1994: 52)

Historic accounts of Cairo's spatiality, such as that of the Iraqi traveller and scientist, Abd al-Latif al-Bagdadi, who visited in 1193, suggest that the city was characterized by wide streets and large spaces around palaces that were 'at times a daily market, where meat, pastries, and fruits were sold' (Alsayyad 1994: 73). In contrast, the lack of an agora-like space in many Islamic cities is argued to be a fundamental difference from the urban spaces of Christian Europe which evolved from Greek and Roman antecedents: those 'in direct line of descent from Mesopotamia ... had no need for either democratic political assembly or communal leisure activity in public' (Morris 1994: 42). This meant that market forms and placement differed, yet a rich history of the *suq* or covered street market within the labyrinthine medina have contributed important food market spaces across the Arab world, as described in Tripoli (Von Henneberg 1994: 144). Until its displacement by more commodified urban forms, Istanbul's Grand Bazaar offered a centre for trade in food and other commodities, by way of numerous small traders (Tokatli and Boyaci 1999). The souks of Casablanca and Fez, meanwhile, remain not only a central element in Moroccan food culture but outstanding examples of Moorish architecture.

Classical Rome provides a fascinating example of a complex food system in which the centrality of the food market was supported by food distribution and retailing forms recognizable in urban conurbations today. As a city of apartment dwellers, where only the very rich could afford a house of their own, Rome was sustained by vast imports for a population, in which up to a quarter were receiving food from the state. Rome's food markets were supplemented by public warehouses that stocked imported commodities and acted as a form of classical 'hyper-market' (Morris 1994: 65). Small food shops, meanwhile, traded from the ground floor premises of ubiquitous flat buildings, and the major concentration of shops was found around the Roman fora, especially the Forum of Trajan (Morris 1994: 65). Already, specialist wholesale food markets existed for vegetables, beef, pork, sheep and fish among other items:

> The holitorium was centrally located between the Capitoline Hill and the Tiber, with the boarium some distance downstream between the Palatine and the river. Near the latter, again centrally located, was the suarium. A third meat market, the Campus Pecuarius for sheep, has not been located. Lanciani describes how these three markets were all used for actual trade, on the spot, and suggests that to avoid yet further congestion in the street the oxen came to the boarium by the river, by barge-loads. However, cattle once bought were driven down city streets to the individual butcher's shops for slaughter. (Morris 1994: 65)

Urban pre-eminence as a market for food could be demonstrated in stark terms. As Tokatli and Boyaci (1999: 182) note, at one point in 1577 all slaughter of mutton and lamb was forbidden in the Balkans, as this was reserved for the 'privileged residents of Istanbul', while the region around the Sea of Marmara and beyond were 'organized around the needs of the capital' for grape syrup, raisins, pickles, meat, bread corn, butter, salt, olive oil, sugar, spices and coffee, among other foodstuffs. More broadly, in medieval times, cities grew out of markets where traders and buyers congregated, often outside a castle or abbey for protection, near a river or harbour for transport, or a river ford for traffic (Sarasúa and Scholliers 2005). As in the Roman city, housing and trade remained closely connected, in a form of vertical mixed use, with shops below and dwellings and storage space above. In the Italian city of Florence, for example, which started as a Roman settlement, fell into decay during the barbarian invasions, and was revived in the ninth and tenth centuries, its former Roman forum became the city's market place, and the local bishop protected trade in his role as temporal territorial as well as spiritual ruler. Similarly, Verona's Piazza Erbe was located in the decayed remains of a much older Roman Forum.

Over time, fairs moved from open-air market places accommodating commerce of all kinds to permanent covered structures constructed over the existing market square, some of which eventually became houses with shops on their ground floors. These sheltered market spaces could take the form of two or more stories, with an open arcaded space on the ground level for market activities and rooms above for government business. Out of these architectural arrangements grew enclosed halls, mainly for marketing, that were often commodious and impressive to attract trade: the Flemish cloth halls, Spanish *lonjas*, Antwerp butchers halls and more modest Cheapside selds are all well-known examples (Girouard 1985; Schmiechen and Carls 1999). Local rulers were always keen to build market halls and establish a monopoly position in order to derive the most rent from these food-related trading activities and spaces, and income could be obtained by renting stalls, taxing sales and letting land on which stall holders could build houses (Girouard 1985: 18; Schmiechen and Carls 1999).

FIGURE 3.2: Market hall entrance, Porto.
Photo: Susan Parham.

In most British towns, food markets also took place in the open air, on particular days during the week, and sometimes in a designated marketplace (Schmiechen and Carls 1999: 3). Many towns were known as market towns and their markets specialized in one or more regional products. In the closed economy of the Middle Ages the town market was a precious asset and its privileges were strictly maintained. Market tolls levied against non-resident users were the town's main income, which acted as a break on growth. Meanwhile, artisans and shopkeepers' livelihoods depended on their nearness to the market place, with the most desirable locations in the market area and the streets towards the town gate (Kostof 1991: 48). In urbanism terms, these towns were already developing what might now be described as a walkable catchment around their markets, with a density gradient centred on the market as an activity node.

Often, the circulation pattern relating to the central market was rationalized to suit new urban conditions: for instance, Roman grids became modified, with new streets pushed through at the weakest points (Kostof 1991: 48). Clearly, the connectivity of the market to its surrounding area was vital, and improvements were made to increase permeability and legibility. Demonstrating the conscious importance given to connectivity were the *bastides* of south-west France. These planned towns were an early development in terms of explicitly designing in markets. Configured as a rigid grid centred on an arcaded market square, in design terms, the *bastide* was a very well-connected urban form, providing many ways through. Its market form was also robust in the sense of being long lasting. In the *bastide* of Monflanquin in Gascony, for example, the weekly market is reputed to have been held in the central square and surrounding streets every week since 1260.

With city growth, sometimes market places themselves were enlarged, or separate markets were established for fish, corn, butter or other foodstuffs. This was not always a positive process. In Nuremberg, for example, the wholesale expulsion and burning of Jews in 1349 enabled the market place to spread out over the former Jewish quarter (Girouard 1985: 69). Buildings constructed over the original market square often had two or more stories with a ground-level market and houses and city government on top. Open arcaded buildings of one storey were a variation on the theme, and many of these remnant halls can still be seen (and are still used) in French towns today, while market halls themselves became increasingly commodious and impressive to attract trade (Girouard 1985: 69). It is worth noting that the close connection of market and government was a sign of the centrality of markets and their potentially anarchic nature. Civic control could never be too far away.

In design terms, the market squares and quarters of medieval European cities achieved a series of outdoor rooms providing a sense of enclosure for public space that promoted diversity of interaction. By the Renaissance, Alberti was able to argue for both private architecture and public infrastructure that would maintain handsome and healthy towns (Rykwert 1988), and markets could be considered a critical element in this infrastructure. City squares (the natural home of food markets) were also the centrepieces for idealized fortified cities, such as Scamozzi's Palma Nova started in 1593 (Morris 1994: 136). Before the physical connection of housing and trade was broken in the modern city, people lived as near as they could to the centre (Frost 1991). As seen in ancient examples, houses above food shops and warehouses became a common urban typology, and for the merchant and artisan classes this meant trade conducted on the lower floors, living space above and storage in the attics. From the fourteenth century, food shops became more common through the conversion of market stalls (on sufferance or by concession) in other parts of town. Such shops had a similar form, with goods sold from an open counter letting on to the street, which was shuttered when the shop was not in operation: a form that can still be seen functioning in the medieval centre of York and other historic towns in the United Kingdom today. Food shops are discussed in more detail in the next chapter.

Continental Europe was well ahead of Britain in developing permanent market structures from the medieval to the early modern period (see Figure 3.2), while British experience demonstrates both the longevity of traditional arrangements and the influences of developing capitalism on market spatiality. By the eighteenth century the typical food market in Britain still consisted of a ramshackle selection of stalls around a market cross (Schmiechen and Carls 1999: 4). The market came to be considered a public nuisance because 'by its very openness' it encouraged lower-class lawlessness, particularly in regard to food (Schmiechen and Carls 1999: 4). Given there was by this stage an urban underclass permanently close to starvation, the market became the locus for fraud, begging, hawking and theft, as well as drunkenness and disorder. Food riots became endemic, and there was also the problem of overcrowding of stalls, congestion, often-malodorous shambles and complaints about animals driven or escaping through the streets: the market increasingly outraged bourgeois sensibilities (Schmiechen and Carls 1999: 11).

Some towns decentralized markets to different locations to deal with overcrowding and traffic but, in fact, often generated more street trading 'obnoxious to respectable town dwellers and town officials' (Schmiechen and Carls 1999: 24). Although street trading was never entirely stamped out, in many towns authorities attempted to draw marketing into single buildings, and, by 1886, two-thirds of Britain's principal boroughs had prohibited marketing on streets, and about half of all towns had developed some kind of enclosed space (Schmiechen and Carls 1999: 24). These issues reinforced the desire of city authorities to gain control of the management of markets and to organize them in ways that would exert more control over anti-social behaviour and make them safe for middle-class women to visit (Schmiechen and Carls 1999: 24).

By the nineteenth century, there were numerous examples of semi-covered structures and some open-sided market halls under enclosed town hall buildings, as part of a wider distribution infrastructure required to support the food needs of increasing numbers of urban dwellers (Scola 1975; Scola and Scola 1992). The market halls built in new industrial towns, for example, formed similar functions to market places in country towns (Cherry 1972: 52). Similarly, as Schmiechen and Carls (1999) point out, by the same period, in many towns, open market places, street markets and semi-covered markets in town squares were clearly suffering from issues of illicit behaviour and poor food quality, as well as sites that had become too small, while attempts to decentralize had shifted congestion and other problems to new areas. As city councils gained control over markets they used the chance to replace problematic market areas with (to paraphrase) well-regulated and designed building where all classes could mix in peace, elegance and comfort (Schmiechen and Carls 1999: 31).

This process occurred in a wide range of English towns and cities, often resulting in the development of fine market buildings. Here and elsewhere, technical innovations in materials and construction methods allowed large, light and airy spaces to be constructed. European market halls (like Baltard's extraordinary iron halls, Les Halles in Paris, of which more below) began to display these innovations in an explicit way: their architecture and urban centrality also grandly reflected the civic virtues they were intended to symbolize. Schmiechen and Carls (1999) argue that the market hall was also implicitly progressive because it cut across social class and gender divides, providing a place where women could be in public space, to work as food sellers and form the majority of buyers of the fresh food that the market celebrated. Ebenezer Howard's proposed Crystal Palace at the centre of his diagrammatic illustration of the urban functions required for a new Garden City perhaps demonstrates most clearly late-nineteenth-century aspirations for food (and other) consumption elevated to its most civilized, convivial level. Yet, paradoxically, the food-retailing transformations ushered in by the invention of the market hall as an urban food typology sowed the seeds of its own destruction and brought in new food-based urban forms:

> As the first large-scale, environmentally controlled general merchandize retail space, the market hall paved the way for the department store, the supermarket, and the shopping mall – all of which made the old market halls look shabby and obsolete. Yet the market hall introduced glass roofs, the multiple shop front

façade, several shops under one roof, and large-scale ventilation, heating, and lighting, as well as many modern display and pricing practices. Equally important, the market hall was often the centrepiece of urban renewal. (Schmiechen and Carls 1999: x)

FOOD MARKETS AND THE DEVELOPMENT OF THE MODERN CITY

Despite the integrative purpose and placement of the urban market hall, from the middle of the nineteenth century onwards many plans for new urban areas were explicitly based on a physical separation of activities including food wholesaling and retailing. The preference for *rus in urbe* was especially strong in the English-speaking world, and between the latter part of the nineteenth and across the twentieth centuries, a number of significant trends for food saw its presence decline in the centre of cities. These included the moving of many wholesale markets to peripheral locations, the decline of traditional covered markets, the demolition of wholesale and retail markets, and eventually the reorganization of much consumption in supermarkets, lock-up suburban malls and then to out-of-town supermarkets (as explored in Chapters 5 and 8). Effectively, through these activities, the capacity of public space to make outdoor rooms centred on food was under attack. Camillo Sitte articulated the public space case in the late nineteenth century in reference to the development of the Ringstrasse in Vienna, where he recognized that a significant area of decline was in the role of the public square:

> The lesson of urban history was that public spaces must be viewed in three dimensions, as volumes carved out of the solid of the built fabric. For thousands of years squares and streets have been enclosed units, and served as legitimate urban stages of social interaction. That fundamental social value of public spaces was being sacrificed in the modern metropolis to the functionalist calculations of traffic engineers and the grandiloquent agoraphilia of planners. (Sitte, quoted in Kostof 1992: 138)

Changes in the United Kingdom between 1890 and 1939 offer a test case of the stark choices between decline and modernization that were faced by market halls in many places. Public markets were often allowed to deteriorate by town officials refocusing on other urban infrastructure priorities. As middle-class families moved out from the centre of towns and cities, with developments including the rise of Garden Cities, and the growth of 'metroland' allowed by the spread of rail infrastructure, the customer base changed and declined. So-called modern food wholesaling methods, which broke the connection between farmers and consumers, became ubiquitous. Co-operatives and chain stores grew up and provided greater competition to food markets. Similarly, food consumption patterns changed between the two world wars, and shops again took market trade. Increased access to cars and buses (of which more in Chapter 5) also meant smaller food markets could not compete with larger ones, which could now draw a more distant catchment.

Street hawking and street markets returned, offering still further competition to covered markets.

The Second World War had a huge impact on food supply and thus markets, and despite periodic attempts to revive and redevelop food markets, by the post-war period many had either become derelict or had been demolished. Moreover, post-war urban redevelopment schemes on modernist principles identified markets with an outmoded past, and the advent of supermarkets proved the death knell for many markets across the United Kingdom and Europe (Schmiechen and Carls 1999: 211). The decline of Borough, Broadway and Exmouth Markets in inner south, east and north London in the twentieth century (documented in Parham 2012) are examples of a common trajectory of decay and dilapidation, as a range of social and economic forces came together to transform food wholesaling and retailing preferences away from traditional food markets and high streets and towards the edge-of-town distribution centres and supermarkets visited by car on a weekly basis.

Although substantial market hall complexes continued to be developed and used as food space into the early part of the twentieth century, in the United Kingdom and elsewhere the close physical connection of housing and trade described earlier in the chapter largely disappeared from new urban development. With the decline of food markets, most cities lost the urban design features of enclosed space forming outdoor rooms that had been common to all market places and halls, and the urban areas around them. Instead, cities and towns turned to the low-density, land use-separated urban patterns for food supply and retailing which are the subject of Chapters 5 and 8. The loss was not only spatial. The physical design of markets in cities had given food consumption a central role in convivial public space, and this allowed the spatial intertwining of the material, political and spiritual dimensions of life on an everyday basis. Food markets were critical elements of the urban armature that created and reinforced a sense of civic connection, and their social and cultural role was powerful in supporting conviviality.

Carol Field (1990) has written of the way that in Italian towns, where traditional market arrangements are often still relatively resilient, long-term food traditions in public space – food days, rituals and festivals – are also in evidence. Over their long history, certain foods – especially meat – came to be associated with market places, and even with cities as a whole. Waverley Root (1971: 282–3) reports that 'after the theatre, especially on Saturdays, the Milanese would flock to the restaurants of the *Porta Garibaldi* and *Porta Ticinese* to eat *Busecca*', a dish of caul and curly tripe. This was an early example of elite appropriation of working-class food (a process that is now ubiquitous), but market food was generally the food of the poor who worked there. The onion soup of *Les Halles* in Paris or a bloody steak for meat workers at a pub in the early morning at Smithfield market in London are well-known examples.

It may have been this very symbolism that was one of the reasons why market halls fell into oblivion. Food markets went out of fashion not necessarily because they were no longer functional but because they did not fit into thinking about the nature of public space and economic activity that shaped the design responses of town planners, architects and government officials (Schmiechen and Carls 1999: 203). A preference for massive, unitary town rebuilding schemes required the removal of 'undesirable' and 'inefficient' pre-war shopping streets and markets (Schmiechen and Carls 1999: 203) to usher in a more vehicle-friendly spatiality and, over time, shopping based on supermarkets. As modernist ideology was predicated on sweeping away past attitudes and practices to city building (Trancik 1986), one of the crucial actions was to separate out food wholesaling and consumption-related activities from other land uses. In so doing, cities also lost some of the critical design elements that reinforced social interaction and thus conviviality through food. New development intentionally avoided building housing over food shops, arcaded edges to streets, human-scaled enclosed squares, lively centres based on food trading or in close physical proximity to market, city government and housing. A Corbusian-inspired vision of the city as a set of towers in a green park replaced earlier approaches where food markets had been central to constructing the enclosed outdoor rooms that made up public space. As designers note, though, what came to replace these elements did not offer the same social diversity, pleasure or conviviality (Alexander et al. 1977: 247).

Just as retail food markets declined, the loss of inner-city food wholesaling functions was also a structural trend that could be seen across developed world cities. In perhaps the most famous example of civic vandalism on a grand scale, Baltard's wonderful iron market halls at *Les Halles* were demolished in the 1970s, leaving a large and problematic hole in the centre of Paris, filled by an unsuccessful subterranean mall that wrecked the urban fabric and undercut the connectivity of one of the liveliest, most historic *arrondissements* of the Right Bank. In the mid-1970s United States, James Rouse's Faneuil Hall development in Boston, meanwhile, became the precursor to other 'festival markets' based on suburban mall typologies which were inserted within historic buildings and spaces, thus beginning a broader trend for suburbanizing the food-related spatiality of traditional cities that can be seen at flagship developments like Covent Garden and other more local shopping streets. 'Rouse took malls from the suburbs to historic or scenic urban locations as part of a larger strategy of urban revitalization. ... He reconfigured the merchandizing mix to feature food and souvenir shops alongside noncommercial tourist venues such as museums and historic ships. Most importantly, Rouse blurred the boundaries between the mall and the urban setting' (Crawford 2002: 32).

In the Australian city of Adelaide, meanwhile, the redevelopment of the East End Wholesale Market in the late 1980s for luxury apartments and boutique shops, cafés and restaurants demonstrated that when the produce market was moved out of the city centre to a remote outer suburb, there were two connected food results. First, the city lost a significant element supporting its food-centred spatiality, and, second, the space for direct relationship between food growers and sellers was destroyed, undermining the economic opportunities for growers at a smaller scale who tended to be more gastronomically inclined than large-scale wholesalers. Like Spitalfields Market in inner east London, redevelopment commodified the food space, maintaining a gloss of food-related activity that made design references to the now defunct market, while undercutting its gastronomic, cultural and economic diversity. In contemporary London, similarly, arguments over the future use of parts of Smithfield Market (one of Europe's longest surviving central city meat markets, and including a disused wholesale building) continue to rage. A proposal was recently rejected which intended to gut some of the historic buildings to their façades and refit them with a land use mix in which the possibility of the return of any food wholesaling functions would be excised:

> Rebranded, inevitably, as 'Smithfield Quarter', the £160m development, by the architect John McAslan + Partners (who recently completed the well-received new concourse at King's Cross station) proposes turning the fish market, poultry market and general market, which sit at the western end of Smithfield and have been disused for a number of years, into boutique shops and restaurants. The plan also introduces a series of 20m high office buildings within the market's existing footprint. (*The Londonist*, 2013)

FOOD MARKETS MAINTAINED: FOOD MARKETS REVIVED

Although decline and removal has been the predominant trend, in a number of cities, the late-twentieth- and early-twenty-first-century period has also seen a revival in the fortunes of urban food markets (O'Neil 2014; van Odijk 2014), with certain market areas transformed into new kinds of food-centred spaces conceptualized as urban food quarters (Parham 2005, 2012: introduction) and associated with both gentrifying spatial tropes (Zukin 1982; Parham 2005, 2012) and increased conviviality and sustainability (Robinson and Hartenfeld 2007; Parham 2012: 10). Social movements such as Slow Food and Slow Cities (Paxson 2005; Petrini et al. 2005; Pink 2008), the burgeoning of farmers' markets and hybrid market forms like Borough in London, L'Enfant Rouge market in Paris (Parham 2012) and Torvehallerne in Copenhagen all speak to the changing design perspectives and social attitudes to market space, while fieldwork interviews suggest a longing for urban food space to create convivial outdoor rooms (Parham 2012). Focusing on the complex, paradoxical social space created by the architecture and infrastructure of a revived public food market in Copenhagen, one blogger explains in relation to the new Torvehallerne:

> Copenhagen's IsraelsPlads was once home to a vibrant outdoor market, which took place daily in the plaza until the mid-20th century, when a wholesale market was established in Valby, on the outskirts of the city. After decades of neglect, the site at IsraelsPlads has been revitalized as a public food market, conceived and designed by Danish architect Hans Peter Hagens. Torvehallerne opened in 2011, after 14 years of political negotiations, design and construction processes, and is almost instantly one of the busiest public spaces in a city full of them. (archinect.com)

The writer articulates their sense of ambivalence about its charms, because as at Borough and Broadway Markets in London (Parham 2012), this space is also understood as a form of habitus production for hip individuals prepared to pay for being close to stylish people, places and things, in which the market's proponents invite all the chain shops around Copenhagen to squeeze in a miniature version of themselves into the new, trendy starchitect-built glass halls so the tourists and Copenhageners who don't know better will have yet another place to buy a latte, a box of frozen dinner or an overpriced premixed cupcake. ($5.50 for a small cupcake, anyone?)

So, that is state of the art 'farmer's market' anno 2011. As you can probably tell, yours truly is slightly underwhelmed. But now that I've gotten all this out of my system I have to admit that it's better than nothing. Even though things are overpriced with discrete [sic] hints of tourist trap and snobbish wannabe hip city culture, I actually have fun there. (umamimart.com)

In certain urban cultures, too, more long-term, traditional markets have, to varying degrees, been able to maintain or even increase their use in ways that are very convivial. Interesting visually based work on Hong Kong's wet markets suggests that they are an unacknowledged success story in a neoliberal urban context, places that allow poorer residents access to fresh meat, fish and vegetables, offer employment to those unable to find jobs in other sectors; and provide ordinary, everyday 'spaces that are contrary to the hyper-globalized world city that Hong Kong strives to be' (Goldman et al. 1999; Blake 2013). As Blake explains,

Finally, these ordinary market sites offer a glimpse at urban values that extend beyond that of profit to include community, interconnection, and support for one's fellow humans. As such they are sites, which as Marcus argues, are a force of multi-dimensionality working against a neoliberal tide driven by greed. It is these aspects of the market spaces that are illustrated in the photographs presented in this photo essay.

FIGURE 3.3: Fruit stall and cart, Alexandria.
Photo: Matthew Hardy.

In certain Western cities, in particular, such market spaces have been sustained or redeveloped despite the depredations of modernist built-form approaches. Increasingly, the social and economic forces driving customers away have been balanced by gastronomically informed cultures that emphasize quality, freshness and daily shopping practices, and design approaches that support these (Parham 1990, 1992, 2005, 2008, 2012). In Stockholm, for instance, the historic late-nineteenth-century Östermalm food hall still operates very successfully albeit as a kind of *slow food* food court as much as a regular food market (pers.obs.), while in Helsinki a combination of outdoor market space in summer and enclosed food halls in winter reflects the interplay of a gastronomic culture and extreme climatic conditions. Thus, 'Helsinki's open-air markets are lively gathering places in summertime, but when autumn comes local gourmands head indoors to shop at the city's fine market halls. Hakaniemi Market Hall is filled with food stalls, many run by small family businesses selling unique delicacies' (VisitHelsinki).

Both redevelopments and newly created market spaces are emerging in a variety of places. As Lee (2009) has noted in relation to Stockholm, these offer qualities of authenticity and atmosphere that twentieth-century food-retailing forms have failed to match. O'Neil (2014: 27) provides examples from Spartanburg, South Carolina and Flint, Michigan, of the development of new farmers' markets in which community-based design is argued to be central to the process of moving from a transaction-based to a place-based market experience. In Spartanburg, the street market was shifted from its 'mediocre sidewalk location' to 'a place designed for and by the community' in a historic tram depot. In Flint, again involving the community in the design process, the market was moved from a site on the town's outskirts to a downtown building that formerly housed the city's newspaper. In both cases it is salutary to note that downtown space became available through the loss or displacement of other civic infrastructure. In Egypt (see Figure 3.3) and as van Odijk (2014: 28) reports from Moroccan experience, despite hard economic times in North Africa, 'markets are still very much the centre of a town or city's daily life, and ... they are growing, demanding new and better space, and questioning the historic format of the *souk* as they do'. Here, the winners of an architectural competition to design a sustainable market square focused on developing both a food market and an open public square as a centre supporting 'strategic objectives for social cohesion' and have employed traditionally inspired, climate-adapted canopies to deal with intense heat (van Odijk 2014: 28).

Similarly, in central Granada, the relatively new Mercado Central San Agustin, despite using an architectural idiom for its built structure that is a stylistic contrast with its location, relates sensitively in scale, grain and materials terms to the surrounding urban fabric. The new market hall helps form one side of an enclosed outdoor room in an urban square which provides a foreground to the hall's façade, but is more than just an aesthetically pleasing setting for the food market, acting as a bustling space for small-scale food trading activity in its own right (Parham 2005). The market maintains urban continuity in social practice and physical form terms in which food is the basis of an active place used for produce deliveries and market stalls, supported by simple, elegant landscaping and space to sit or promenade (Parham 2005).

Of course, not all food halls are this uncommodified: there are equally examples where food market halls have been redeveloped by supermarket chains, and while giving the appearance of a diverse range of stalls and food options, are actually in one ownership (Parham 2001). In such spaces the reality of vertical economic integration is obfuscated by spatial design to suggest diversity and a convivial basis for the space. However, this should not detract from the wider point that the outdoor market and indoor food hall examples discussed above, and others like them, are not romanticized or nostalgic places, as has been argued, but reflect in their food-centred spatiality an ongoing urban reality of food retailing for many across Europe and other regions (Parham 2008, 2012). Such food spaces have suffered in architectural and urbanism theory terms from mistaken perceptions about what constitutes an appropriate modernity in urban place shaping (Parham 2008). In Paris and most other French towns, for example, urban *arrondissements* still maintain a strong tradition of local open air and covered markets. A *quartier* like the Tenth in Paris, for example, boasts the covered Marché St Quentin and Marché St. Martin, as well as the open air Marché Alibert on Sundays. However Paris, too, has seen a burgeoning of *supermarché* of the 'fake local' variety (Oram, MacGillivray and Drury 2002: 2). Evidence from elsewhere suggests the cultural preference and immigrant shopkeepers are keeping small food shops alive among minority ethnic communities (Collins et al. 1995; Goldman and Hino 2005). It is also clear that food consumption practices are shifting away from being solely market based, at least for some. At the same time, as an account of French shopping practices reported by Friedland (2005) suggests, while market shopping is increasingly mixed with other forms of food buying it still garners support for reasons that seem both gastronomic and convivial:

> CD: I think this is mostly a myth, based on a truth from a long, long time ago. What a lot of French people do nowadays is diversify their sources for food: They will buy most things at the grocery store, some from their local shops or at the market (bread, meat, fish, produce, cheese, wine), and some from local producers even (if they live in a region where there are any). But since most of these things are available from the grocery store (and some supermarkets have a very good selection of fresh products), many people do all of their shopping there because it's more convenient.
>
> I do some of my shopping at the grocery store, but I try to support my local food shops by buying from them what they are specialized in. For instance, I would rather buy cheese from the fromagerie, meat from the boucherie, and produce from the open-air market than buy them at the grocery store: It is more time consuming (several stops vs. just one) and usually more expensive (though not always), but the products are better quality and I enjoy the advice and human contact.

DIVERSITY AND ADAPTION: LONDON FOOD MARKETS AS AN EXAMPLE

Although, as traced above, many traditional markets declined to the point of oblivion in the twentieth century, the strength of traditional working class and minority ethnic consumption practices in maintaining long-standing food markets should not be underestimated. In Berlin, for example, the Turkish weekly market (*pazar*) in the Kottbusser area of Kreuzberg was reported in the 1990s to wind several blocks along a canal and offer dairy products, meat, flowers, vegetables, fruit, bread, spices, olives, pork-free sausages and Turkish tea (Metcalf 1996: 152). In London, too, although many food markets are now again at the point of transformation owing to structural forces including gentrification, they are also thriving. Market places and streets like East Street, Southwark; Inverness Street in Camden; Ridley Road Market in Dalston Kingsland; Walthamstow Market in Walthamstow; Queen's Road Market in Newham; and Shepherds' Bush Market in west London have continued to adapt as their ethnically diverse working class (and now middle class too) catchments prefer them. Site observations suggest that their users mix their market consumption with supermarket-based food buying. Interesting work on such ordinary places suggests that they can be sites for convivial everyday practices and relations among diverse social groupings (Duruz 2005; Hall 2012).

These London markets tend to have minimal spatial provision made for them apart from location on or abutting a high street (the relationship between East Street Market and Walworth Road in Southwark is fairly typical). They tend to comprise dismountable structures that are removed after each day's trading, although there may also be semi-permanent and permanent stalls like those at Ridley Road and Shepherds Bush. Exceptions include the covered markets in Brixton and Queen's Road, which are located in fully developed, long-standing market halls. Often, such street and covered markets are grubby and down-at-heel. Sharp practices such as customers being given rotten fruit and vegetables are not unknown, some markets are known to sell illegal foodstuffs such as so-called bush meat, and the chaotic, sometimes aggressive atmosphere may deter would-be buyers, as well described by Howard Jacobsen in his Manchester-set novel *The Mighty Waltzer* (1999), which depicts a trader's traditional spiel. Just as prior to the development of the market hall, such markets may be considered exotic, other, lower class or even dangerous. As one visitor to Ridley Road Market explains,

> Its difficult to fully describe the riot of colour and noise that is Ridley Road market for anyone who hasn't been there. Places it reminds me of are the medina in Marrakesh, Electric Avenue in Brixton, and the old Moore Street market in Dublin – not sure if thats any use. Its basically a market where you can get your hands on mostly anything, with the likelihood being you'll pay significantly less for it than in a regular shop.
>
> The central 'avenue' of stalls are mostly fruit and veg at the top but then turn into other types as the market thins out towards Dalston Lane. ... In the real shops behind the stalls, there is an even wider variety of goods on offer. These include: hairdressers, Indian food, a 24 hour bagel shop (a reminder of the street's Jewish history), the grimiest greasy spoon caff I've ever witnessed, lots of Halal butchers and fishmongers with boiler (sic) chickens hanging by the neck and bowls of spotty Tilapia for £5, fabric rolls, electrical kitchenware, and heaps of west African suppliers, including crates of live giant land snails.

I generally do more staring than shopping here. But when I do buy things I go for the fresh herbs, the egg stall, the £1 bowls of cherry tomatoes or sweet peppers and the avocado and mango stall at the very top. I also shop for fresh bread in the Turkish supermarket TFC down towards the east end. (Derry 2008)

FARMERS' MARKETS AND HYBRID MARKETS

Alongside these extant market forms, farmers' markets have long been a ubiquitous urban element in parts of Europe, including in Italy and France, although dying out elsewhere prior to a renaissance since the 1990s (Vecchio 2009) in new, more self-conscious forms. With a contemporary practice often situated within Alternative Food Networks (discussed at length in Chapter 9) farmers' markets and similar 'direct from producer' markets have sprung up in many Western cities and towns, although as Markowitz (2010) reports from Kentucky these have been more difficult to establish in low-income communities. Research findings also reinforce that in North America at least they attract a disproportionately white visitor base (Campigotto 2010). There are some thousands now operating in the United States; with over 450 in the United Kingdom in 2004 (Kirwan 2006) and more than 750 in 2012 (Bardo and Warwicker 2012). There are estimated to be more than 130 farmers' markets in Australia (Willson 2010), and those in New Zealand are argued to be providing new opportunities for small-scale food entrepreneurialism (Guthrie et al. 2006). A recent study of thirty-three farmers' markets in the Canadian province of British Columbia, meanwhile, found that they were injecting over $170 million Canadian dollars into the local economy annually (Connell 2013). Work on the attitudes of farmers' market customers in the Stour Valley in the United Kingdom suggests that short supply chains and organic, genetically modified (GM) free food are important elements in attracting buyers (La Trobe 2001). Farmers' markets' move into the mainstream has been reflected in the US government's Agricultural Marketing Service (2013)'s recent description of them as 'an integral part of the urban/farm linkage [which] have continued to rise in popularity, mostly due to the growing consumer interest in obtaining fresh products directly from the farm'. According to the Agricultural Marketing Service, this is a growing sector that allows

> consumers to have access to locally grown, farm fresh produce, enables farmers the opportunity to develop a personal relationship with their customers, and cultivate consumer loyalty with the farmers who grows the produce. Direct marketing of farm products through farmers markets continues to be an important sales outlet for agricultural producers nationwide. As of National Farmers Market Week (the first full week in August), there were 8,144 farmers markets listed in USDA's National Farmers Market Directory. This is a 3.6 percent increase from 2012.

Consumers are seeking out local foods for reasons 'ranging from its quality, freshness and taste, to more intangible attributes such as its benefits in contributing to local economy, preservation of cultural heritage and environmental issues, and higher food security' (La Trobe 2001: 3). Fresh food, contact with farmers and a sociable atmosphere that is not simply about monetary exchange all play a part (La Trobe 2001: 6), demonstrating the importance of such markets as sites for expressing conviviality. One intriguing finding of Vecchio's (2009) case study fieldwork in two farmers' markets in Washington DC and Naples was how, in the latter, obtaining cheap fresh food was the aim of a larger proportion of buyers (which makes Naples a convenience market in his view) rather than the desire to respond to more amorphous ethical, social or environmental considerations (which situates Washington DC as more of a community market). Farmers' markets in the mainland European context might be thought to be largely continuing culturally normative food-buying traditions, rather than expressing a radical, reflexive departure from these norms as does experience in the United States and other Western countries. As in the hybrid markets studied by Parham (2012), the issue of price is an ongoing area for contestation between farmers' markets and more conventional supermarket-based consumption. Yet commentary from market vendors themselves suggests that the actual experience of farmers' market consumption makes price comparisons a somewhat complicated matter rather than straightforward like-for-like comparisons:

> Fernleigh Farms' Fiona Chambers says customers often comment that her free-range meat and seasonal organic vegetables are cheap. '[But] If you say expensive compared with supermarkets, I would say probably, yes, but they are operating within a completely different food distribution system.' Fiona's husband Nicholas says it's about putting a true value on food: 'I mean 90 per cent of meat in supermarkets is put through feedlots and whacked up on hormones.' Bronwyn Cowan talks pragmatically in terms of selling to 'people who can afford to make food choices'. 'But we say to people, just eat less meat – a smaller portion – and really enjoy it. My pork is in the paddock on Monday and sold in the market on Saturday. You couldn't get that under any other system.' (Willson 2010)

There are now at least twenty formally accredited farmers' markets operating in London, and their role in consumption is complicated. For instance, buying organic food at such markets is understood in popular discourse as an obvious way of undertaking conspicuous consumption. The particular style and atmosphere of both farmers' markets and hybrid markets (discussed below) have provoked journalists into making comparisons between such shopping and buying a Mulberry handbag (Rayner 2011). This is situated as an act of narcissism with 'a greasy veneer of self-righteousness' (Rayner 2011); and, invoking a well-worn cliché about markets' supposed nostalgia, it is suggested that they are expressions of a bygone era that never existed. In other words, shopping at such markets is both about luxury food retailing affordable only to a self-regarding elite and pandering to a deluded view from market shoppers that they can change the world. Such perspectives are informed (although this is perhaps not recognized by those holding them) by broader assumptions about food's relationship with nostalgia and authenticity. By contrast, as prefigured in the Introduction, certain theorists have sought to redeem these contested terms and concepts from their normatively negative use (Parham 2008; Wills 2011).

More evidence-based research findings suggest that alongside alternative food consumption that reflects a stylish individual habitus, which, it has been argued, ushers in area gentrification (Zukin 2008), other more positive trends have emerged (Watson 2009; Parham 2005, 2012). As Watson (2009: 1578) notes following detailed empirical research in a number of London markets, markets offer unparalleled spaces for sociability that are often not recognized and 'contradict pessimistic accounts of the decline of public space'. Other primary research findings suggest that the physical layout of market space strongly supports this inclusive social quality (Watson 2009: 1578; Parham 2008, 2012). In inner London (and other places), the street markets and newer farmers' markets discussed above now coexist with hybrid food markets that cater to a very diverse community of interest not dissuaded by arguably higher prices for organic, speciality and bespoke produce from small-scale, local and artisan food producers (Parham 2008, 2012). Argued to offer a distinct break with past food market practices, these hybrid markets are not farmers' markets (nor do they wish to be) but share some of their ethical and sustainability concerns, with a gastronomically informed stress on food quality, origin, production ethics and freshness (Parham 2008, 2012). Their traders and some users argue that invidious price comparisons discount the way that market buyers forgo bulk for quality and recognize price encompasses aspects of sustainable production that the mainstream food industry externalizes onto others and the future (Parham 2008, 2012).

Established in traditionally shaped, formerly moribund market streets and spaces not only, most famously, in the historic setting of Borough Market, but also in market streets such as Broadway Market in east London, their design qualities have been central to a process of rapid food-centred renewal. One of the elements they may now incorporate is space for artisan food processing, as at Borough, and as is described in relation to Toronto's food scene (Biggs 2009: 32; Donald and Blay-Palmer 2006). These consciously reduce the spatial scale of some food processing and distribution in gastronomically informed ways. Another is a focus on extending food skills and education, with market-related spaces designed or retrofitted for such activities. Such places can now be conceived of as food quarters which although shading into gentrification, exhibit both sustainability and conviviality benefits that are based on a strongly positive interplay of new food practices and traditional urban forms (Parham 2012).

MARKETS AS DESIGNED SPACES – LESSONS FROM PRIMARY RESEARCH

Having explored a range of contemporary food market forms, the next part of the chapter briefly reviews examples from primary morphological and design research on food markets in Italy, France and Australia, to explore the shaping of markets that have resisted modernist dismantling, or have reconfigured city space in new ways to revitalize food's outdoor room (Parham 2001, 2005, 2012). Covering both traditional and hybrid markets that incorporate farmers' market-like elements, research findings suggest that despite their geographically divergent locations and trading styles, these markets are urban elements that have 'a family resemblance' in design terms (Lozano 1990: 55). They can be typologically grouped into four categories: open markets in urban squares with space for open stalls; semi-covered market hall structures in urban squares or streets; enclosed market halls in urban squares; and enclosed halls within a perimeter city building block (Parham 2001).

The morphological shifts at a market-like Campo dei Fiori in central Rome are long term and slow changing, with food remaining central to socio-spatial use day-to-day by local food buyers, although it serves now as a prime site for tourism as well. The Campo is reputed to be Rome's oldest food market and is a rectangularly shaped outdoor room that forms an urban square in the central city. Important elements in the design of the Campo include five-to-six-storey buildings with frontages to street alignments on all sides, and narrow openings into the square from surrounding streets, giving the quarter's streets appropriate solid-to-void relationships, and the Campo itself a proper degree of spatial enclosure. There is sufficient local population density to provide a walkable catchment in a highly permeable area where the market is an activity node towards which various paths direct the walker. Although its path complexity might create legibility issues for those unfamiliar with the area, this also allows exciting serial vision opportunities and vistas. Once at the market square, the rows of produce stalls provide 'streets' for pedestrians, demonstrate a comfortable height-to-width ratio and have human-scaled stalls. The stalls also allow significant opportunities for visual richness and personalization and are supported by local food shops including the famous Antica Norcineria Viola (devoted to pork products), as well as bars, *enotecas*, *osterias* and restaurants. All these qualities and food elements contribute to giving Campo dei Fiori a strong sense of place.

Produce is brought into the square by small vehicles, which, by virtue of their modest size, avoid undercutting human spatial scale and the pedestrian-friendly atmosphere in the way that larger truck servicing areas do. Boxes on the back of APE vans add to the lively character of the space and also reflect the predominance of very small-scale food production and retailing, a characteristic of markets historically, implying a fine grain of land use. The daily market helps to support a highly diverse and fine-grained set of food-related frontages on the square and in surrounding streets, such as cafés, grocery shops, wine merchants, bakeries, restaurants, bars and cafés on ground floors, with offices and housing above. As noted previously (Parham 2005), a market-like Campo dei Fiori also attracts one of the most common criticisms levelled at markets of high-quality foods in historic locations; that they are gastronomic tourism zones trading on nostalgia about a lost way of life and pandering to the obsessions of wealthy food-literate tourists. It is certainly true that such markets are attractive to affluent tourists. The Campo, in the centre of historic Rome, draws a global community of interest, yet fieldwork observations of local shoppers suggest it is also a much-loved and well-used local foodscape too (Parham 2005).

In Trastevere, meanwhile, a traditional working-class area of inner southern Rome, a small daily food market takes up one end of a triangular square, the Piazza san Cosimato. In this case, the market reinforces the square as a strongly enclosed outdoor room and is supported by a range of local food shops, cafés, bars and restaurants. The market's permanent semi-covered stall structures are found at the well-enclosed northern end of the square where they contribute to a comfortable, human-scaled place. Trastevere market provides substantial opportunities to see, smell, taste and touch fresh, affordable, seasonal food, which in turn helps market produce to consumers. Like Trastevere, a number of outdoor food markets successfully rely on simple construction materials, a simplicity made possible by a supportive urban design context. In the French town of Villeneuve-sur-Lot, for example, stall elements including trestles and umbrellas are readily set up and dismantled while giving a strong structure to the food market located in an arcaded *bastide* square that forms an outdoor room. Buyers congregate at cafés under the arches on all sides. In terms of the scale of the catchment and the market's connectivity, on two site visits it was noted that the buyers coming to the market predominantly arrived on foot or by bicycle. While there is car parking on peripheral boulevards near the river Lot, the streets around the square are pedestrianized and the high-density, mixed-use fabric of the town reinforces this walkable catchment.

A typological variation on the covered market hall is the semi-covered hall in an urban square, or encompassed by a net of streets, as at Borough in London. The semi-covered market can encompass a roof without walls, or a less physically enclosed, partly roofed structure or series of structures. Some such markets are architecturally exquisite, as in Foix, in southern France. Others, like the recently demolished Testaccio market, in an inner southern quarter of Rome, are very ordinary in design terms, but bustling and vibrant. Testaccio's recent history is an illustrative example of the way that food market spaces and related food land uses are being renegotiated in the contemporary city. Testaccio ('the ugly head') took its name from the mound of broken terracotta pots that grew up in the area as workers drained imported foodstuffs from large amphorae into smaller pots for sale, tossing the empty vessels on to a growing rubbish heap. Terracotta was found to be a good insulator; tunnels and grottoes were dug into the mound for storing wine and food and thus 'this practice began a tradition of gastronomic activity in the area' (Roman Urbanism, undated).

Fieldwork observations undertaken in the early 2000s showed the food market attracting a large number of local pedestrians, as well as shoppers on bicycles, scooters, motorcycles and in cars. A major tram and bus network in the neighbouring Via Marmorata connects the quarter to central Rome to the north of the Tiber, and the regional train station of Ostiense is located nearby, meaning that shoppers could visit the food market from further afield by these means. While the food market building itself did not have any architectural distinction, it took up a city block in an area configured by a strong rectilinear grid of streets demonstrating a number of positive urban features. These include a fine-grained land use mix supporting the area's food-centred qualities, with medium-to-high housing densities that provided a thriving local population catchment for the market. The local built fabric in Testaccio is generally of vertical mixed use, comprising cafés, restaurants, shops and offices on the ground floors with housing above. Buildings are contiguous on regular city blocks and have strong edges to the street. These outdoor room 'walls' contribute to good height-to-width ratios in streets abutting the (then) market; making human-scaled public space reinforced by an absence of large servicing and delivery vehicles. Food shops and cafés in the area, including Panificio Passi, Sicilia e duci,

Gatti – Pasta all'uovo and ca Palombi, supported the market's economic vitality; with one of Rome's best food shops, Volpetti, found on nearby Via Marmorata.

The food market operated at this location until early 2012 when issues with local drainage and sanitation were reported to have led to a decision by the municipal authorities to knock down the market and rebuild as a new, privately managed hall nearby. After a lengthy process of negotiation with existing stallholders and the local community, the market was demolished and relocated, to mixed reviews. One unwelcome result of the move was the designing in of a more commodified suburban retailing model (as discussed earlier in the chapter) that included an underground car park built below the market, the appearance of a supermarket as part of the new development, higher rents for stallholders and the jarring presence of food out of season in stallholders' displays. On a trip to the new covered market, one commentator also noted a loss of vitality and sense of community that might also be described as a decline in its convivial quality:

> The thing that struck me the most was that it was much cleaner and brighter than the old market. However, there weren't any throngs to speak of and, because it is so much larger than its predecessor, the atmosphere didn't seem so 'communal' or cosy – that's if 'cosy' is an adjective that can be applied to a market. It is early days yet and I am sure it will grow into its new spaciousness over time, and redesign a character all of its own. For nostalgic old-timers like me, however, there is a nudging feeling that yet another bastion of 'romanità', i.e. the romanness that made Rome and its home-grown Roman citizens what they were, has been directed to the 'to be filed' archive. (My home food that's amore 2012)

Testaccio's market redevelopment also needs to be contextualized in relation to the abandoning of the traditional slaughterhouse complex of Mattatoio, designed by the architect Gioacchino Ersoch in the 1890s, which gave Testaccio its particular flavour as a food quarter. Slaughterhouse workers were given the less saleable meat such as offal, heads, and tails, and took this 'fifth quarter' to local taverns to be cooked in return for wine; spawning particular dishes like *coda alla vaccinara*, which became not only a part of Testaccio's character and sense of place but also a famous dish in Roman cuisine. Like Parc de Villette in Paris, once the slaughterhouses had been turned into a multi-use complex including an art gallery, museum and music school, fundamental qualities of the quarter were lost. Reports of the new market and the redeveloped slaughterhouses appear to support the sense that Testaccio's food-centred character has been challenged and possibly undermined by these intertwined spatial transformations and processes of economic commodification.

Turning to the Australian case study examples, the Queen Victoria Market in central Melbourne is nineteenth century in origin, developed shortly after the founding of the city. Comprising a combination of enclosed and semi-covered halls in a location next to the city centre, the area is characterized by a strong grid of streets in which the axis has been shifted, making a number of triangular city blocks. The market's catchment constitutes a community of interest, drawing buyers from across inner Melbourne, as well as tourists, for food quality and atmosphere. A weekly night market operates, and plans for a gourmet food hub influenced by Borough

Market were mooted at the time research into the market was undertaken. The residential area around the market is medium-to-high density, close to universities and the central business district. Historically, this is an area that has housed the left-leaning middle classes: artists, activists and academics who favour market shopping for cultural and political reasons, but the market has a diverse visitor catchment not restricted to those with a particular habitus. The fresh fruit and vegetable area of the market is found in its semi-covered spaces while its enclosed halls include an early-twentieth-century Art Deco dairy produce hall (also known as the deli hall) which now houses a range of dairy produce, small goods, groceries and food stalls. The fixed stalls in the dairy produce hall interior have high-quality finishes and have shown considerable robustness as their uses have changed over time, without compromising the human-scaled stall structures. Café seating on the footpath outside provides places for people to eat the food and drink the coffee purchased at these stalls, with traffic intrusion from local streets mitigated by a glass barrier at the pavement edge to screen market users from passing cars.

A perennial issue in food markets is how buyers carry their goods around the market and then home. Traditionally, markets were held on enough days of the week to lighten this burden and in some of the European fieldwork sites it was apparent that buyers would be taking their purchases home in the capacious straw baskets and cloth shopping bags they had brought with them. In the Queen Victoria Market's case, like many markets, there is a tacit acceptance that at least some buyers will be arriving by some kind of private vehicle rather than on foot, so the issue then becomes more about moving purchases around the market space in a convenient, comfortable way. Here portable trolleys are used to carry goods around the market and transport these to nearby car parks. It is worth noting that young and old, men and women alike, use their own trolleys at Australian markets without the embarrassment they seem to engender in the United Kingdom, where they are associated with elderly people in an often disparaging way. In fact, trolleys have stylistic conventions and fashions, with 'retro'-inspired fabrics of the 1940s–70s as particular favourites.

Like the Queen Victoria Market, the Adelaide Central Market, in the centre of South Australia's capital, was also established in the mid-nineteenth century shortly after the city was founded. The central city square mile is laid out as a grid with five squares breaking up the uniformity of the city blocks, and the food market abuts the major central city square, Victoria Square, to the west. On its eastern side it is hidden behind a high-rise hotel and a low-rise court building redeveloped from a defunct department store. Also on the eastern side, but running north-south, are a series of food-related spaces making up a cross shaped arcade specializing in butchers' shops and fish and delicatessen counters at the southern end, and providing an internal entrance to a supermarket on the northern end. On the southern and northern frontages the market is enclosed behind speciality food shops, butchers, fishmongers, cafés and small restaurants. On both the western and eastern sides the market runs into internal malls of small food-related shops, food courts, cafés and two supermarkets. The area to the immediate west of the market houses Adelaide's Chinatown, with Asian shops and a food court with Vietnamese, Burmese, Chinese (various regions), Laotian, Cambodian, Malaysian, Philippino, Indian, Sri Lankan,

Japanese, Korean and Thai food available, reflecting various waves of immigration to the city, especially in the post–Vietnam War period.

The Central Market is an important land use element and social space in Adelaide's urban form. Its exterior has landmark status, reinforced with tower elements and a strong, relatively active edge to the street on northern and southern sides. Connectivity to surrounding areas is highly developed for both pedestrians and public transport users, with a number of major bus routes and Adelaide's last and recently extended tram line (the city pulled up an extensive network in the late 1950s) terminating or passing through Victoria Square, Gouger and Grote streets. The market and environs provide a large number of ways through, making the spaces highly permeable on foot although both permeability and legibility is less strong especially where 'big block' suburban typologies like large floor plate supermarkets have been allowed to disrupt obvious connection points and pathways.

The market's visitor catchment constitutes a community of interest rather than solely a geographically based community, although buyers are predominantly from the medium density, middle-class residential areas of the city and inner suburbs to the east, north and south, as well as the more mixed income, ethnically diverse areas to the west. Buyers at the market are attracted by a very wide variety of speciality foodstuffs, organic and artisan produce, and low prices for fresh fruit and vegetables, but this is not simply a utilitarian space. The Central Market has also long been a convivial centre for social life, with cafés such as Lucia's and The Athens Continental Delicatessen next door well-known Adelaide landmarks where market frequenters can be sure of experiencing unplanned meetings with friends and acquaintances. Changing market consumption habits are reflected in the increasing number of stalls offering food to eat in situ, shifting the focus to the market as a kind of slow food court as well as a more traditional produce market.

Despite the market's design strengths, Adelaide's very wide street grid means that major cross streets like Grote Street (on the north side of the market) do not provide the right height-to-width ratio to effectively support market-related uses like food shops, cafés and restaurants, except on the market-side frontage. With Grote Street's absurdly overgenerous eight-lane road, the scale is by far too large and traffic flow too intrusive. On the southern side of the market, by contrast, Gouger Street's height-to-width ratio is considerably more human scaled, and these more sympathetic design conditions have provided a sufficient sense of enclosure to allow a thriving café and restaurant scene to develop on the perimeter and across the road from the food market. Tree planting and wide verandas also serve to reinforce Gouger Street's outdoor room status. On the southern side of Gouger Street, in particular, there are a number of cafés and restaurants, again with an emphasis on fresh fish cafés (mostly Greek), cafés (mostly Italian) and Asian restaurants (Thai, Malaysian, northern Chinese, Vietnamese, Korean and Indian), using produce from the market. All in all, the food market and its environs have a sense of place that is widely perceived as a special asset to be protected as the symbolic heart of the city.

It is worth reflecting that more recent work on farmers' markets as designed spaces in a North American context, emanating from landscape architecture (such as Francis and Griffith 2011: 261), similarly define a range of spatial design characteristics that support such markets as a 'socio-spatial ecology'. These markets are based on 'mixed life' space that reinforces health, community, and social and environmental justice, and are understood to reflect demands for new kinds of civic space (Francis and Griffith 2011: 263). Francis and Griffith (2011) identify four types of space at different scales: the promenade, the working market, the market landscape and the market neighbourhood, as relevant in design terms and suggest that the promenade constitutes the primary social space of the market, as it does in the examples discussed above. It is interesting that the promenade is picked out as the key spatial element: this kind of self-conscious promenading space in the syle of the *passegiata* was also identified as a critical spatial and performative element in food quarters studied in London (Parham 2012). A key difference is the association of these farmers' markets with green park spaces rather than the more urban squares and streets that the European and Australian examples showcase. What they share, however, is a concern for sufficient permanency and sense of place in their spatial design to allow for ongoing use over time as robust, convivial, social and economic space based on food.

FOOD'S OUTDOOR ROOM IN REVIEW

Food markets have been at the centre of urban space for thousands of years, giving rich vibrancy to cities and putting food-centred conviviality at their heart. Yet, since the twentieth century in particular, markets have been threatened by dominant modes of place shaping largely tied to modernist architecture, design and economics. The spatial logic that had previously offered public space primacy and created enclosed outdoor rooms was largely abandoned, and food markets lost their pre-eminent place in the urban armature. The market street, square and market hall declined in many places, despite the last often representing a pinnacle of civic architecture. As the market disappeared, its attendant townscape of fine-grained, vertical mixed use of food shops on the ground floor and storage and living space above also went into decline. Yet in certain places daily food market shopping practices and spaces have been maintained, revived or developed anew, as part of particular food cultures and spatialities which have resisted or renegotiated these transformations. Both traditional and revived markets reflect a resurgence of interest in the food market as a sociable design setting for everyday conviviality, and fieldwork case studies offer some notable examples of highly successful market-centred urbanism from a diversity of urban locations. In establishing the centrality of the market, the discussion of food's outdoor room provides a basis for Chapter 4's focus on a wider, linked set of food spaces that, along with the urban food market, contribute to making the gastronomic landscape of urbanism.

CHAPTER FOUR

The Gastronomic Townscape

INTRODUCTION

This chapter expands the scale of Chapter 3's discussion of food's outdoor room to explore some of the other elements that contribute to making a 'gastronomic townscape' (Parham 1993) of convivial food-related land uses that also broadly support sustainable urbanism. Many of the cities and towns that appear to work well in a food sense exhibit the urban design qualities discussed in previous chapters, reflecting their settings and reinforcing their surrounding food-growing context. This chapter considers whether it is fair to conclude that townscape which is based on the limitations and traditions imposed by agricultural and building technology, materials, climate and soils allows for more appropriate and convivial design responses than do places that have abandoned or ignored these connections. It investigates how location-responsive social and economic practices interweave with design qualities to contribute to a food-related sense of place that also underpins sustainability. In so doing, attention is given to food shops and food streets, street food and eating out, café society, public dining and institutional food spaces, food festivals and feasting. These and other elements in the gastronomic townscape are considered not only in relation to their convivial contribution to the civility and intellectual life of cities, but also in connection to aspects of social exclusion and incivility that have particular spatial expressions.

As is evident from Chapter 3's discussion, food markets have an extremely long-term history of centrality in shaping urban space, having contributed richly to both the material and symbolic life of the town until these connections were first broken and then began to be reimagined and rebuilt in certain places in the late twentieth and early twenty-first century. The case study examples touched on in Chapter 3 demonstrated how some food markets have maintained or revived close spatial, social and economic connections to food shops and food streets in both historic and contemporary practice. In this chapter, those food shops and streets' roles are explored, as is the corresponding increase in street food in the growth of eating in public space as a widespread socio-spatial practice with particular design expressions and implications.

SITUATING FOOD SHOPS AND FOOD STREETS

Food markets have traditionally supported a range of food-related shops on the squares and streets adjoining and near them. Evidence from Classical cities shows that the Greek agora was a place where specific goods were sold, e.g., ... the fish market, the meat market, etc. Markets for perishable and nonperishable goods were separated. ... Around the main agora the government of the town often leased shops and stalls to merchants ... and as more and more space was taken over by business ... eventually stoa and halls were erected exclusively for marketing purposes and for exchange, often presenting just a glorified facade for rows of modest shops behind a magnificent portico. (Zucker 1959: 36)

In Pompeii, where the poor lived in dense multi-storey blocks, these *insulae* of small apartments had no kitchens, and fast-food shops at the ground floor level supplied foods which were cooked and kept hot in large stone bowls let into stone counters (Davey 2008: 100). Archaeological sources offer similar examples of the plentiful street corner bars found in Herculaneum, with pitchers built into their counters that housed takeaway foods, while in Rome, there were also *kapeleia*, *popinae* or *tabernae*, 'shops and bars which offered food and drink on a commercial basis, either to eat in or take away' (Wilkins and Hill 2006: 53). Much more recent evidence demonstrates how urban food provision exhibited a very long-term coherence in design terms with these classical antecedents. For instance, in Paris of the 1100s, Sennett (1974: 194) notes that food was an integral part of the street's function as an economic space, represented through 'window architecture' of fold-down counters:

> People went into the streets to shop before or after they had finished their own labors; the dusk as well as dawn became hours of consumption, the bakery at dawn, for instance, and the butcher shop late at night, after the butcher had purchased, prepared, and roasted his meats, during the day. The counter stayed down and the courtyard stayed unlocked as long as there were people in the street.

Lively food-centred streets could also be found in other urbanism traditions. In Song period (960–1127), Kaifeng in China, accounts speak of how

> the many taverns, tea houses and restaurants in the capital added gaiety and colour to the streets, with their richly decorated entrance scaffoldings, or cailou. Many of these food and wine establishments had an upper level: some were even three stories high ... numerous street hawkers set up tables and stands selling all sorts of snacks to tease the appetite of passersby. Corner eateries boasted tables and benches set outside, under light mat awnings or huge parasols. (Kiang 1994: 54)

Such food spaces in the form of small bars, shops and stalls form a central design thread in the development of urban space up until the early modern period (Girouard 1985; Stobart et al. 2007: 5). As was described in the previous chapter's case studies, local food shops and other food-related land uses such as cafés and restaurants have traditionally been an integral part of the whole experience of shopping at the market (Morrison 2003). In the United Kingdom, for example, an economic and spatial structure of small food shops remained a feature until at least the middle of the nineteenth century when 'one master, one shop was the rule and the opening of a branch was rare' (Cherry 1972: 52). Already though, by the nineteenth century, technical innovations including the advent of rail transport were shifting some food-retailing arrangements away from this model, as in the case of 'railway milk' which was shipped in from country areas to towns overnight, in the process destroying the trade of urban dairymen who had kept their cows in the shop cellars or at the back of their shops (Cherry 1972: 52).

As noted by Atkins and Oddy (2007), increasing urbanization added complexity to food's spatiality. By around 1900, 'Europe contained several large urban agglomerations, notably London, Paris, Berlin, followed in size by Moscow and Vienna. Their size meant that the supply and distribution of foodstuffs was complex, creating problems that required the state, the municipal institutions, and also the town-dwellers themselves, to make numerous adaptations' (Atkins and Oddy 2007). Although new retailing forms including co-operatives, multiples, department stores and bazaars all began to emerge, it was still possible by the late 1930s for the design qualities of the traditional high street to be celebrated in Richards and Ravilious's *High Street* of 1938. Through text and superb lithographs, food was shown as artfully yet pragmatically displayed in each individual shop frontage to the street, to attract customers and for the convenience of selling. Thus, each shop

> has acquired its own characteristic appearance through custom and the habit of arranging its goods in a particular way ... the style of the butcher's shop, for example, with its big window with the door at one side, and rows of joints hanging up, arranged so that customers inside can point out which one they want ... and the fishmonger's and greengrocer's with their goods displayed on one big shelf in the front of the window, so that people passing can see what is in season. (Atkins and Oddy 2007: 7)

This physical shaping was reflected in characteristic servicing arrangements so that 'each trade has its own kind of cart for delivering goods – a baker's cart is quite different from a milkman's; and the men in the shops have their own costumes – you never see a grocer in a butcher's blue apron' (Atkins and Oddy 2007: 7). Moreover, these arrangements contributed significantly to social richness in public space. Richards and Ravilious (1938: 7) point out that nothing could be so exciting or satisfying as the ordered arrangement of foodstuffs 'and nothing could look better ... than the whole shop window packed tight with cheeses or the shop interior lined with rows of wedding cakes in identical glass cases' (Richards and Ravilious 1938: 7). The artistry of food shop windows remains a strong feature of the gastronomic townscape, as any stroll through an Italian town will demonstrate, and is explored more systematically in Parham (1998, 2005). In certain places the dominance of small shop food retailing remained strong well into the latter part of the twentieth century, as in Istanbul (Tokatli and Boyaci 1999).

Food retailing has undoubtedly undergone significant changes in the latter part of the twentieth century, including a long trajectory of decline (Ellaway and Macintyre 2000; Smith and Sparks 2000). This has been claimed to represent only a remnant historical food-retailing form that is now in a 'death spiral' (Press Association 2012).

The traditional high street in many places has been situated as under threat through so-called clone town developments, out-of-town supermarkets, malls and e-commerce that are discussed in Chapters 5 and 7 (Forsberg 1995; Conisbee and Murphy 2004; Kjell and Potts 2005; Portas 2011). By contrast, the link between shopping and vitality has been well documented (Schiller 1994; Bromley and Thomas 2002). A number of strategies to reinvigorate town centres and main shopping streets have been reported on (Dokmeci et al. 2007; Docking 2009). Food streets that mainly comprise small food shops still exist as vibrant food hubs in many places (Hamlett et al. 2008). These are often at the centre and act as the spatial and social focus of urban quarters, sometimes with a strong minority ethnic presence, or in resurgent neighbourhoods, which are shading into gentrification, as explored in relation to the changing nature of community life on Abbot Kinney boulevard in Los Angeles' Venice neighbourhood (Deener 2007).

In London's inner north, the small high street of Highbury Barn, for example, has a split personality in terms of its spatial food economy, with fast-food shops, a convenience store and a down-market bakery chain interspersed with an expanding range of more upmarket small food shops of high quality. The latter include a highly regarded cheese shop, an organic butcher, a revived greengrocery, a newly established fishmonger, a busy delicatessen specializing in Italian products, a wine store, an independent public house and dining room and a well-stocked health-food-cum-grocery shop. At the most 'gastronomic' end of this continuum, high quality and high prices are reflected in the products offered by the cheese shop which advertises 'regional pairings and food boxes' to its customers by way of a professionally designed leaflet. These offer cheeses and other products from Italy, France, Spain and the British Isles, such as a regional French box with cheese and wine from the Loire, Haut-Doubs, Franche Comte, Auvergne and Haut Garonne (£100), a regional French and Italian box with products from Piedmont, Lombardy, Veneto, Savoie and Haut-Doubs (£195) or from October to March, Vacherin du Haut-Doubs and the shop's own rye biscuits (£20).

The food-retailing changes to Highbury Barn could well be situated within Bridge and Dowling's (2001: 95) arguments about the microgeographies of gentrification. Drawing on earlier work by Crewe and Lowe (1995) they suggest that 'based on the particular mix of shops ... a local area can reinforce a distinct geographical retail identity that then attracts further similar forms of investment. Such microgeographies of retailing also create distinct locales with which individual consumers can identify and in some senses develop their sense of identity'. The range and nature of these local food shops may well reflect this kind of identity formation tied to neighbourhood gentrification, but neighbourhood change in food terms could also be connected to preferences for food-centred elements to be available at a small, individual, specialized scale as was once ubiquitous on high streets. In food quarters and other ordinary places studied in London, findings demonstrate some positive implications in terms of sustainability and conviviality as well as more critical judgements, and this somewhat complicated picture could also be the case in relation to such high streets. The transformation of high streets like Highbury Barn would benefit from more in-depth fieldwork to tease out some of these issues.

FIGURE 4.1: Traditional food shop frontage, Paris.
Photo: Susan Parham.

In any case, preferences for shopping at small, individual food shops are not just the province of the well-to-do, aspirational or self-conscious food consumer. At shops around London's traditional markets, such as at Dalston Kingsland's Ridley Road in the inner northeast of the city (briefly discussed in Chapter 3) buyers can expect cheap prices and a diverse range of foods from West Africa, Turkey and elsewhere that reflects their gastronomic and cultural preferences. Similarly, South Asian food shopping patterns in Britain demonstrate that hubs of South Asian shops emerged in the 1960s, and dual shopping practices were established by which shopping from local stores was mixed with travelling further afield to buy spices and other specialist ingredients (Hamlett et al. 2008). Similarly, in the streets and squares adjoining food markets in many European (and other) cities and towns, food buyers can be observed visiting local food shops in conjunction with trips to market halls (Parham 1998, 2001, 2005, 2006).

The closure of larger food shop chains on high streets has even been welcomed as offering opportunities for more quirky start-ups (Hemingway 2012). In certain examples, like Marylebone High Street in London, a small food-shop-led revival has paved the way for the outstandingly successful economic growth and social resurgence of a formerly moribund high street area (Parham 2006). Meanwhile, in the market street of Broadway Market, in east London, a number of the individual small shop frontages that escaped demolition post-war have again been taken up as food shops, cafés and restaurants, setting up a positive interplay with the revived market in social, economic and physical design terms (Parham 2012). In the capital city of Adelaide, in South Australia, Gouger Street, next to the main produce market (discussed in Chapter 3), has enjoyed a kind of renaissance based on small food shops, cafés and restaurants. In other places, support for traditional shops has been situated as part of efforts to combat food poverty and obesogenic environments (of which more in Chapter 8). Thus, in 2006,

> the New York City Public Health Department, produced the report Eating In, Eating Out, Eating Well: Access to Healthy Food in North and Central Brooklyn, which documented Brooklyn residents' difficulties in finding affordable, healthy food. As a result of that study, the public health department – in collaboration with state agencies and nonprofit groups – developed the Healthy Bodega Program, in which 15 bodegas (convenience stores) in the Bronx, Brooklyn, and Harlem increased their offerings of low-fat milk and healthy snacks, such as two-ounce packages of sliced carrots and apples. (Pothukuchi 2009: 357)

Shopping for food at small shops sometimes reflects the resilience of long-term cultural practices that support conviviality (see Figure 4.1). The strength of a Parisian culture of small food shops, for instance, can be seen in areas such as Montmartre, where despite the overwhelming numbers of tourists, local food shopping streets remain healthy and thriving (pers. obs.). In fact, a recent example in Paris showcases some of the intriguing trends in food shop and food market interconnections that are reshaping traditional high streets. At the redeveloped L'Enfant Rouge market in Paris's third *Arrondissement*, the local streets around the market, such as the Rue de Bretagne and Rue de Beauce, boast a significant diversity of food shops that work in synergy with the revived hybrid market (discussed below in relation to 'slow' food courts). As a food blogger notes,

> The whole of *Rue de Bretagne* is actually super concentrated with some of Paris's finest foods. Leaving the market, you run right into the excellent cheese shop, *Jouannault*. There are the famous rotisserie *Stevenot* chickens at number 31, the bakery and pastry shop *Fougasse* at number 25 (where there's always a line) and a fishmonger's with fresh oysters from Breton at number 19, called the *Marée Du Marais*. *Rue de Bretagne* is a veritable parade of food shops, bistrots, boulangeries and pâtisseries. It's impossible to leave the neighbourhood hungry. (finedininglovers 2013)

STREET FOOD TRANSFORMED

Food streets are not only sites to buy food to prepare at home; a certain amount of eating in the street is implied by the profusion of food shops and other out-of-home food sources. Eating outside the home 'on the run' became increasingly integral to modern life over the twentieth century. Examples such as *Avtomaticni bife* and street foods in pre-war Slovenia described by Godina-Golija (2003: 133) and the *automatiek* hatches between fast-food kitchens and their young buyers on the street in Dutch cities of the 1960s discussed by de la Bruheze and van Otterloo (2003: 321) suggest this has taken particular cultural expressions and resulted in supporting infrastructures over the relatively long term. Although street food may contribute substantially to nutrient intake, street trading and hawker opportunities have equally become a significant focus for concern about health and nuisance including litter, congestion and harassment in both Western and developing world contexts (Oguntona and Kanye 1995; Mensah et al. 2002; Von Holy and Makhoane 2006; Mikkelsen 2011: 209–10). Food safety is often seen as open to improvement (Lucca and Torres 2006). At the same time, street food hawkers, as part of the informal food economy, may be subject to harassment by a predatory rather than a regulatory state, as Anjaria (2006) reports from Mumbai, where bribes demanded by, and violence from, the police are commonplace:

> To those who are familiar with the conditions of public spaces, street markets and the sale of vegetables in the city, these restrictions are a bit odd; the rules banning handcarts and limiting licences to one per family contradict the historical practice of street vending in Mumbai; and railway stations, colleges and municipal markets are precisely the places where the majority of the population requires the food and goods provided by the hawkers. (Anjaria 2006)

Yet in Western cities, eating quickly in the street has become increasingly normative, and the notion of the 'self-indulgent street' may help explain how and where eating in public has become socially acceptable and understood as a civilized response to time compression in relation to food consumption, when formerly such consumption was not seen as appropriate (Valentine 1998: 198). Both reluctance to eat in the street and culturally sanctioned spatial exceptions are clearly connected to changing mores about how the public domain should be approached as a food space in which to perform civility in Norbert Elias's (1982) terms (see Figure 4.2). Conversely, in some other cultural settings the enjoyment of a convivial, public, street-based food culture appears to be less self-conscious. 'In Mexico, where so much of life takes place outdoors-in bustling squares, open-air markets, and restaurants whose seating spills

FIGURE 4.2: Street food, Whitecross Street, London.
Photo: Susan Parham.

out onto the sidewalk – "the street" is a surprisingly expansive concept' (Santibanez 2012: introduction).

Although street food is not a new phenomenon, its expression is currently being reshaped in certain ways, such as changes described in relation to Toronto's gastronomically informed 'snacking' street food culture (Duffin Wolfe 2009: 28). Offerings of street food are ubiquitous in the history of urban development, with accounts previously noted from the earliest cities of cook shops for buying takeaway foods (Bray 2003; Alcock 2006). There are more recent examples from medieval street space onwards (Tinker 1997; Mennell et al. 1992; Albala 2003). These reflect street food's longevity as an element in urban food space. Reporting on research from Bangladesh, the Philippines, Egypt, Nigeria, Thailand, Jamaica, Senegal and Indonesia, Tinker (1997) found that street food buying and selling offers much needed sources of income to the micro-businesses of vendors, and the home-based women who supply some of the street stall, table and food cart items. These vendors offer sources of cheap and (sometimes) nutritionally rich food for poor urban dwellers, a finding echoed in other works (Oguntona and Kanye 1995; Ekanem 1998). Meanwhile, home-cooked food for sale, like the *tiffin* boxes of urban India, the *rantangan* of Indonesia or the *pin-to* of Thailand, that are sold directly to customers on contract, can be defined as 'invisible street food' because they are sold 'through the streets but not on the streets [and are an] important part of urban feeding patterns' (Tinker 1997: 181).

In design terms, too, street food spaces, infrastructure and specific dishes contribute richly to urban vibrancy and sense of place (and thus conviviality), providing a heady mix of tastes, smells and sounds, as Fernando (2005) has reported from Chinatown and Little Italy in Manhattan. Certain examples have all but disappeared, like Adelaide's much-lamented, late night to early morning pie-carts with their pie floaters – meat pies served upside down in thick green pea soup, often topped with tomato sauce or London's historic jellied eel and cockle, mussel and whelk stalls. Recent accounts of Tubby Isaac's jellied eel stall at Aldgate in inner east London suggest that it remains the only one of its kind, serving its long-term regulars, some of whom are now in their eighties and nineties:

> It is a testament to Paul Simpson's tenacity and the quality of his fish that Tubby Isaac's is still here, now that this once densely populated former Jewish neighbourhood has emptied out and the culture of which jellied eels was a part has almost vanished. Tubby Isaac's is a stubborn fragment of an earlier world, carrying the lively history of the society it once served now all the other jellied eels stalls in Aldgate are gone and the street is no longer full with people enjoying eels. (SpitalfieldsLife 2013)

Yet other forms of street food have remained resilient. From the *Imbiss* stands and kiosks of urban Germany, to American hotdog stands, the street corner food sellers of Nigerian cities (Ikpe 1994: 157), the hawker food stall markets of Kuala Lumpur, Singapore and other Asian cities and the *ekibentō* (known as Ekiben) *bentō* lunchboxes of Japanese train stations (Noguchi 1994), provision has to be made for eating on the move in the public domain, or sending out household members with lunch (Allison 1991). Although food stalls are sometimes contested as a legitimate use of street space (as explored by te Lintelo 2009), particular street food dishes can come to represent a national cuisine or city, like the *chivita al pan* (steak sandwich) of Uruguay, the *batata vada pav* (potato burger) of Mumbai, the *pho* (soup) and *banh mi* (Vietnamese baguette) of Saigon, the *kelewele* (fried plantain) of Ghana or the *Yangrou Chuan* (mutton kebabs) of the hutongs of the cities of north-west China. Often, the narrative of street food is about place-based authenticity in the gastronomic landscapes of travel (Parham 1996). Writing about Thailand's *som tam* exemplifies this approach:

> Som tam is a tangle of crisp, unripe papaya, peanuts, and dried shrimp, tossed in a lip-tingling dressing of fish sauce, palm sugar, and lime juice, then crammed, to-go style, into a plastic bag. You'll find it all over Bangkok, but the quintessential version is found just off Phaholyothin Soi 7, a busy street in the Soi Ari neighborhood packed with vendors seek out the cart whose window flaunts stacks of shredded papaya and tomatoes, plus a coiling bunch of long beans. More daring chowhounds should seek out the style of som tam popular in Isaan, Thailand's Northeastern region, where many think the dish originated. Stop by the open-air haunt called Foon Talop, in the Chatuchak Weekend Market, where the salad is made with *pla ra,* a supremely funky, murky fish sauce whose flavor you won't soon forget. (Concierge.com 2012)

Equally, as Gilman (2011) describes, in his discussion of street food in Mexico City, the *puestos* (street stalls serving *antojitos* or 'before the eyes' dishes), the *comedores* (eating stalls in markets) and *tianguis* (temporary market-based eating stalls), are both very appetizing and a source of anxiety about food safety, a concern well-documented globally in the food microbiology, food safety and urban health literature (Mosupye and von Holy 2000; Mensah et al. 2002; Estrada-Garcia et al. 2004; Von Holy and Makhoane 2006; Rheinländer et al. 2008). In Gilman's (2011: 20) view,

> Some of the most down to earth and truly exquisite food in Mexico is found in street stalls and markets. It is common advice not to eat ANYTHING on the street: even some Mexicans adhere to this rule. I don't find it necessary to be so strict, as many stands are simply micro-restaurants whose ingredients are as fresh as could be and handlers are as spic-and-span as a hospital kitchen ... I don't eat dishes that look like they have been sitting outside for a long time. I avoid certain kinds of street food, such as seafood during very hot weather. I search out crowded stalls that have been discovered by locals: they have already selected the good ones.

FOOD VANS AND PODS: POP UPS AND NIGHT MARKETS

A resurgence of interest in the gastronomic quality of street food has seen the rise of 'gourmet' food vans and pop-up food spaces, which their 'devoted fans' will take the bus or train to find, 'or walk that extra block to snag their fare, just like they would for any popular restaurant' (Spens and Gilland 2012; Square Meal 2013). In London, for example, locations vary but include the fashionable and popular Whitecross Street Market, Exmouth Market (Parham 2012) and 'The Real Food Market' held weekly from Friday to Sunday on the Southbank Centre Square behind the Royal Festival Hall. Street food operators have sometimes been attracted in by developers or even city government (see Figure 4.3) to enliven particular street spaces, (often leftover, marginal or redeveloping ones), such as former railway land near the relocated Central St Martin's College of Art and Design within the renovated King's Cross Station quarter. In other examples, they have been moved on by city authorities, as in Kerb's food van's short-lived presence outside the 'Gherkin' high-rise building in the financial quarter of the City of London. Night markets and pop-up spaces on roofs and in derelict spaces have also emerged, such as the one organized by Street Feast, which in late 2013 was to be found in a former factory in inner east London, and operating in the style of a hawker centre.

These food forms generally have a website attached, and their proponents will tweet, blog and post online about them, providing information as to their location and food on offer. They tend to be supported by a lively social media community. The gastronomic quality and supposedly revolutionary, guerrilla like or just hipster nature of this shifting food scene is emphasized by some media accounts (while the link to the guerrilla gardening movement is discussed in Chapter 6). As one online writer explains, 'Forget greasy kebabs at 3am; thanks to a slew of adventurous chefs and entrepreneurs, London's street food has gone gourmet. [Name] scouts out 20 of the best guerrilla grillers and breakaway bakers driving London's street food revolution from their trucks, trailers and minivans'

FIGURE 4.3: Food truck, Leigh Street, Adelaide.
Photo: Susan Parham.

(*Evening Standard*, 20 July 2012). London's experience offers a useful example of the nature and economic trajectory of this emerging food form:

> Food vans first started causing a commotion in London when the former ballet technician and guerrilla burger chef Yianni Papoutsis began cooking juicy burgers and hot dogs out of The Meatwagon mobile in Peckham, tweeting daily about what was on the menu. Soon his updates were being retweeted, a local blogger turned up to review the van and Papoutsis suddenly found himself serving crowds of hungry Londoners. Hot in his wake came Pitt Cue Co, a silver trailer positioned beneath Hungerford Bridge on the South Bank, that began selling out of its homemade barbecue food in boxes following a similar Twitter storm. Both vans continue to serve street food around the city but have also since opened permanent bases: MEATliquor near Oxford Circus, MEATmarket in Covent Garden, and Pitt Cue Co's tiny restaurant in Soho. (*Evening Standard*, 20 July 2012)

In the United States, meanwhile, food trucks (Mclaughlin 2009; Shouse 2011) and 'pods' of food carts, as in Portland Oregon, have become equally popular, with guides to developing such a food business (Weber 2012; Myrick 2012), cookbooks of food truck recipes (Edge 2012) and smartphone applications to help the cognoscenti track the path of their favourite food offerings (FoodCartsPortland 2013). The food truck, cart and pod movement is widely associated with, but not limited to, the cities of the American west and northwest including San Francisco, Seattle, Los Angeles and the 'culinary renegade zone' of Portland, and is generating its own parodies and in-group cultural references (Mclaughlin 2009; Tway 2011; Brooks et al. 2012). In Los Angeles, taco trucks are understood to have particular economic and gastronomic implications:

> Los Angeles is no stranger to street food or food trucks, but the new *gourmet* food truck trend is popularizing a historically low profile industry. Much like the food guilds of post revolutionary France, restauranteurs [sic] today are wary of the fresh competition these new trucks present. Los Angeles consumers, however, are happy about the array of new choices, as well as the reasonable prices. Lower costs of operation have allowed the trucks to introduce higher-cost items such as duck confit and truffle-infused whatevers to a larger audience. (Geller 2010)

GASTRONOMIC TOWNSCAPE AS DESIGNED SPACE

Like fixed food shops, the placement of food stalls, stands, kiosks, carts, vans, trucks, pods and pop-ups is a townscape design issue with conviviality and sustainability implications. In spatial design and planning terms, researchers have noted how food trucks and pods are drawn to areas with big lunchtime populations and a few 'sit down' dining choices and support both vibrant public spaces and local economic development (Geller 2010; Newman and Burnett 2013). Moreover, there are well-documented 'wars' between food truck vendors and regulatory regimes in places such as Los Angeles (Hernandez-Lopez 2011). Like food vans in the United Kingdom, this food space movement in the public realm has been situated as a form of 'tactical urbanism' that

> the city can encourage at relatively low cost and low risk. No large scale infrastructure is required since the bulk of the monetary investment will come from private entities, people who want to operate food trucks in the city. However, if successful, the city can realize economic, social and environmental benefits that more than justify any expenditure. Citizens, whether those who own and operate food trucks, those who patronize them, or the general public at large can also derive economic, social and environmental benefits from development of a vibrant food truck culture. (Tanenbaum 2012)

In design terms, successful spaces for street food are not just happy accidents. Such food streets and spaces tend to share a number of characteristics which reflect well-grounded design principles. As an example, the physical design of enclosed public space to support food uses is exquisitely represented in Andō Hiroshige's woodcut, *Night Scene at Sarawakacho*, one of the hundred views of nineteenth-century Edo which shows how 'the street facades of the shops and residences lining the street comprise a membrane that changes the degree and location of permeability throughout the day and night' (Treib 1994: 42). Food stands can not only be part of the 'stage set' for an active, lively and well-used public realm (Davies 2001: 99), but also be carefully located within that public space to maximize their impact, being compressed in to certain areas rather than spread out over the whole space (Whyte 1980: 53). In fact, Whyte (1980: 50) argues that if 'you want to seed a place with activity, put out food. In New York, at every plaza or set of steps with a lively social life, you will almost invariably find a food vendor at the corner and a knot of people around him eating, schmoozing, or just standing'. These vendors, Whyte (1980: 52) notes, have a 'good nose for space' in a situation in which 'food attracts people who attract more people'.

The props needed can be minimal, and most such temporary food structures, like the food stands described by Alexander et al. (1977: 454), will be located near the edges of an open space, reflecting that such building edge space is a place or zone with volume to it and should offer an active frontage with 'depth and a covering, places to sit, lean and walk; particularly at points along the building perimeter which look onto interesting outdoor life' (Alexander et al. 1977: 752). Paraphrasing Jan Gehl, such placement reflects the design understanding that people prefer being at the edges of open spaces where they feel protected and enclosed, and will usually place themselves near something: 'a facade, pillar, or street furniture if these are provided' (Alexander et al. 1977: 754, from Gehl 1968). Gehl and his colleagues have long advocated attention to the details of place design to support a vibrant public realm, proposing soft edges to public space, as these offer prime opportunities for sitting and standing. With urban life growing 'from the edges to the middle ... [soft edges] make a vital contribution to spatial experience and to the awareness of individual space as place', in this way having 'a decisive influence on city life' (Gehl 2010: 75).

Thus a common design feature of food streets from ancient times onwards is the use of the arcade or portico as edge space (Alexander et al. 1977; Geist 1983; Yegul 1994; Parham 1998, 2005). Rudofsky (1964: npr) calls arcades 'altruism turned architecture – private property given to an entire community' and notes they were once ubiquitous in many places including Spanish towns about which he offers a series of beautiful examples in illustration. Robbins (1994: 170) describes the arcade's use in medieval Rome's Trastevere district, providing a space that linked private and public domains, and 'in many cases, a place of commerce, with counters and stalls extending out from an interior shop'. Famous typologies such as the shophouses of Singapore's Little India include a porticoed walkway that by offering a semi-enclosed edge that feels pleasant and protects against oppressive weather supports lively food streets. In contemporary Bologna, meanwhile, miles of porticoed streets provide protection from rain and sunlight and places to comfortably browse and visit the numerous small, specialist food shops that make up a rich gastronomic townscape, while contributing to height-to-width ratios that support an excellent level of street space enclosure (Parham 1998). Places suffering climatic extremes have understandably tended to take the requirements for shelter from the elements more seriously than have milder regions. Thus, in southern Italian, North African and Middle Eastern cities, narrow and covered streets and canopied spaces, using canvas, trellises, mats, nets and vines, have helped support street and market life despite intense heat (Rudofsky 1964: npr). In future, with global warming, the design techniques traditionally used by cities in hot climates to make food space livable will become more important models for others. In terms of conviviality, environmental and psychological protection for the pedestrian is a useful physical support to public interaction including that in relation to buying and carrying food and drink.

SLOW FOOD SPACES AND COURTS

Like the food van and truck movement, a common feature of the areas around hybrid food markets that have emerged from the ashes of derelict traditional marketplaces and market streets is the advent of what might be designated 'slow' food spaces and courts. These incorporate the fine-grained design features of long-term market areas rather than the coarser spatiality of food courts associated with suburban and post-suburban shopping malls, which are discussed in Chapters 5 and 8. The Slow Food and Slow Cities movement is explored in Chapter 9; it is worth noting that its gastronomic principles appear to neatly correlate with the designed spaces of 'slow food court' markets such as Borough in London and Marché L'Enfant Rouge in Paris. Food bloggers describe the experience of eating at the latter in the following terms:

> Located in the chic northern part of the Marais, Le Marche des Enfants Rouges is a compact indoor market with a small iron-gate entrance that's easy to miss if you aren't looking for it. I arrived in the late afternoon and was starving for lunch. The nice thing about the market is that many of the stands sell fresh prepared food that you can eat on tables with cheerful plastic tablecloths spread out through the market. The heavenly aroma from an Asian food stand seduced me and I chowed

down on a delicious dish of Asian-style chicken nuggets with seaweed salad. (I prefer Paris 2013)

A ubiquitous development in such food spaces is the eclectic diversity of what is on offer. At Whitecross Street and Exmouth Market in inner north-east London, for example, the 'slow' food stalls offer lunchtime eaters a variety of food options and form a kind of informal food court space supported by the tight height-to-width ratios of their former market streets (Parham 2012: 206). In the previously mentioned L'Enfant Rouge market in Paris, where the market spaces are set back from the street within a block interior,

> Today, neighborhood locals still congregate to shop for produce and fresh products, to have a coffee and to converse with other locals, old-timers and merchants. What is interesting about the marché now is the diversity of its different merchants and products. An Afro-Antilles café serves up delectable bites to hungry shoppers; the friendly owner of the Italian booth proposes Illy café on its year-round terrace and will advise you exhaustively on his wines (for which he has a secret passion); a Breton will whip you up a crêpe (but don't photograph him without asking, he might squawk at you); Moroccan delicacies tempt you from one corner while Portuguese and South American products beckon from the other. ... You can even browse an antique shop or park yourself on a bistrot terrace in the back. People even actually still come here for their weekly produce runs, for goodies from fish to flowers to figs. (Ladd 2010)

Such food courts need not be so self-consciously gastronomic in nature and may form parts of interiorized or below-ground shopping malls, reflecting a suburban spatial food consumption model explored in Chapter 5. In the tighter space of the inner city, however, examples of such internal, sometimes subterranean, food courts can also be found in downtown areas across the world. In design terms these reflect new ways of shaping food space focused on interior, privately owned and managed walkways rather than exterior public streets. As Treib (1994: 39) notes, 'as the combined shop and dwelling of the traditional town has opened to the street, so the new commercial spaces open to the promenade: shops with walls of sliding glass, restaurants with terraces'. In Tokyo, for instance, food courts are famously connected to that city's vast underground rail network, while in Ōsaka, the Umeda Station offers a case study of underground commerce, including many food spaces (Treib 1994). In Sydney, similarly, labyrinthine underground shopping spaces, including food courts and cafés, form a second city that provides respite from hot and humid summers. At the other end of the climatic spectrum, Canadian cities, including Toronto with its PATH network, have extensive underground shopping and food spaces in contrast to an apparently empty snowbound central business district at ground level (pers.obs.). As Jackson (1996: 1112) notes from Montreal's La Ville Souteraine experience, 'since 1962, for example, Montrealers have been able to survive their harsh winters by working, shopping, and living, often for months at a time, underground-or at least inside glass and concrete. Large parts of the core city are now linked by miles of subterranean walkways, all lined with shops, restaurants, snack bars, and theaters.'

In Singapore, mostly ground-level food courts are often located in covered, but open to the breeze, spaces beneath or between apartment buildings, as can be seen in a neighbourhood like Changi Village where observational visits show bustling food court spaces comprising small, individual stalls surrounding communal seating areas. These are successful because they offer good-quality, well-executed, hygienically prepared and inexpensive traditional and hybrid dishes that may be enjoyed as part of everyday urban food practices (Choi et al. 2010), suitable even for those employing 'defensive dining' strategies to fulfil their religious obligations in relation to food (Nasir and Pereira 2008). Of course, the notion of ethnicity in terms of Singaporean food has been critically interrogated (Huat and Rajah 2001). The artful restaurant representations and commodification of its Peranakan past and Nyonya cuisine have attracted critiques (Wong Hong Suen 2007). As elsewhere, there is resistance to ethnicity being reified through cuisine as a natural or an uncritical given (Klopfer 1993). However, the city's public, hawker-style food spaces do clearly reflect the multiple ethnicities and foodways of Singapore's diverse population. These sites unselfconsciously accommodate a cosmopolitan, multicultural public dining experience (Nasir and Pereira 2008). Similarly, in London, bloggers reported with great sadness the closure of a hawker-style food court, Oriental City, at Pacific Plaza, in Wembley in north-west London, that offered 'good, unpretentious food, at a good price, cooked by someone who loved what they did'. Recently reopened in a new incarnation, a blogger reported that he

> had the pleasure of visiting the Pacific Plaza tonight, its well worth a look in, only about a quarter of the size of Oriental city food court, and a few boutique shops down stairs at the moment, there about dozen food stalls in the court, Thai, Malaysian, Chinese, Korean and Japanese downstairs. Beer is available from the main chinese [sic] stall which dominates the back wall, there is a bar stall under development along with a number of other vacant stalls/kiosks well worth a look for all you entrepenuers [sic]. (Thailand Food Forums 2013)

CAFÉ CULTURES AND SPACES

Of course, informal, impermanent eating structures are not the only way people have traditionally interacted in a convivial way with food in city space. Bars, cafés and the cultures that surround them are also critical elements in the gastronomic townscape (Rudofsky 1964, 1980; Whyte 1980, 1988; Parham 1990, 1996; Haine 1992, 1996) (see Figure 4.4). Notwithstanding this centrality, even in France cafés are thought to be in decline because of changing social mores, losing their status as 'a kind of public living room' (Erlanger 2008: npr; Vix 2008). Affected by the economic downturn, the smoking ban and tighter controls on drunk driving, such spaces are additionally suffering from 'competition from American-inspired coffee chains [which] with their easy sofas and Wi-Fi connections are also drawing younger people away from traditional cafés' (Hird 2010: npr).

FIGURE 4.4: Street café, Cluj, Transylvania.
Photo: Susan Parham.

Historically, the rise of the café in pre-revolutionary France and the coffeehouse in post-Restoration London have been cited as central to the creation of a novel form of the public sphere in cities (Sennett 1974; Habermas 1989), although these readings of the development of civic space for political engagement have been challenged in relation to coffeehouse history (Cowan 2001, 2004). While the coffeehouses that emerged at the beginning of the eighteenth century did not look very different from taverns or ale houses, they offered exotic drinks not found elsewhere and were 'notable centres of news culture', in which 'the peculiarly "virtuosic" emphasis on civility, curiosity, cosmopolitanism, and learned discourse' made them 'a distinctive space in the social world' (Cowan 2005: 87–9). In nineteenth-century Paris, informal *cercles* performed similar functions in regard to providing space for male exclusivity and food consumption (Rich 2011).

Equally famously, feminist analysts of city space have noted the gendered, exclusionary nature of the coffeehouse and café as the supposedly democratizing centre for public social life (Wilson 1991; Ellis 2004). While there were some women owners and women working at coffeehouses, 'there was no need formally to exclude [women] because it was assumed that no woman who wished to be considered virtuous and proper would want to be seen in a coffee-house' (Ellis 2004: 66). While women were both more present and perhaps less remarked upon than some have argued in the restaurant culture that grew up in the urban nineteenth century, a strong strand in the development of dining rooms that emerged out of coffeehouses and clubs was their tradition of masculine sociability (Rich 2011: 155). At Simpson's in London, for example, women were not admitted until about 1910 (Rich 2011: 155).

At the same time, from the eighteenth-century onwards, with the new social practice of tea drinking in England, upper-class women were, in certain places, able to make a daytime visit to a tea shop, as depicted in Anne Elliot's visit with Mrs Clay to Mollands (in Jane Austen's novel *Persuasion*). This access though was rather asymmetric in its results, as it 'produced no social revolution in public culture equivalent to the coffee-house'. Teahouses themselves were very limited in spread to spa and resort towns until their wider development in Victorian times when the teahouse as a form did become extremely popular (Ellis 2004: 210). The teahouse's trajectory can be traced in relation to the Lyons Teahouses and Corner Houses in the United Kingdom and the adoption of the form as Le Tea-Room in Paris (Rich 2011), while the tradition of the *konditorei* of Germany and Austria and the *konditori* of Sweden similarly had a female-friendly character.

In mainland Europe, too, café culture was famously both the centre of intellectual life and one that generally excluded women except as servers and often as prostitutes. The Paris Commune 'marked the peak of women's participation in cafe politics and the Belle Époque registered the greatest incidence of cafe prostitution' (Haine 1996: 179). In Madrid, cafés like El Parnasillo, which was considered the centre of a publicly engaged culture that grew out of the earlier *botellerîas*, kept women out until at least the 1920s. Famous nineteenth-century examples of café culture include *fin de siècle* Vienna's grand Kaffeehaus; Taine's description of Café Florian in Venice; the Cafe Royal in London, and Manet and Degas' paintings of Parisian café life (Ellis 2004: 216). By the early twentieth century, in Naguib Mahfouz's (1994) novel, *Palace Walk*, meanwhile, the coffee shops of Cairo are at once places for political agitation and male exclusiveness. The protagonist, Yasin, uses the coffee shop as place to drink and to pick up lower-class women whose need to be on the street as traders or shop assistants, unlike the strictly secluded women in his own more upper-class family, make them more vulnerable targets. By the 1950s, in the United Kingdom, though, the espresso bar

> had exploded like a bomb ... importing an idea and some technology from Italy, it was transformed by London's particular forms of urban sociability, by jazz and skiffle, by teenage fashions and migrant expectations. Repudiating conventional politics in favour of consumerist hedonism, the espresso bars were surgeries of the new politics of later decades. (Ellis 2004: 244)

Thus the café presents a somewhat mixed picture in urbanism terms. On the one hand, as a sociological reading would suggest, 'the modern city, with all its bazaars, street shopping, cafés and other delights for the voyeuristic urban flaneur, is also Engels' city of cut-throat competition, survival, poverty, racism and stark inequalities of wealth, lifestyle, life-chances and living conditions' (Amin and Graham 1997: 13). Work by Roseberry (1996) and others demonstrates that modern 'coffee culture' is grounded in extremely unequal relations between producers and consumers. Various critical perspectives tease out aspects of the café's complicity in providing spaces of exclusion based on race, class or gender, and both resting on and obfuscating extremely exploitative economic relations. At the same time, in early twenty-first-century cities, cafés do offer convivial opportunities for social interaction focused in part on food.

This is a role that has become increasingly important, as work on a wide range of places including fast-food outlets suggests (Cheang 2002). Cafés are understood to function as third spaces or places (Oldenburg 1989, 1997; Thompson and Arsel 2004b; Rosenbaum 2006; Rosenbaum et al. 2007; Farnham et al. 2009). They offer necessary and desirable ordinary places, contribute to quality of life and provide a sense of care in low-income communities (Knox 2005; Jeffres et al. 2009; Warner et al. 2013). Cafés can operate as adjuncts to convivial everyday life more broadly (de Certeau 1984, 1986, 2000; Miles 2000; Jarvis et al. 2001; Maitland 2007; Bell 2007). Thompson and Arsel (2004b) make the point that while 'Oldenburg argues that corporate chains are inherently antithetical to genuine third-places, the success of Starbucks is due to its skill at creating a third-place ambiance on a global scale'.

Similarly, other spaces including libraries that are wireless internet enabled also play this third-space role (Lawson 2004; Hampton and Gupta 2008). In London, the British Library café and restaurant space is a case in point, both in terms of its attractiveness to third space users and the increasingly contested nature of this indeterminate spatiality, somewhere between public and private realms. Here, as elsewhere, this in-between quality is reflected in both the territory capturing strategies of users and the gatekeeping responses of food business operators (something Laurier et al. 2001, also noted in their work on café ethnography). Café visitors stake out areas through carefully casual placement of their books, computers, coats and bags, often making minimal, if any, food purchases, and such digital nomads can be seen working alone or carrying out client meetings, tutorial sessions and discussions with project teams (pers.obs.). Food business management has responded to these incursions by placing signs on tables which request 'guests' to match their time at the café or restaurant to the amount and nature of their consumption, in what they describe somewhat euphemistically as 'seating windows' (see Figure 4.5):

> Tables Reserved For Diners. Monday to Saturday 11.00-15.00. Due to the high demand for tables in this area Peyton and Byrne work on the following seating windows prior to requesting guests to move on – For a coffee and pastry 30 min. For lunch 45 min. No picnic lunches to be consumed in our restaurant. Thank you for your cooperation.

It is not only café patrons who employ informal strategies to control space and interaction. Café work can require wearing emotional labour, and those employed in cafés are often subject to exploitative conditions, but they can fight back in a minor way through subtle (or more obvious) tactics of incivility in relation to those they serve. An absence of conviviality is foregrounded by withholding smiles, avoiding eye contact, ignoring verbal interaction beyond the minimal and offering required mantras of welcome with an ironic edge, while further driving home the point to their customers by speaking to colleagues in a more authentic way during these interchanges or playing music very loudly. Yet, equally, café space can play a crucial role in supporting convivial communities of practice and settings for particular meals including breakfast (Laurier and Philo 2004; Laurier 2008). Drawing on evidence about the rapid rise of the use of cafés in the United Kingdom in the 1990s, these spaces are defined (perhaps rather portentously) as 'generative nodes,

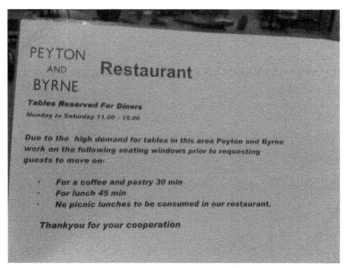

FIGURE 4.5: Negotiating 'third space' use in a public building, London.
Photo: Susan Parham.

as heterotopias, where economic, political and cultural matters run up against one another and are mutually transformed' (Laurier and Philo 2004: 3–5). Work on hospitable spaces similarly suggests that cafés can play a crucial part in developing "a convivial, hospitable ecology … through which hospitality and commensality are woven into new patterns of urban living (and eating and drinking)' (Bell 2007: 7). Reflecting on Latham and McCormack's (2004: 1719) insights, Bell (2007: 7) suggests that this can have quite broad urbanism effects:

> the conviviality, the commensality, the hospitableness of commercial venues is seen by Latham to spill out into the streets, generating 'new solidarities and new collectivities' and a greater sense of belonging … the ways of relating that are practised in bars, cafés, restaurants, clubs and pubs should be seen as potentially productive of an ethics of conviviality that revitalizes urban living.

It is worth noting that café or restaurant work also sometimes offers a route into economic activity for migrant communities often otherwise excluded from productive life (Hirvi 2011). Thus, ethnographic analysis of Sikhs employed in restaurants in Finland demonstrates that these sites have been places for asserting and maintaining cultural identity (Hirvi 2011). Similarly, in an Australian context, through the process of remaking Australian foodways and developing café culture, such communities are thought to have resisted the dominance of café chains and revived retailing at a small scale (Humphery 1998: 102). With reference to post-war experience, it has been noted that 'the delicatessen or the café were to provide migrant communities with a way of resisting cultural assimilation, within the dominant culture these commercial activities were constructed as a means of gaining citizenship' (Humphery 1998: 102).

CONVIVIALITY AND THE KOPITIAM

Although food is not always central to the accounts of café space, particular ethnographies provide intriguing insights into the role of café-based food interactions in cementing social relations in public urban space that appear very convivial in nature. Lai Ah Eng's (2010: 3) detailed exploration of the kopitiam (Chinese dialect for coffee shop, known as *kedai kopi* in Malay) found throughout Singapore demonstrates that this 'quintessential' feature of Singapore's everyday life is 'one among several institutions and spaces in Singapore within which are embedded dynamic aspects and processes of migration and social-cultural diversity, set within the larger context of rapid changes and globalization throughout its history' (Lai Ah Eng 2010: 3). As Lai Ah Eng (2009) notes elsewhere, such authentic everyday experience occurs in both planned housing developments and more traditionally shaped urban environments in Singapore. Serving drinks, snacks and sometimes meals, and open from dawn until nearly midnight, the kopitiam features drinks and dishes originating in Chinese, Malay, Indian and 'Western' cuisines (among others) that have been hybridized over time. In the 1990s, kopitiam food experienced further fusion (i.e. beef carpio ramen), with 'Singlish' (creole) names for beverages such as the 'milo-dinosaur' (thick Milo topped with Milo powder) and the 'Michael Jackson' (mixed soy bean milk with grass jelly drink), reflecting these cross-cultural flows (Eng 2010). The kopitiam has acted as a close reflection of both spatial and gastronomic changes over time in Singaporean urban space and food culture:

> Individual items such as chicken rice, noodles, roti prata and nasi padang, once identified as ethnic and introduced by/for immigrants, is now iconic of Singapore food and readily available in most kopitiam as basic and popular everyday foods. The Hainanese-Western half-boiled eggs, kaya toast and coffee set is now standard breakfast fare in the kopitiam. The 'economy rice' stall that once offered a cheap combination of rice and dishes to poor Chinese immigrants retains its status in the kopitiam as its major stall, but with an expanded range of traditional home-cooked style foods, while the other main stall, the *tze char* Chinese kitchen is now in effect a small restaurant serving a wide range of cook to order Chinese dishes, including favourite seafoods and steamboat, still at affordable prices. ... It is now common practice for an individual to rotate different ethnic dishes among his or her meals at the kopitiam, and common for members of the same family to be eating different dishes at the same time, for example, father eating pig's organs soup, mother eating nasi padang and children eating pizza. (Lai Ah Eng 2010: 12)

Spatially, hundreds of kopitiam are found sprinkled across Singapore's vast public housing estates where many local communities were resettled in the post-war period, as well as in 'historical ethnic enclaves and in business districts, streets and downtown shopping malls' (Lai Ah Eng 2010: 12). In some instances their popularity has meant a spillover of the coffee shop space into a fully fledged food court, and they have also transformed socially, economically and gastronomically as the city-state has itself changed in the era of post-colonial globalization (Lai Ah Eng 2010: 3). While the kopitiam and hawker centres are spaces which are inhabited by elderly residents (Lai 2009), Singaporean foodies will seek out a particular kopitiam to taste especially good dishes. These kopitiams' 'claims to fame and sometimes ownership of a particular dish [are] based on "first setup", originality, authenticity, cultural heritage and tradition, special skills, styles and ingredients, and sometimes simply being the offspring of an original stall' (Lai 2009: 13) although pressures of competition and low wages have made running or working at a kopitiam a more cut-throat, uncertain business. Emphasizing its civility, and reflecting its important role in the production of everyday conviviality, coffee or meal time in the kopitiam is open to all, and, 'fully displays the public culture of eating, drinking and talking that is considered by some as the national pastime, well beyond the basics of the necessity to eat' (Lai Ah Eng 2010: 23). Its very openness and egalitarian nature has meant that the kopitiam has a uniquely important role in gluing together the diverse strands of Singaporean everyday life:

> The komitiam stands out as a unique institution with its particular local-global nexus of economic, social and cultural inputs, ingredients and infusions, through the generations of diverse people who inhabit it and through the foods and activities that they bring and partake of. At the same time, the kopitiam within a local community serves as a natural and spontaneous gathering point for many residents and their activities, making it the most significant public site that manifests aspects of local community and its everyday life. (Lai Ah Eng 2010: 4)

SNACK BARS AND GASTRONOMIC NICHES

Another, rather unglamorous, element in the gastronomic townscape is the form of the snack bar, which can be considered as a kind of fixed relation or ancestor to the food stalls, carts and stands discussed earlier in the chapter. The snack bar is also strongly connected to the post-war survival strategies of migrants seeking economic opportunities in their new locations, and may be an adjunct in marking identity, as Rabikowska and Burrell (2004) explore in relation to Polish food shops in the United Kingdom. The snack bar has a particular place in urban public space and in discourses of diaspora. In Adelaide, South Australia, for example, the grand Classical names and allusions employed by local snack bars in the post-war period confirmed the mostly Greek immigrant origins of these fledgling small businesses at a particular moment in the diaspora's history: with the Olympic, Gladiator, Spartacus and Athens snack bars, matched by the boldness of the Paragon, Matador and Kontiki. Similarly, ethnography from a more recent period of the 'NIL' Sudanese snack bar in Berlin demonstrates immigrants to Germany developing an income through 'ethnic' food provision within a hip urban quarter:

> In my further visits it became increasingly apparent how embedded the little Sudanese snack bar was in the multicultural eating, dining and consumer culture of the Friedrichshainer Kiez: Arabian snack shops compete with Turkish doner kebab stands, the streets are filled with alternating Indian restaurants and German beer bars, there is a grocer's with the charming name 'Esperanto' which offers

Spanish, Italian, Turkish and Lebanese antipasti. This is the sort of neighbourhood culture I know from Ottensen in Hamburg, the sort that produces its own atmosphere, and which is particularly sought out and created by young people. (Weisskoppel 2004: 98)

Similarly, work on the development of Turkish restaurants in Brussels shows how in certain areas they have shifted over time from such an 'ethnic minority niche' to 'exoticism' and finally, economic assimilation (Kesteloot and Mistiaen 1997). In design terms a notable aspect of such niche businesses is their need to access small scale and cheap sites with substantial pedestrian catchments. City form that is fine grained, highly mixed and relatively dense, with active building frontages, is much more able to accommodate these requirements than are single ownership, monofunctional, large floor plate buildings, set back from and with few openings to the street. These overly large and expensive buildings, which are a feature of twentieth- and twenty-first century city shaping, are aimed at maximizing the commodification of urban space and where they are the dominant spatial mode, effectively design out the possibility of including such start-up businesses. They are also often associated with private, institutional food spaces that turn their back in the street. It can be argued that the post-war shift away from the intricate human-scaled urban design of the city was part of what caused the decline of such individually operated snack bar spaces, and presaged both the site-swamping tactics of food chains and the rise of street-based food as an alternative way for individual food businesses to gain a foothold in urban space.

EATING OUT: INNS, TAVERNS AND MODERN RESTAURANT SPACES

Turning to restaurants as a significant aspect of the gastronomic townscape, eating out is now considered central to the urban 'experience economy' (Pine and Gilmore 1999). Yet this is not just a phenomenon of plenty but often about resilience, as reflected in the *bukas* of Nigerian cities, small restaurants which offer cheap food on a daily basis to urban workers (Ikpe 1994: 157). City design has incorporated dining out from antiquity onwards. As noted earlier in the chapter, accounts from early and Classical cities demonstrate hot food was available away from home, with bars and taverns supplemented with street vendors' offerings (Albala 2003; Alcock 2006). Greek *kapeleion* (taverns), Roman *taberna*, *thermopolia* (cook shops), *popina* (somewhat dubious eating and drinking places) and *caupola* (hotels) were all in evidence (Alcock 2006: 126). During Rome's imperial period, *mansiones* were found along Roman roads, sited a day's journey apart to give imperial messengers accommodation, food and drink, while the somewhat later development of *palatinates* to accommodate peripatetic courts was associated with boarding houses, again offering food and rooms (Klauser 2008: 112).

Inns have, for millennia, been a feature of urbanism, as O'Gorman (2009) traces through the analysis of Mesopotamian, Greek, Roman and Middle Eastern sources, with Mesopotamian hostels and inns evidenced from at least 1800 BC, and *katagogion* or inns described in Greek accounts. Evidence of *hospitiae* from Pompeii similarly shows inns or taverns which supplied food, drink and accommodation, and in one case can be seen to include a separate restaurant room (O'Gorman 2009). A form of *hospitiae* called 'stabulae' was located at city fringes: 'close to the city gates, often described as the ancient equivalent of modern motels or, more romantically, coaching inns; they had an open courtyard surrounded by a kitchen, a latrine, and bedrooms with stables at the rear. Businesses within city gates were smaller than those in the countryside, due to pressure of space' (O'Gorman 2009: 6). From the seventh-century AD there is also substantial written evidence of the establishment of caravanserai, in which travellers could stay, as part of the Islamic tradition of hospitality (O'Gorman 2009: 6). While medieval monasteries offered guesthouse accommodation to pilgrims or crusaders, these were hospitable spaces reserved for particular categories of traveller or diner, whereas 'caravanserais could also be used as commercial centres for merchants' (O'Gorman 2009: 9).

Mennell (1985: 136) reports that 'in London, Paris, and every other city in Europe, it has always been possible to purchase cooked food' and accounts emanating from the twelfth century describe urban cook shops where townspeople could both buy cooked dishes and take their own food to be cooked (Carlin 2008). In the early modern period, cook shops were associated with dishes including braised ox shanks and turbot, and equally allowed other dishes to be brought to bake on their premises (Albala 2003: 112). By the late seventeenth century, some urban cook shops had gradually evolved into the coffeehouses discussed earlier in this chapter, and themselves sometimes transmogrified into dining rooms serving meals 'almost to the exclusion of coffee' (Mennell 1985: 137). Paris's traiteurs or 'public cooks', meanwhile, were sustained by regular local customers, making 'cooked foods available to the vast population of urban workers and artisans who did not have their own kitchens or cooking equipment' (Spang 2000: 31).

Inns, too, were a feature of European towns, catering to upper-class travellers and offering *table d'hote*. A dense network of these existed throughout Europe by the late Middle Ages, selling food, providing banquets and offering meals to takeaway (Kumin 2003: 71). This was of very uneven quality. As Godina-Golija (2003: 125) reports, the food in Slovene inns was 'bad, tasteless and expensive', whereas in eighteenth-century London inns, the 'ordinary', a fixed price and fixed menu dinner was reputed to be of high value and quality (Mennell 1985: 137). Inn food was generally more elaborate than that found at taverns and might include soups, stews, roast poultry and fish (Albala 2003: 113). However, the tavern also grew up as a form of restaurant, with its origins as a place for men to drink wine but also serving food that was often salted or preserved: hams, sausages, anchovies and other preserved fish (Albala 2003: 113). By the eighteenth century, at least in London, 'many taverns in the capital were noted eating-places and centres of social life [for a] superior clientele' (Mennell 1985: 137). The London Tavern, The Crown and Anchor Tavern, The Globe in Fleet Street and the White Hart in Holborn, all became known through cookbooks produced by their cooks, and Samuel Johnson celebrated the tavern's conviviality (Mennell 1985: 137).

From early-to-mid-nineteenth-century London, though, the best cooking was to be found in small private hotels (resembling early twentieth-century 'service' flats) where the boundary between public access and private space was again rather blurred, rather than the taverns where cooking standards had declined (Mennell 1985: 154), or the chop houses where more basic meals were served to the working classes. Similarly, the proliferating private clubs that grew out of earlier coffeehouses produced good, average food (Burnett 2004: 75). Tavern culture, of course, was not just a London phenomenon and did not entirely disappear from the more modern urban space of the late nineteenth century, as Orhan Pamuk's (2006: 123) discussion of Istanbul makes clear. Reflecting on the work of Ahmet Rasim 'who made the city his subject', Pamuk notes Rasim's writings about 'the beauties of the towns along the Bosphorus to its rowdy taverns and meyhanes' (Pamuk 2006: 124).

In France, meanwhile, the rise of the modern restaurant has been argued to be a result of changes brought about by the Revolution (Rich 2011: 136), but restaurants serving 'rather ordinary food in not very elaborate surroundings' were already opening in the preceding decades, with the Boulanger case cited as an early example of the loosening of the grip of the guild of *traiteurs* on the provision of cooked food outside the home (Mennell 1985: 138). As Spang (2000: 2) notes, in the last twenty or so years of the Ancient Regime, 'one went to a restaurant ... to drink restorative bouillons, as one went to a cafe to drink coffee', and these spaces could be distinguished from inns, taverns or cook shops by 'their individual tables, salutary consommés and unfixed mealtimes'. By around 1804, there were over 500 'establishments in Paris in which wide-ranging menus could be enjoyed', with the emergence of the bistro, a result of returning French soldiers introducing the Russian fast-food restaurant to the Parisian scene (Klauser 2008: 112–13).

By the 1820s, it is argued, the version of the Parisian restaurant with which we are familiar today had emerged: 'a public setting for private appetites' (Spang 2000: 2). This move to restaurant dining blurred the distinction between public and private space, affecting both sociability and food consumption patterns (Rich 2011: 138) and ushering in new forms of spatial design, including in some cases separate dining rooms for men and women. In fact, their *cabinets particuliers* (small private rooms within restaurants) became notorious as venues for sexual liaisons (Rich 2011: 138). For foreign visitors the preponderance of women in restaurants was a great surprise (Spang 2000: 201). Rich (2011) describes the sometimes substantial lengths proprietors went to spatially manage the possibility that women (with their supposed lack of self-control) might misbehave socially or sexually, by allowing them to sit only at certain tables and in certain less conspicuous areas. Women alone in restaurants are still subject to suspicion and discrimination, demonstrating the continued ambivalence that exists about womens' 'right to the city' and suggesting a distinct lack of conviviality.

The rise of 'modern' European restaurants has equally been associated with the development of bourgeois culture, allowing the middle class to enjoy what had previously been the luxurious preserve of the aristocracy (Mennell 1985: 142; Finkelstein 1989: 13). The restaurant as a form of semi-public space is thus predominantly understood as a democratizing influence on public food culture, in which the number and variety of options for eating out steadily increased through forms including bouillons, brasseries, dining rooms and teahouses (Rich 2011: 153). For example, there was a burgeoning of a culture of restaurants, cafés and bars from the period of late nineteenth century in German cities, resulting from regulatory and political changes, economic boom conditions (later replaced by hyper-inflation and instability) and the physical and social transformations in the nature of urban life (Klauser 2008: 113). Hotels and restaurants emerged close to theatres and railway stations, and perhaps reached an apogee in terms of the interplay between art, food, music and performance, at the enormous Haus Vaterland café in Berlin's Potsdamer Platz adjacent to Anhalter Bahnhof (Klauser 2008: 113).

Yet, drawing on the work of Goffman (1959) and Elias (1982) on the civilizing process, Finkelstein (1989) has suggested conversely that far from being a benignly convivial activity, restaurant dining in the city was and is uncivilized because diners are playing out roles of conspicuous consumption of private pleasure in public space, which reinforces a lack of reflection about, or questioning of, conservative social norms. In language that brings to mind Bourdieu's habitus formation, Finkelstein (1989: 2) argues that 'at the more exclusive bistros, pleasure may accrue from the diner's use of the event to suggest the personal possession of culturally valued characteristics such as wealth, fine taste and savoir faire'. Dining out is situated as a way of expressing cultural capital in the public domain of the city, and the restaurant in this reading is 'a popular arena in which skirmishes over social location and claims for prestige can take place because it provides the tools through which claims of cultural capital are made, namely, artifice, esoterica and fashionability' (Finkelstein 1989: 122).

RESTAURANTS AS DESIGNED SPACE

In terms of the spatiality of such dining, Finkelstein's (1989) typology of restaurant spaces denotes different kinds of dining out experiences. Finkelstein (1989) begins with the 'fete special', in which the restaurant itself is the attraction, drawing diners who may be rich tourists who demonstrate more economic than social capital in their gastronomic taste, and employing staff who openly show their disdain for these diners (Finkelstein 1989: 29). In the second category are restaurants to which diners go to be amused and entertained: these encompass the bistro mondain in which ambience and food fashionability are important, and the parodic restaurant in which diners are 'offered a pre-constructed material fantasy' in relation to setting and food styles (Finkelstein 1989: 29). The third category is the convenience restaurant, where diners eat because they are not inclined or able to cook for themselves, and includes the café mundane (like the bistro mondain but much cheaper), local ethnic restaurants and fast-food chains. Suen (2007: 115), meanwhile, considers design from a slightly different angle, offering an analysis using examples from Singapore of the historically themed restaurant, which, he argues, provides the 'external props and tangible reminders' we need now that tradition is no longer woven into the fabric of everyday spaces. The restaurant's design provides 'a commodified historical narrative' to make up for this absence (Suen 2007: 115).

Finkelstein's work has been criticized as 'curious' and 'poorly substantiated' (Burnett and Ray 2012: 148). In fact, in urbanism terms at least it has the strength of offering a typological analysis that is suggestive of spatial design considerations, unlike more recent work on eating out, which offers a 'service provisioning' framework for its analysis (Warde and Martens 2000). Thus both Finkelstein's typology and Suen's perspectives seem relevant in situating the role of design features in contemporaneous examples. In the pre-constructed fantasy category, for instance, is the following from New York. 'Do you like CNN and bison meat? Then perhaps you might enjoy Ted's Montana Grill, the turn-of-the-century Western-themed megarestaurant in the Time Life Building from mogul extraordinaire Ted Turner. The décor is McCabe and Mrs. Miller meets Restoration Hardware, the menu is a mix of burgers, big-ass salads and things cooked on cedar planks' (eater.com 2013). At a more upmarket level, a bistro mondain offers its dinners the chance to express a stylish habitus, in which design and architecture are central to this capacity (Riordan 2006; Ryder 2007). In Toronto's Blowfish Restaurant and Sake Bar, for example, there is

> A melding of yin and yang, a blend of Eastern and Western architecture, with fusion cuisine that melds divergent ingredients and cooking styles, Blowfish has a hip, stylish interior inside a structure with a classical exterior ... the paramount design objective was to create a brand that is a study in contrasts yet works together to create fusion between the French-inspired design elements and the Asian-oriented cuisine ... the unisex, four-compartment restroom incorporates a cantilevered hand-washing sink that divides the space in half and provides a must-see for Blowfish patrons. (Baraban and Durocher 2010: 181)

Intriguingly, conviviality is invoked in some descriptions of such 'bistro mondain' restaurant styles; yet the details given suggest something more artful and manufactured:

> Public's dining areas include wide-open rooms for convivial dining, while the adjacent Monday Room (added three years later) is a small, intimate space with a library-like feel intended for a comfortable drink, flight of wine, or light meal. Invested with a hip yet convivial atmosphere that feels like it has always been there, Public fast became a neighbourhood icon known for great food: the restaurant was awarded a coveted Michelin Star in the 2009 New York Guide for executive chef Brad Farmerie's inventive Australasian cuisine. (Baraban and Durocher 2010: 212)

In socio-spatial terms the restaurant holds a somewhat ambivalent place in the gastronomic townscape. Work on Vieux-Quebec suggests that particularly desirable historic locations can lead to an over-abundance of restaurants and the loss of housing stock and other urban services, as well as the sense that the space has become a kind of Disneyfied one (Gazillo 1981). In an extreme example from Cracow's 'Old Jewish Quarter', restaurants emerged in the area where the film *Schindler's List* was based, with 'Jewish style' cafés and restaurants offering tourists traditional Jewish food but in fact entirely staffed by Polish Catholics (Shaw, Bagwell and Karmowska 2004: 1987). The use of design-based gimmicks in restaurant interiors, which perhaps reached their zenith in *Carême's enormous pièces montées* (Mennell 1985: 145), may be matched in contemporary practice by the stylistic overload of the parodic restaurant interior identified by Finkelstein (1989: 38). Yet particular elements in restaurant interior architecture have also been read as the pursuit of a national style designed to express something quintessential about the place, as in the work of Japanese designer Takashi Sugimoto (Locher et al. 2006). Less bold in their claims, but still suggesting restaurant design says something fundamental about urban dwellers' tastes and spatial design preferences over the long term, Ryder (2007: 5) argues that 'there is no abatement in the hunger for elegant, unusual and spectacular dining spaces' and currently fashionable design tropes include 'global views', 'new baroque', 'modern classics' and 'high concept' spaces. Restaurant design becomes 'a cohesive design experience' with 'places of high theater that feed not just the appetite but the soul as well' (Riordan 2006: foreword).

In a design sense, the restaurant offers transitional space between the public realm of the street and the exclusive, private space of the building. Design may be used in more or less subtle ways to give out messages about who is welcome and who might be excluded. Fieldwork interviews in London suggest that restaurants designed to 'give something back to the street' are seen as more convivial than entirely interiorized examples that operate as private worlds, and this openness to the public realm is thought by diners to signal a 'European' food sensibility and lifestyle (Parham 2012: 233). Restaurant design could thus be seen to have the capacity to either promote or undercut spontaneous convivial contact of the kind which is not solely about economic exchange (Lefebvre 1991). Such dining can thus contribute to the townscape's gastronomic richness rather than simply drawing energy from and draining it of vitality.

The emergence of pop-up restaurants in cities in the United Kingdom, United States, Australia and elsewhere has been associated with taking over private, transitional and leftover public urban spaces at gastronomy's 'high end' (Mclaughlin 2012). Clearly tied to a particular kind of habitus formation, being conversant with both the latest food and urban spaces, the pop-up restaurant phenomenon has been described as an underground restaurant scene (Square Meal 2013: npr) and celebrated, critiqued and parodied. In locations including private homes, car park rooftops, warehouses, old tram sheds, disused or underused restaurants (and even in Finland, a former mine), these pop-ups 'are morphing into a multipurpose tool, used by different strata of the restaurant industry to test concepts, market new brands, engage with a younger audience, or prove to landlords, lenders and investors that they are worth the risk' (Square Meal 2013: npr). An example from London encompasses both typical elements and trajectory:

> A cooking collective formed by ex-Loft Project chefs James Lowe, Isaac McHale and Ben Greeno, The Young Turks use other people's venues – often other supper clubs (including the Loft), and often top-secret – as their base for their fabulous multi-course dinners. Dishes could be anything from beetroot, goats milk and pickled elderberries, to pheasant egg, ramson and snails. Their latest venture, Upstairs at The Ten Bells, was so successful as a pop-up it has since become permanent.

In more formal spaces, restaurant design may be approached as a purely technocratic activity (Baraban and Durocher 2010) in which architecture is utilized to maximize profit. Using a systems approach and introducing the topic as 'where design begins', the authors seek to define and describe design solutions for every aspect of a restaurant's operation, noting that 'the secret to a good relationship of concept, menu, and design is to conduct a careful market study and menu analysis before determining specific design elements in either kitchen or the dining area' (Baraban and Durocher 2010: 1). These attempts to systematize and control spatial design processes, though, are always some steps behind the untidy reality of dining trends and preferences. The rise of the big table, although not an entirely new phenomenon, is one example that could be relevant in increasing the level of interaction, and by implication, the conviviality of restaurant dining. Having to share a large table space has become a commonly employed design technique, although much of the concern in this area is about increasing table turnover and spend-per-minute (given technical legitimacy through the use of the acronym 'SPM'), and by careful attention to table shape and configuration (Kimes et al. 2004), rather than any interest in the experience as a pleasurable one for the diner.

Technically driven design analysis, for instance, shows that booth designs are associated with slightly higher spend-per-minute than are banquettes (Kimes et al. 2004). Diners' perceptions of physiological comfort are influenced by seating configurations, while norms of seating behaviour can change according to cultural background (Robson 2002). As in the discussion of third spaces, power over personal territory is the key here, with diners attempting to establish and defend sufficient space for themselves to ensure privacy, regulate contact with others and express power. They may use architectural features as props to help them do so (Robson 2002). While restaurant ambience is cited by diners as important to their repeat business, other service factors also seem to play a part (Soriano 2002). As noted in relation to cafés, there are more or less submerged tensions and hierarchies created between diners themselves, and between diner and server (Finkelstein 1989), while

> increasingly, restaurants are recording whether you are a regular, a first-timer, someone who lives close by or a friend of the owner or manager. They archive where you like to sit, when you will celebrate a special occasion and whether you prefer your butter soft or hard, Pepsi over Coca-Cola or sparkling over still water. In many cases, they can trace your past performance as a diner; how much you ordered, tipped and whether you were a 'camper' who lingered at the table long after dessert. (Craig 2012)

It is worth noting that particular forms of urban dining space, including the French *brasserie*, the German beer hall and the British public house, reflect strong cultural traditions connected to consumption of wine, beer and spirits. The Italian *enoteca* and the Spanish tapas bar, for instance, are forms of drinking space which have sparked the production of numerous cookbooks about the simple dishes served, with the latter's gastronomy seen as part of national branding (de Lera 2012). Institutional forms of dining, too, are an area of intersection between private and public access to urban dining space that has taken on a range of intriguing forms in contemporary urbanism. The *mensa,* or student dining room, is often found in Italian and German towns and local people (and tourists, if they can find them) can also dine there on affordable food of reasonable quality (Rebora 2001). As one blogger in Germany notes: 'Although Mensa cafeterias are commonly found on campus at German universities and packed with healthy, athletically-inclined students, the dining halls are open to the public, and you'll often find yourself dining alongside area businessmen and women. For the adventurous tourist, the cafeterias offer a unique and budget-friendly opportunity to get a taste of the real Deutschland' (Meyers 2008).

Similarly, in Berlin, the backdrop to present-day office dining rooms is of a long history of industrial canteens originally dedicated to 'people's rational feeding' in which the worker was required to 'match his bodily physical needs and functions to those of the industrial process' (Thoms 2003: 351). As industrialization advanced, the disconnect between home and office grew spatially greater and food at work became both more necessary and more sharply differentiated in relation to restaurant quality meals for managers and more basic offerings for staff (Thoms 2003: 363). Contemporary office canteens and business dining rooms are still known for their cheap, excellent food and are visited by aficionados of particular dishes (Evans 2012):

> In Berlin, countless works canteens are open to the public. Anybody can wander in, pick up a tray, queue of course, and then buy a mountain of a meal for the price of a snack. And then return your tray and dishes because that's what you do in a canteen. You might choose the chichi surroundings of the Universal Music canteen overlooking the River Spree. Or you could negotiate your way through the maze of theatre trucks and stage sets stacked in the yard behind the theatre on Schiffbauerdamm. ... At the very smart canteen for the combined Nordic embassies, they serve fresh fish every day – roasted fillet of hake, for example, on lentils in balsamic vinegar with vegetables and new potatoes, at a price of 5.20 euros.

The open dining room or canteen is a much more unusual form in Anglo-Saxon cities, in part because these spaces are habitually made more private and physically inaccessible to the public, although particular examples, like the Indian YMCA in London, have famously cheap, accessible and tasty food. As accounts of architect David Chipperfield's temporary 'canteen' in Berlin (Burrichter 2011), Richard Rogers' 'employees café' in west London, now a very upmarket restaurant (Garnett 2012) and the Rochelle Canteen, in London's inner east, demonstrate, the humble works canteen has been transmogrified, allowing only an insider's access to the gastronomic townscape:

> I first heard about Rochelle Canteen through a food blog friend of mine and thought it sounded charming. then i heard about some pop-up hidden spot that was run by the wife of the chef of St John. and then by (now-ex)boyfriend randomly took me there. You need to be available at odd hours (ie lunch only on weekdays), and not too fussy an eater. in a predominantly residential neighborhood, just on St Arnold circus there is a school, and in the massive brick wall around it there is a big blue door with a panel of buzzers next to it. Almost unreadable one says 'canteen', but you would never think to buzz it on your own. in you go, across the

once-playground into a redone shed with huge glass doors. inside is essentially an open kitchen and several long bare tables. The walls, white, stand bare except a row of hooks along on side (anyone been to St John Bread & Wine? you'll find these familiar). (Hayley 2011)

School dining rooms meanwhile have been portrayed as somewhat conflicted spaces (Metcalfe et al. 2011) and undoubtedly the site in some cases of very poor-quality food which has led to vociferous public policy debates, campaigns and programmes of reform (Campbell 2011; Gilchrist 2012) including to connect students to local farmers and seasonal and regional produce (Vallianatos et al. 2004). As Sonnino (2009: 426) notes from her research into the radical reshaping of school meal arrangements in Rome, these can offer a profound challenge to an unsustainable, unconvivial food system, through a focus on quality:

> The public sector is emerging as a powerful actor in the food chain-one that has the capacity to reconnect producers and consumers through a process of qualification that extends beyond the market and the food products alone. By also acting upon the less visible aspects of the food system-including service, transport, labor, eating practices-procurement policies such as those implemented in Rome are designing an 'economy of quality' that has the potential to deliver the environmental, economic, and social benefits of sustainable development in and beyond the food system.

Mainstream rather than 'gastronomic' office canteen dining similarly has a mixed reputation, including being associated with food poisoning cases such as outbreaks of norovirus (Showell et al. 2007). What seems to link these canteens and office dining rooms to the conviviality of the townscape is their capacity to maintain a relationship to the public domain in their design arrangements, or at least to allow easy accessibility from public space, so that outsiders can visit them to eat. Reflecting on personal experience, as a guest at a design organization's staff dining room, atop an architecturally prized modernist building in central London, the dining room offered reasonable food, a lively atmosphere and great views of the city, but did not engage with the public realm from which it was completely hidden. It was possible to speculate that this conferred a feeling of exclusivity and privilege that those 'on the inside' enjoyed (of course, some might argue they simply preferred the conveniently located, affordable food on offer). Yet the lack of access by the public to the benefits of affordable food in a desirable location, and the loss for the most part of the possibility of sharing food with others outside the organization, seemed to demonstrate the gap between exclusionary preferences and more convivial practice that would contribute to the gastronomic townscape.

TAKING OVER STREETS FOR DINING

Another significant contribution to the gastronomic townscape comes from the practice of taking over public space for dining (Parham 1990). Within various urban traditions, at least two different kinds of convivial use of the street can be discerned.

In some instances, local restaurateurs rope off or otherwise commandeer sections of the footpath or road and set up tables for their patrons, which may be organized at the owner's discretion and with the authorities' connivance. Yet this kind of appropriation of street space for restaurant tables and chairs can be a contested use of public space because it may suggest conspicuous consumption in the public domain in the face of those less entitled (Whitelegg 2002; Parham 2012). It can also offer physical impediments to others using street space: something planning authorities have often frowned upon.

In other instances, town dwellers have themselves appropriated street space for dining, with special occasions allowing public and private space to be merged (Parham 1990; Franck 2005). Field (1990) makes links between festivals held in Italy's urban public domain and the rhythms of life determined by agricultural production and food's seasonality that still hold sway, especially in village- and town-based urban life, a theme further explored in Chapter 9's consideration of the critical food region. Although sometimes with a religious cast, traditionally, these celebrations of abundance have acted as the counterpoint to *magro* (poor food) experienced for most of the year; making up for austerity and want as the usual condition. Representing the topsy-turvy world of *carnevale* (Salvalaggio 1984), public occasions like the orange festival in *Ivrea* and *La Tomatina* in Spain arguably dissipate the potential for real violence with symbolic conflict, and use food as their mode of expression. Roden (1989) has noted the way that food festivals can transcend deep political divides: a point nicely made in Bertolucci's 1962 film of Stendhal's *Charterhouse of Parma*, where the protagonist attends both religious and communist festivals in order to eat the best sausages. Such *festas* have supported the renewal of gastronomic traditions, as in the case of the revival of balsamic vinegar (Roden 1989: 9). Recent London-based fieldwork demonstrates a trajectory of food-centred area regeneration, from holding a one-off food festival to developing a thriving regular food market or food street, as evidenced at Borough Market, Whitecross Street and elsewhere (Parham 2012).

Food festivals based in public space have also been critiqued, including being defined as ways of commodifying place and identity (Adema 2006). Often used to celebrate aspects of ethnic, cultural or regional identity among diaspora communities, it is argued that their foodways may also express ambiguities and contradictions through food stereotypes that present a 'collective ethnic fantasy' (Van Esterik 1982: 209; Shaw et al. 2004). However, these perspectives do not seem to entirely capture the meaning of examples where food festivals reflect more place-based connections, which is in turn reflected in particular foods. The *jambalaya* of Cajun festivals in Louisiana discussed by Esman (1982), or the Italian specialties of Adelaide's *Carnevale* (including *salcicce con sugo e polenta, arancini Siciliani* and *pollo alla brace*) reference historical and contemporary migration links; in the former between French-speaking Acadian settlers from Canada, and in the latter case between South Australia and the Calabria and Campania regions of Italy in particular. While the *Carnevale* festival outgrew its on-street location, many other examples remain street-based. The *Glendi* festival on Lonsdale Street in Melbourne, for instance, and the Notting Hill Carnival in London showcase foods of various countries of origin, while those origins may stretch back some generations. Most critically in the context of this analysis, they are public space focused: *Glendi* occurs in an inner urban area of Melbourne known as the 'Greek precinct', and the Carnival in an area of west London traditionally home to immigrant communities from the Caribbean although now an upper-middle-class gentrification hotspot.

These expressions of gastronomy within the townscape might be considered in part as urban feasts, a form that has a very long pedigree. We know that cooking and sharing food is an ancient practice: with archaeological evidence from close to 30,000 years ago of early meals cooked and eaten (Jones 2007). From myth come stories of the feast promised to Apollo by Theseus (the *Pyanepsion*, its name derived from boiling beans); from drama the resolution of action in a classical Greek comedy; and from gastronomic history the medieval feasts held by the Milanese aristocracy (Root 1971: 264). Similarly, certain cities are associated with particular foods (such as Parma ham and Roquefort cheese), while there are equally close links with foods served on particular days. The *ceci con la tempia di maiale del giorno dei Morti* (pig's head meat boiled with chick peas, onions, carrots, celery, sage and rosemary) served on All Souls Day in Milan, the *ossa dei morti* (bones of the dead biscuits) made for the same day, northern Italy's *polenta e baccalà* (salt cod and polenta) on Christmas Eve and *cotechino con lenticchie* (boiling sausage and lentils) on New Year's Eve all spring to mind. These spatial expressions of street-based feasting appear more effectively convivial than the spectacle of the lavish civic banquets that formed an important historical form of 'public yet exclusive' dining in which women were largely excluded both practically and symbolically from connection with the heart of the city (Rich 2011: 172). This might in the contemporary city relate most closely to the exclusionary nature of some private members' clubs in which female diners have up until very recently been admitted on sufferance, as portrayed by William Boyd in his novel *Restless* (2006).

THE GASTRONOMIC TOWNSCAPE IN REVIEW

Cities have traditionally maintained a very close spatial design and cultural relationship with gastronomic land uses, and while these connections can be seen to be shifting as socio-spatial practices change, the focus on the primacy of the public realm and public engagement in city space in relation to food remains strong. Often spatially connected to the food markets, as explored in Chapter 3, the food-centred land uses of street foods, shops, inns, cafés, snack bars, restaurants, canteens and festivals all contribute richly to the gastronomic townscape. While it is clearly not possible to reduce this survey of such townscape elements to the question of whether there are 'perfect' gastronomic cities or urban spaces, certain elements seem to recur in this analysis. The examples discussed here include a sufficient density of housing and people to act as catchments for these land uses, appropriate height-to-width ratios of streets and squares to support their use as outdoor rooms and attention to the design richness of the edge and transitional spaces between the public and private realms. Perhaps most important are the ways that culturally supportive approaches to food production and consumption patterns are played out in sustainable, life-affirming ways in which convivial traditions put food at the centre of social and economic life, which is in turn reflected in particular forms of urbanism. In the next chapter, the exploration of food and city form continues at a larger scale, with a focus on food-related elements that have contributed to making the more recent – and gastronomically ambivalent – urbanism of suburbia.

CHAPTER FIVE

Ambivalent Suburbia

INTRODUCTION

This chapter acts as both an extension and a counterpoint to Chapter 3 and 4's discussions of food's outdoor rooms and the wider gastronomic townscape. The scale of the analysis expands to consider ways that food has interconnected with the rise of suburbia before suburban food space was itself refashioned by the growth of sprawling urban regions, a megalopolis whose food elements are the subject of Chapter 8. Just as the two previous chapters considered how various food-centred land uses, spaces, designs and practices support (or undercut) a rich conviviality in the public realm and contribute or otherwise to making a sustainable townscape, this chapter moves to the larger scale to explore some of the ways suburban development has shaped connections to food in the nineteenth and twentieth centuries.

The story is a complicated one in food terms but not entirely negative; although, broadly speaking, early suburban design can be judged as often more capable of supporting gastronomic conviviality than post-Second World War models have achieved. The most significant spatial and cultural shift this chapter explores in food terms is the advent of supermarkets and shopping-mall-based food consumption, food courts and fast food that aligned with car-dependent sprawl and prefigured the imposition of suburban food typologies in more urban places, largely to their detriment. The consideration of suburban design in food terms foreshadows the discussion in Chapter 8 of the landscapes of food poverty and social exclusion, the contested spaces of food deserts and the increasing prevalence of obesogenic environments (Parham 1998; Lake and Townshend 2006; Lake et al. 2010).

This exploration of suburbia's foodscapes starts by briefly sketching out the interrelated history of suburban development and outlining some of its food-related expressions. This is not as straightforward as might first appear. As Harris and Larkham (1999) have noted, there are a number of myths about suburbs. They have suffered from over-simplifying accounts of both their origins and later development, and gaps remain in understanding their nature. While acknowledging this incomplete context, this chapter focuses its survey of the *sub urbe* in terms of its complicated interplay with food in both historical and contemporary spatial practice.

FOOD AND THE BURGEONING OF SUBURBIA

Just as cities have historically been understood as centres of economic, social and cultural excellence, suburban development has contained a connotation of inferiority, spatially inhabiting 'land immediately below hilltop walled towns, the houses, gardens, workshops and other built forms literally beneath-and beyond-ramparts and gates' (Stilgoe 1988: 1). In the western European traditional, suburbs have been a feature of many post-Roman towns, identifiable as places where noxious trades were banished or where water was required for industrial activity, and equally shown to thrive around medieval towns (Harris and Larkham 1999: 4). It was when wealthy households sought to escape from urban centres that were increasingly dirty and congested (Harris and Larkham 1999: 4) that a new kind of suburb began to emerge in the eighteenth century. By the nineteenth century, the advent of trains, trams, buses and bicycles let people live further out from where they worked, and the geographical spread of towns became a 'marked urban feature' of the second half of the century. This was spectacularly so in places like London where 'the leapfrogging suburban cycle of selection, settlement and later partial rejection by a widening band of middle class people, was the process by which a growing urban population was redistributed' (Cherry 1972: 62).

While the modern term suburbia has only been in common use since the 1890s, it has tended to be employed in a polarized way, either to celebrate or decry forms of low density, mono-functional living space (Webster 2002: 5). In a switch from the suburb understood as an inferior kind of space, ever since Olmstead and Vaux's Riverside suburb was developed for 'the more intelligent and more fortunate classes' (Kostof 1991: 74), suburban development has been argued by some to be the highest form of settlement, representing 'the most attractive, the most refined, and the most soundly wholesome forms of domestic life, and the best application of the arts of civilization to which mankind has yet attained' (Olmstead, quoted in Fishman 1987: 198). On the other hand, the idea that suburbia is a lesser form has continued to be held, with suburbs thought to represent a boring and tedious way of living that both constrains and represses their inhabitants (Fishman 1987). Whether loved or loathed, the spatial forms that suburbia encompasses should be understood as the 'most fundamental realignment of urban structure in the 4500-year past of cities' (Jackson, quoted in Frost 1991: 10). Many residents have continued to find an escape in suburbia from the problematic inner city (Wilson 1991). Thus, although suburban landscapes are not a spontaneous expression of consumer demand in a rational market system, as has sometimes been claimed, over the course of the nineteenth and twentieth centuries, they have been overwhelmingly desired by many urban dwellers (Murrain 1993).

By the early nineteenth century, Cobbett (1830) on his rural rides was already lamenting the relentless growth of the Great Wen of London, with its disease, poverty, overcrowded slums and dark, narrow streets. London's unhealthfulness (understood as both moral and physical) provoked social reformers with the desire to build more low-density areas suitable for London's working and lower middle classes, and developments like Bedford Park for modest habitation in urban areas made accessible by the growth of the railway network, were a late-nineteenth-century precursor of later 'streetcar suburbs' (Frost 1991: 14). However, the so-called true suburbs were not for the less affluent in the intermediate areas around the centre of cities, but found on the very edge of the metropolis: 'here were the houses of size and quality, invariably inhabited by upper-income groups. Large, landscaped lots and open spaces provided seclusion and exclusiveness and thus the appropriate marriage of town and country' (Frost 1991: 15).

Similarly, in the suburbanizing United States, not all suburban development was the same, with the 'borderlands' of larger, set back residences distinct from more modest 'front lawn suburbs', and from 'the scruffy, new, open, poorly built zone of mixed residence and manufacturing that hung about the edges of large cities' (Stilgoe 1988: 9). Recognizing suburbia's considerable diversity, a useful morphological definition of suburbs identifies the presence of individual houses as a salient feature and emphasizes how urban space changed from streetscapes to landscapes in which dwellings 'mostly detached or semi-detached, are separated from the road, and often from one another by private gardens' (Whitehand and Carr 2001: 1). In an American context, Dolores Hayden has offered a chronology of suburban development from 'borderlands' (starting about 1820) through picturesque enclaves (1850) to streetcar buildouts (1870), mail-order and self-built suburbs (1900), mass-produced 'sitcom' suburbs (1940), edge nodes (1960) and rural fringes (1980)' (noted in McManus and Ethington 2007: 318). Despite assumptions that suburbia was solely residential, this was not just a dormitory domain but also encompassed a process of industrial expansion from an early period (Walker and Lewis 2001).

In the 'new urban frontier' cities of America, Canada, Australia and New Zealand, among others, householders who could afford to rejected the higher housing densities of European cities, associated with overcrowding, slum blight and unhealthy air, in favour of detached single-family cottages (Frost 1991: 10). Food was already part of this spatial shift. It was cheaper food costs enjoyed by the inhabitants of these new cities that provided the surplus income households used to purchase suburban dwellings (Frost 1991: 74). Once living in this outer urban space, food distribution arrangements made both a virtue of location and reflected these lower densities in their shopping patterns, as can be seen in relation to borderlands' suburban development: 'housewives expected almost every purchase to be delivered into their kitchens, sometimes put away. Icemen delivered blocks of ice directly into iceboxes; milkmen, often operating from local farms, placed milk on top of the ice. Along the curving roads, according to set schedules, came the competing butchers, fishmongers, vegetable sellers, and other vendors of food' (Stilgoe 1988: 209). Food retailing was sometimes designed in as the centrepiece for suburban development. In the United States, the settlement of Lake Forest, laid out in 1857 on the shores of Lake Michigan in Illinois, was intended to provide a peaceful, green and secluded environment for family living in middle-class style. Designed by the architect Howard Van Doren Shaw, and completed in 1916, its Market Square is thought to be the first planned shopping centre developed in the United States, prefiguring the suburban retailing spaces that later blighted outer urban areas in the post-war period. Yet this design demonstrated Shaw's European sensibilities and showed both that food-centred space could be an essential element in such a suburban design and way of life, and that it could be organized as outdoor rooms at a human, walkable scale.

FOOD AND THE THREE MAGNETS

In the United Kingdom, Ebenezer Howard's *Tomorrow: A Peaceful Path to Real Reform* (1898) proposed Garden Cities on the basis of the 'three magnets' of town, country and town-country, 'in which all the advantages of the most energetic town life, with all the beauty and delight of the country, may be secured in perfect combination' (Howard 1902: 46). Garden City designs based on Howard's proposals, such as those by Barry Parker and Raymond Unwin at Letchworth, north of London, offered their inhabitants the possibility of suburban living, with areas for housing and industry around a cultural and social centre, encircled by a productive agricultural greenbelt (see Figure 5.1). These ideas were extremely influential across Europe, in parts of Asia, the Middle East, Australia and the Americas (Garnaut and Hutchings 2003; Goldshleger et al. 2006; Hutchings 2011; Gallanter 2012; Parham 2013). Socialized food features proposed included some of the houses having common gardens and co-operative kitchens (Howard 1902: 54). It encompassed the improvement of land not in use for building, where fruit trees could be planted or a dairy set up (Miller, in Parsons and Schuyler 2002: 106). The Garden City programme (written and diagrammatic) also proposed allotment areas around settlement edges, within a broader, productive agricultural periphery that would return both food and farm rental income to the town, as well as deal with its food waste. These ideas owed a debt to nineteenth-century utopian models including Fourier's *phalanstère*, Godin's *familistiere* at Guise and the workers' model villages

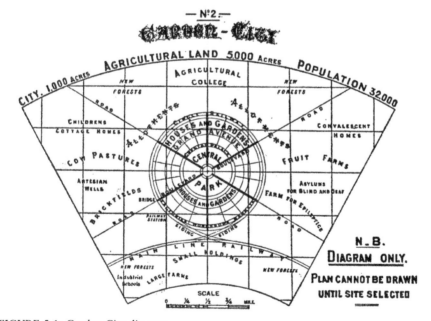

FIGURE 5.1: Garden City diagram.
Diagram by Ebenezer Howard in Garden Cities of Tomorrow (London: S. Sonnenschein & Co., Ltd, 1902).

of Cadbury's Bournville, Sir Titus Salt's Saltaire and the Lever brothers' Port Sunlight, among others. However, Howard's integration of food into his Garden City vision and practical plans was notably holistic.

Howard resisted the blurring of his city ideals in suburban interpretations (Dentith 2000: 20), and unlike many later suburban developers, he was particularly concerned with the possibilities for agricultural production in close vicinity of such settlements. These were intended to be of benefit to the farmer in producing a local food market and to inhabitants in lower food costs. 'Every farmer now has a market close to his door. There are 30,000 townspeople to be fed ... and this is a market which the rent he contributes will help to build up' (Howard 1902: 12). It was argued that Garden Cities would advance healthy living not just because houses would be well sited but 'because the gardens and surrounding agricultural belt will supply fresh and pure food and milk in place of the transit-soiled articles to which the average dweller in an ordinary city is condemned' (Purdom 1925, in Cherry 1972: 136). The productive agricultural greenbelts proposed by Howard were not just a setting for his Garden Cities but also a practical component of the economic base underlying these settlements, and it is notable that Howard saw urban food waste (and human waste) going back into the countryside to enrich the soil (Howard 1902: 13–14). Presciently, Howard proposed electrically powered transit arrangements which would keep the 'smoke fiend ... well within bounds' and allow produce to be sent to more distant markets (Howard 1902: 6). Howard did not expect Garden Cities to be completely self-sufficient in food rather, townspeople would be

> perfectly free to get their foodstuffs from any part of the world [but] consider vegetables and fruit. Farmers, except near towns, do not often grow them now. Why? Chiefly because of the difficulty and uncertainty of a market, and the high charges for freight and commission ... [but by] placing producer and consumer in such close association ... the combination of town and country is not only healthful, but economic. (Howard 1902: 12)

Fishman (2002: 58) has pointed out that Howard's idea of the 'boundedness' of a city reflected significant depth of thinking about urbanism and was central to his conceptualization of the Garden City, yet was an area in which imitation often failed to properly reflect the original model. Building on the neglected work of nineteenth-century economist Alfred Marshall and the American reformer Henry George whose anti-urban focus articulated a view that city slums reduced men to 'below the level of the brutes', Howard challenged the idea that continuous growth and increasing scale should be the measure of a city's success (Cherry 1972: 98). This was a food issue. For Howard (1902: 9), the greenbelt around the Garden City helped ensure its bounded quality, and as noted above, was of immense practical use in maintaining a productive spatial, environmental and economic relationship in food terms with the town it served. Of course, the problem of sprawl, loss of agricultural land and food security are still issues that confront us – undoubtedly more so given climate change. In looking at how Howard's ideas were applied in other places and renewed interest in Garden Cities today, how to respond spatially, socially and environmentally in food terms to the ever-growing megalopolis provides a critical context to which we will return in Chapters 7 and 8.

Although, as Hall (1988: 88) notes, many later critics have confused Howard's Garden City 'with the garden suburb found at Hampstead and in numerous imitations', aspects of Howard's legacy undoubtedly influenced many suburban and New Town developments (Parsons and Schuyler 2002). In the process, this created a new urban movement and term, variously the *cite-jardin*, *gartenstadt* and *ciudad-jardin* (Cherry 1972: 121). Approaches to settlement that referenced the Garden City may have misunderstood, misapplied or simply ignored some of Howard's extremely ambitious intentions such as the development of high-density, populous conurbations, the creation of self-governing communities and the reconstruction of capitalism (Hall 1988: 88), but were employed in both the United Kingdom and the new urban frontier where there grew 'a broad functional region of *truly* new cities, which coped with very rapid rates of population increase by spreading outwards through the replication of dispersed suburbs of single-family houses' (Frost 1991: 19).

It does seem clear that rational, local food-growing and shopping arrangements (and ethical approaches to food) were central to Howard's conceptualization of the Garden City and influenced later examples, yet were reflected in suburban developments in sometimes perverse ways. For instance, Howard's (1902: 4) notion of a Crystal Palace, which was conceived as a large covered circular arcade ringing the City's central park and in part given over to 'that class of shopping which requires the joy of deliberation and selection', is thought to have contributed to the idea of the 'covered collective retail space' of the regional shopping mall. The historical connection between the Garden Suburb and vegetarian ideas has also been noted (Spencer 1996; Miller 2010). In the United Kingdom, notable Garden City examples such as Parker and Unwin's plan for Letchworth Garden City (thought to be the first Garden City development) included shopping 'parades' built in the town centre while Louis de Soisson's 1920s master plan for Welwyn Garden City showed a town divided into four by railway lines, with each area boasting its own local food shops (Miller, in Parsons and Schuyler 2002: 125). Barry Parker's 1927–9 plan for the municipalized Garden City of Wythenshawe in Greater Manchester, meanwhile, included neighbourhood shops and plans for a major town centre (Miller 2010: 84).

These Garden City ideas also spawned a large number of more modest garden suburbs, such as the Hull Garden Suburb opened in 1908, which was sponsored by the manufacturers Reckitts (Cherry 1972: 123). Unwin's 1905 plan for an artisans' quarter in Hampstead Garden Suburb in London, proposed and promoted by Henrietta Barnett (Miller, in Parsons and Schuyler 2002: 113) offered 'a garden suburb for all classes ... where the innovation of the Garden City Movement [was] allied to a broad social purpose' (Miller 1992: 22). Unwin's proposals had included two shopping parades but these failed to be translated into Edward Lutyens plan: instead, a 1930s development called 'The Market Place' was built 'astride a diversion to the major roadway of the A1' (Miller 2010: 65) to make good this absence. In fact, a salient characteristic in food terms of the near ubiquitous garden suburbs of post-First World War Britain was the loss of a relatively direct spatial relationship between housing and food shopping areas, although food-growing opportunities on allotments were sometimes more accessible, as is discussed in Chapter 6. Morphological work on the suburbs of the interwar period in the United Kingdom suggests rather that

> residential areas were frequently laid out with little or no thought having been given to the provision of social services. ... Despite the fact that movement within suburbs was still mostly on foot, town-planning schemes prescribed the concentration of shops within a limited number of designated areas, often at main road junctions or near railway stations. This was in marked contrast to the individual corner-shops at both major and minor road junctions and the shopping ribbons along main roads that were characteristic of the majority or residential areas constructed before 1914. (Whitehand and Carr 2001: 44, based on Jackson 1973: 129–30)

This food-related spatial shift towards an increasingly coarse grain of food shop provision was evident in a number of areas in London and Birmingham suburbs in 1939. These showed a pattern of shop front buildings in which 'no sample square contained more than one shopping area and nearly one-half of squares lacked even a single shop-front building, in marked contrast to the numerous end-terrace shops in most areas developed before the First World War' (Whitehand and Carr 2001: 76).

FOOD AND POST-WAR SUBURBIA – RISE OF THE 'SMOKE FIEND' AND OTHER INFLUENCES

In the United States, by the late nineteenth century there were a number of examples of towns with Garden City–like attributes that had arisen independently of Ebenezer Howard's work and demonstrated some of the strengths of urban containment, including in relation to food. Yet over the course of the twentieth century both the rise of mass car ownership and changes to the structure of food production, exchange and consumption overtook such approaches in most urban development shaping. So keen were Garden City–New Town proponents to distinguish their models from the traditional city that went before them that they under-emphasized the Garden City's bounded nature, with its rings of allotments, agricultural land and concentrated food shopping arrangements, and emphasized instead 'loosening urban textures in favour of models drawn from the small town or the suburb' (Fishman 2002: 64). The attempts to find space for cars from the 1920s onward saw the development of neighbourhood units with Radburn layouts designed by Clarence Stein and Henry Wright. These separated out pedestrian and vehicular paths, effectively making those models even more anti-urban (Fishman 2002: 64). In the United Kingdom, anxieties about the spatiality of suburban development were already apparent by the late 1930s, as a hyperbolic description of sprawl by architect Clough Williams-Ellis makes clear:

> We are making a screaming mess of England ... the jerry-built bijou residences creep out along the roads. Beauty is sacrificed on the altar of the speeding motorist.

Advertisements and petrol stations and shanties ruin our villages. The electric grid strides across the hill-sides. A gimcrack civilization crawls like a gigantic slug over the country, leaving a foul trail of slime behind it. (quoted in Cherry 1972: 153)

New Towns, in particular, failed to achieve either the density or walkable scale that would allow them to work in food terms in the way that the Garden City did (at least in principle terms). Thus, while the New Towns movement and establishment post-war was a direct result of the success of Garden Cities and Garden Suburbs, the twenty-eight New Towns built in the United Kingdom, especially the latter ones (the so-called Mark Twos and Threes), were very much shaped around private car use, and this was for many urbanists part of their problem (Hall and Ward 1998: 53). There is a deep irony for food and urbanism scholars that Howard's ideas have often been called upon to defend suburban sprawl when he so ardently planned what Mumford had called 'a compact, rigorously confined urban grouping' (Fishman 2002: 62). Within this spatial schema, sustainable approaches to food were very much to the fore. As Southworth and Owens (1993: npr) explain in relation to their morphological work on American urban fringe locations,

The form of the contemporary suburb is not a completely new invention but has roots in early models for ideal suburban communities such as the Garden City schemes of Ebenezer Howard and Raymond Unwin and the neighborhood unit of Clarence Perry and Thomas Adams. However, the Garden City ideals of a communal, self-sufficient satellite city set within a greenbelt of parks and farms seem to have been completely lost, as have notions of walkable, transit-supported living within convenient mixed-use neighborhoods and communities. High land values and pressures for development have eroded or completely obliterated most greenbelts. What has survived from these models is the residential district of primarily single-family homes, set on a green plot of land, within a short distance of an elementary school. ... The problems of design and development at the urban edge are apparent at all scales, from individual house lot to entire subdivision or community.

Post-war suburbia has drawn the most concerted criticism from designers and planners for well-documented problems associated with its exclusionary practices, and lack of sustainability and conviviality in relation to food and other elements of everyday life. While revisions in thinking about pre-war American suburbs highlight their social and income diversity (Harris and Lewis 2001), recent scholarship has traced how post-war suburban development across the United States (including the famous Levittown on Long Island) was also the site of restrictive covenants designed to keep out black and minority ethnic Americans (Fogelson 2005: 76; Loewen 2005). Similar practices excluded Jewish households in suburban Toronto (Dennis 2008: 191). Suburbia has similarly been critiqued by feminist geographers for the circumscribed and blighted lives its spatiality and mores imposed on women (Hayden 1981, 2009). Urban sociologists have explored the way that it excluded poorer households (Gottdiener 1994). Designers have decried its parodying of nature as space that is both 'over processed and denuded of vitality' (Relph 1987: 216).

As noted above, the design break point between earlier and later suburban development is widely argued to be related to the rise of motorized transport: including from connected streets to the imposition of road hierarchies ending in culs-de-sac. These spatial shifts played out differently according to location but have had relatively similar effects on food. Broadly, walking and public transport trips to grow, buy or consume food became much more difficult and time consuming than they had been in pre-war suburbs. Meanwhile, car-based consumption was expedited both in newly developed areas and in existing suburbs which were sometimes reshaped to be more car focused in food consumption terms. In the United Kingdom, for example, Whitehand and Carr's (2001) morphological work gives an insight into profound shifts in suburban food patterns which occurred in the post-war period, as suburban development grew in scale and also altered in spatial terms, driven by transformations including increasing land scarcity, increasing numbers of households (but decreasing household size) and alterations in work patterns and mobility. Rapid growth in car ownership after the Second World War affected the 'geographical pattern of accessibility among house purchasers at the cheaper end of the housing market, combined with restrictions on availability of greenfield sites, helped to stimulate major reevaluations by developers of the most profitable development on various sites that already had houses on them' (Whitehand and Carr 2001: 189).

Key changes in the physical shaping of existing subdivisions (to increase the number of dwellings) were the insertion of culs-de-sac to increase plot densities, the building over of allotment lands and the truncating of individual plots, which in turn reduced private garden space for food growing (Whitehand and Carr 2001: 124–5). All these changes also reduced accessibility to food services for those on foot, while for newly developed suburbs, the imposition of road hierarchies created the same effect from the start. As Gottdiener (1994: 84) puts it, 'In the suburbs we seem to have found a happy marriage between middle class demand-side preferences for housing and living arrangements and supply-side subsidies providing the incentive structure for growth patterns ... the result is a suburban landscape consisting of immense regions of single-family home developments and the hegemony of the automobile culture'.

The analysis of suburban shortcomings in relation to sustainability and climate change is a more recent area for sustained criticism, and this has particular food resonances. The dispersion created by suburban space-shaping and the economic incentives that underpinned it is seen as symptomatic of approaches by both its builders and designers that reflect a disregard for nature and natural processes in pursuing a wasteful, polluting form of settlement that must be reversed (Hough 1984; Fishman 1987; Lozano 1990: 299). Yet recent, rather polemical, defences of suburbia, of which Bruegmann (2005) is perhaps the best known, have discounted sustainability arguments against sprawl and have challenged the notion that 'postwar suburbanization and sprawl were different in kind from what had gone before' (Bruegmann 2005: 43). Rather, in this view, these urban development forms

were different in scale but not really different in kind ... really just an extrapolation of the process visible in London since the seventeenth century or in American cities for more than a century ... specific parts of the story [such as] the rise of the suburban shopping center ... tend to perpetuate stereotypes that were popularized by upper-middle-class anti-suburban writers during the 1960s.
(Bruegmann 2005: 43)

In fact, for food-shaping at least, post-war suburban space tends to show foodscapes that reflect significant changes to urban form rather than mere extrapolations of previous trends, as explored here. The existence of these developments undercuts the 'class bias' criticism or the argument that any such analysis should be discounted on the grounds that it represents an overly narrow preference for more urban places. Rather, in situating suburban food-related spatiality broadly within both its historical urban design and gastronomic context, it is clear from the discussion below that the development, and sometimes decline, of post-war suburbs created a unique, historically specific landscape of food. Suburbanization has been extraordinarily influential in shaping the way urban form and food have connected, and its rise is associated with particular food spaces and practices, which this chapter explores in detail, starting with the advent of the supermarket.

SUPERMARKETS AND SHOPPING STRIPS – RISE AND DECLINE

The dominance of suburban development as a spatial form in the mid-to-late twentieth century has shaped suburban food spaces in often unsustainable and unconvivial ways (see Figure 5.2). Supermarkets have been a central element in the growth of suburban space, so dominant that they are taken for granted. However, the rise of the supermarket in fact reflects dramatic changes in the political economy and cultural norms and practices of urban dwellers in relation to food over the course of the twentieth century. In American experience, supermarkets

> are not a given ... the growth of a retail industry with the political and economic power to dominate food retailing (as well as food production and distribution) was highly dependent on the increased mobility of the upper- and middle-classes, the willingness of the government to relinquish a number of regulatory controls, and the development of technology which vastly improved both communication and information management for those who could afford it.
> (Eisenhauer 2001: 129)

The shift away from shopping in food markets and individual shops on street corners and high streets, to supermarkets as the primary site for food buying was a fundamental change in everyday life. While these shifts occurred in many places, it is difficult to avoid considering American experience as a blueprint for these transformations. Although notable exceptions are referenced above, in suburbia's early incarnations, the United States saw very few suburban developments where food shopping space was designed in from the start or placed intentionally at the

FIGURE 5.2: 1960s shopping 'street' with supermarket, London.
Photo: Susan Parham.

development's centre. 'Rather, suburbanites were left to fend for themselves by driving to "market towns", which often offered the only commerce for miles, or by returning to the city to shop' (Cohen 1996: 1052). There had been attempts though. By the 1950s, while food market structures had become aligned with a society that was both suburbanized and predicated on mass consumption (Cohen 1996: 1052), the first experiments in capturing growing suburban retail markets had already been made by large department stores in New York and Chicago in the 1920s (Jackson 1985: 257). Sears (among others) decided to locate retail outlets in low-density areas because these not only attracted lower rents for space but also allowed potential customers to reach them by car (Jackson 1985: 257).

Such spatial, cultural and economic changes to food consumption were not limited to the United States, with these retailing approaches imported into the United Kingdom (Shaw et al. 2004; Alexander et al. 2009) and elsewhere (Reardon et al. 2003; Reardon et al. 2007) in a number of waves, including into Scandinavia (Sandgren 2009), Australia (Symons 1982), Asia (Reardon and Hopkins 2006; Gulati 2008), Africa (Amine and Lazzaoui 2011) and rapidly in South America (Reardon and Berdegué 2008) in the post-war and late-twentieth-century period (Stanilov 2004: 179). Recent work at the individual country level tends to show that while 'traditional' supermarkets are losing their grip in advanced Western retail environments, they are still increasing market share in developing countries (Reardon and Hopkins 2006). This occurs similarly in places which maintained more traditional food consumption arrangements for longer, including Portugal and Nicaragua (Farhangmehr and Veiga 1995; D'Haese and Van den Berg 2007). The rapid intrusion of supermarkets into Mexican food retailing, with deleterious effects on small farmers and suppliers, is one notable example of the impact on traditional food systems (Schwentesius and Gómez 2002). The imposition of the supermarket as a retail format challenging the dominance of traditional markets and shops in Vietnam is another example, situated by economists with a typical lack of reflexivity as a positive process of retail modernization being rightly encouraged by both urban authorities and relaxation of regulatory regimes (Maruyama and Trung 2007). Yet, in relation to Australian experience (and elsewhere) the rise of the supermarket was not just a process of transference but one of permutation, in which local cultural traditions also shaped the nature of both supermarket and mall-based food shopping forms (Humphery 1998: 3).

Equally, long-term cultural trends in consumption, sometimes traced back to the rise of the department stores of the mid-nineteenth century, were influential in shaping consumption attitudes and practices in Western cities outside the United States as well as within it (Bowlby 1985). The demarcation between marketing as a domestic chore and shopping as a leisure activity had broken down, and by the beginning of the twentieth century, 'shopping of all kinds, even the everyday purchase of food and household goods, became both a woman's "duty" *and* a medium through which women were supposedly able to assert who they were and gain pleasure' (Humphery 1998: 29). The nature of that food shopping process also changed. Techniques like self-service, which were to become a basic attribute of the supermarket's cash and carry offer, had their seeds in nineteenth-century innovations in packaging, such as canning and bottling, that long predated the growth of the supermarket as a pre-eminent retail form and helped lead to the dominance of chain stores from the 1930s onwards:

> Shoppers quickly saw the benefit of buying insect-free flour and other staples in one-pound or five-pound bags instead of relying on their local shopkeeper's cleanliness and honesty. They also liked buying milk that was sealed in glass bottles at the diary instead of ladled out from open cans in the store ... shoppers knew exactly what to expect, and where to find it, when they went inside [the shop]. At the same time, with control over food production becoming centralized in the hands of the big processors ... seasonal and regional differences in diet began to weaken. (Zukin 2004: 72)

Although supermarkets can be traced back to the 1930s in the United States, it was in the 1940s and 1950s that these stores became ubiquitous, growing alongside car-based suburbs, and reflecting technological innovations in freezing, food packaging and domestic refrigeration (Humphery 1998: 71). As Ellickson (2011: 9) notes, 'The post war boom was a period of steady growth for the supermarket industry. There was plenty of virgin real estate on which to build stores and plenty of markets to convert from chain grocery store to supermarket'. Allowing food shoppers to make direct contact with merchandise, and contributing to both food staff reductions and deskilling, America's 'no-frills' discounters 'took bargain culture to a new level of transparency, both simplifying and rationalizing the shopping experience' (Zukin 2004: 77). Zukin (2004: 78) explains that in an increasingly urban society the corner food store had offered both 'convenience and sociability' but supermarkets, with their 'combination of self service, low prices and variety – encouraged shoppers to develop new habits'. By the end of the 1950s, supermarkets accounted for 70 per cent of American food buying for home consumption (Zukin 2004: 78), and although shoppers were initially reluctant, for example, to use wheeled trolleys, they learnt to relate to these new shopping interiors where food produce was carefully placed to maximize sales and profits. Discount stores offered both a new kind of social space for food shoppers in which people shopped alongside one another but did not interact and a new kind of spatiality that was coarser grained, bigger in scale, car based and low density in nature:

> Supermarket shopping created a mobile public who measured convenience in terms of the time it took them to shop rather than the distance they had to travel to do so, and prized bargain culture over personal relations with store owners and clerks. ... Although shoppers still tended to patronize nearby food stores, more of them began to go shopping by car rather than on foot. Supermarket chains bought land wherever it was cheap – building larger stores but fewer of them. To attract shoppers with cars, they surrounded their stores with parking lots. These arrangements created a different, more diffuse kind of city, especially in Los Angeles and the growing cities of the West. (Zukin 2004: 78)

There was a paradox embedded in these social and spatial changes to suburban food space. On the one hand, the supermarket promised speed, cheapness, greater choice of food products and more uniformity and convenience in food shopping, by allowing a weekly one-stop shop by car rather than requiring daily food buying on foot from a multitude of specialized local food shops close to home. On the other hand, as ever-larger, more widely spaced supermarkets replaced earlier more local examples, 'though parking lots, shopping carts, and multiple checkout lines ... made it faster to get in and out of the store, it took longer to escape the supermarket's totalizing environment' (Zukin 2010: 79).

These paradoxical shifts in the suburban foodscape were not limited to the United States: they were happening in other places too. In Britain, despite a slower rate of adoption than in the United States, by the mid-1950s, the supermarket was on its way to becoming the dominant model of food retailing, with almost all Tesco stores self-servicing by 1955 and Sainsbury's in the process of converting its stores according to the same logic (Shaw et al. 2004). Safeways, which entered the UK market in the early 1960s, meanwhile, 'was quick to introduce pre-packed meat and vegetables and imported fruit on refrigerated stands, which was in part a key to its market success' (Humphery 1998: 76). Similarly, in Europe, self-service shops increased from around 1,200 in 1950 to over 45,000 in 1960 (Humphery 1998: 73). Sandgren (2009), for example, traces their rise in Sweden from 1935 to 1955 as a new way of thinking about food shopping. In Australia, meanwhile, packaging and distribution changes analogous to those noted in the United States meant that food retailing was able to respond to the spatial models the new supermarkets developed to take advantage of rising car ownership levels (see Figure 5.3). The first completely corporate supermarkets did not open until 1960 in Australia (Sandgren 2009: 101), and in an intriguing mirroring of female rejection of the 'rational' kitchen designs discussed in Chapter 1, contemporaneous accounts demonstrate that female

FIGURE 5.3: Suburban shopping centre, Australia.
Photo: Susan Parham.

consumers were less enamoured by the new supermarkets than were corporate retailers. 'All the male delegates ... were completely sold on it. They admired all the shining rows of goods, they wheeled the empty trolleys ... about. The supermarket, they agreed, is a wonderful idea' (Humphery 1998: 85).

CARS AND SUBURBAN FOOD SPACE

As foregrounded in the last section, the rise of the supermarket within a shopping centre or mall context demonstrates that a central feature of the development of suburban food space has been the twentieth century's design primacy for the car: as both a means of access and around which food retailing has been shaped. Suburbs and supermarket-focused shopping centres have formed a tight spatial fit. As Jackson (1996: 1111) notes, shopping centres in an American context 'are the common denominator of our national life, the best symbols of our abundance', and the Country Club Plaza in Kansas City, which was constructed in 1923 is cited as the first shopping centre oriented around the car (Jackson 1996: 1111). The work of Clarence Perry and the suburban neighbourhood street layout that came to fruition in Clarence Stein and Henry Wright's designs for Radburn, New Jersey, in 1928–9, set the template for subsequent suburban subdivisions. Mass distributor roads were used to move vehicles round at speed on superblocks, while a network of culs-de-sac gave access to houses; and walkers were segregated onto separate pedestrian pathways (Parsons, in Parson and Schuyler 2002: 134). We tend to see the antecedents of later 'big box' shopping centres as post-war ones, however Clarence Stein's local shopping centre sketches for Radburn from the late 1920s suggest he was already 'working on the problem of modern pedestrian-circulation-shaped commercial center design' in which cars and pedestrians would be strictly separated and cars given primacy (Parsons, in Parson and Schuyler 2002: 135). Over the course of the twentieth century, this car-pedestrian separation was highly influential on shopping centre design in both the United States and elsewhere, and continues to shape contemporary food space design practice in the twenty-first century, about which more in Chapter 8.

Of course, for the privileged few, cars had been a feature of the landscape since the turn of the twentieth century, offering a delicious freedom to explore culture and gastronomy, as accounts of Edith Wharton (and for part of the trip, Henry James's) early tours in provincial France suggest (Wharton 1908). As cars became a basis for mass transportation in the mid-twentieth century, this allowed the melding of rural and urban landscape to be experienced by the majority for the first time as a kind of urban pastoralism (Rowe 1991). Drivers became 'the only fully entitled citizens of edge city' (Kostof 1992). Cities seen from a car were argued to constitute a new phenomenology (Thrift 2004). These changes, in turn, had particular implications for food space, which assumed and fed off car dependence (Newman and Kenworthy 1989). Although a direct causal link is sometimes contested in the rise of cars and the decline of streets as human-scaled space well configured for food buying and other food-related practices, the way food was distributed and consumed dramatically changed in nature when the neighbourhood and high streets' primarily fine-grained pedestrian space was overtaken by its role in moving vehicles. On food distribution, retailing and consumption fronts, streets were perforce reshaped to encompass the larger scale, coarser grain, higher speed, greater noise, increased emissions, bigger volumes and voracious parking requirements of cars and trucks. Design began to reflect and plan for this situation.

FOOD, SHOPPING CENTRES AND MALLS

After the slowdown effected by the Depression and the Second World War, the suburban shopping centre model rapidly grew to achieve spatial dominance in food retailing in the United States' suburban landscape, with the 'malling' of America an omnipresent feature which seemed unstoppable (Kowinski 1985; Feinberg and Meoli 1991). Jackson (1996) has traced the development of the physical form of the suburban shopping centre from its origins in a range of pre-modern shopping forms, and shows how in the post-war period, this suburban retailing model, with large stores surrounded by plentiful parking space, took root across the United States, arguing 'the multiple-store shopping center with free, off-street parking represented the ultimate retail adaptation to the requirements of the automobile' (Jackson 1985: 258). By the 1950s, in the United States, there was a shift from free-standing supermarkets to those located in growing shopping centres (Ellickson 2011: 9). A prevalent trend, alongside the sheer size of supermarket stores, was their suburban location in such shopping centres and eventually the development of large regional malls and post-suburban mega malls and hypermarkets which are discussed in Chapter 8.

Some of these shopping centres began to be designed with enclosed private spaces that mimicked the configuration of the public streets they had replaced. The Northgate Shopping Centre in the Seattle suburbs became the model for what was to come in the design of the shopping mall. 'Two strips of stores were joined to face a pedestrian walkway anchored by department stores on both sides with ample parking surrounding the entire development' (Stanilov 2004: 182). Further refinements were introduced at Victor Gruen's seminal Southdale Shopping Centre outside Minneapolis. Here the mall was enclosed with an arcade, and 'the weekly visit to the mall became an integral part in the life of middle-class suburbanites and a most visible symbol of postwar economic prosperity' (Stanilov 2004: 182). The Southdale Shopping Centre's enclosed, air-conditioned mall was a refinement of the shopping centre retailing formula. 'The concept proved wildly popular, and it demonstrated that climate-controlled shopping arcades were likely to be more profitable than open-air shopping centers. Indoor malls proliferated, slowly at first but with increasing frequency, and within fifteen years anything that was not enclosed came to be considered second-rate' (Jackson 1996: 1114).

At the same time, a second food shopping form, the shopping strip, often located on a bypass road, that was designed to serve car-based traffic rather than those on foot, also began to threaten traditional corner shops and market-based food-buying forms (Jackson 1985). 'These bypass roads encouraged city dwellers with cars to patronize businesses on the outskirts of town. Short parades of shops could already have been found near the street car and rapid transit stops but ... these new retailing thoroughfares generally radiated out from the city business district towards low-density, residential areas, functionally dominating the urban street system' (Jackson 1985: 258). Such strip shopping environments increasingly functioned as a kind of poor relation to the shopping centre, which itself was transformed. Over time, more local suburban shopping centres gave way to much larger regional shopping malls, as design for growth in the scale of food shopping space began to emerge, increasing even further both journey times and car dependence. Again, highway infrastructure and widespread car ownership, both influencing and made possible by suburban spatiality, were instrumental, as were the retailing calculations of shopping centre developers. As described by Cohen (1996: 1052),

> By the mid-1950s, however, commercial developers – many of whom owned department stores – were constructing a new kind of marketplace, the regional shopping center aimed at satisfying suburbanites' consumption and community needs. Strategically located at highway intersections or along the busiest thoroughfares, the regional shopping center attracted patrons living within half an hour's drive, who could come by car, park in the abundant lots provided, and then proceed on foot (although there was usually some bus service as well).

The shopping mix was also altered, with some convenience food shops no longer included, reinforcing a kind of food-related two-speed economy, which is explored in Chapter 8's discussion of food poverty and obesogenic environments. Having undermined more local, walkable food shopping spaces, these huge centres became the only remaining nodes of intensity and environments for engagement. Described as housing public life and yet privately owned and controlled,

> the enclosed mall created a focused atrium space, a zone of urban intensity, energized by plunging elevators and zigzagging escalators. It dramatically reshaped the retail mix by expelling convenient everyday shopping to the strip in favor of stores featuring clothing, gifts, and impulse items. At the same time, its covered, climate-controlled spaces (especially in places like Minnesota) suggested new forms of public and civic life … Southdale vastly expanded the role of the mall as social and community center. (Crawford 2002: 31)

From a base of eight such shopping centres in 1946, by 1984, some 20,000 centres were responsible for nearly two-thirds of the United States' retailing, including for food (Jackson 1985: 259). By the early 1990s, the United States hosted 38,966 shopping centres, and of these, some 1,835 were large, regional malls. 'Increasingly they were featuring the same products, the same stores, and the same antiseptic environment' (Jackson 1996: 1111). Changes to scanning and other technologies in the 1970s saw an increase in the number of product lines carried at supermarkets and thus the opportunity to further increase store size within growing shopping centres (Ellickson 2011: 12). Johnson (2000a: 130) has argued that the initial form of the regional shopping centre characterized the international post-1945 architectural style but that in the late twentieth century it changed both in spatial and economic terms. Just as such regional malls are now part of the edge cities explored in Chapter 8, rather than sites within 'undifferentiated suburbia around a single city centre' they are now also the core of retailing space rather than a peripheral economic sector (Johnson 2000a: 130). This transformation in food retailing produced 'a new landscape of consumption' with very particular spatial design and urban sprawl implications:

> With cornfields and forests obliterated by a grid of interstate highways and subdivisions of suburban housing, big stores and malls imposed an equally rational arrangement of 'repetition and banality within the unpredictable patterns of consumption'. Regions of the country that had been wilderness, as far as shopping was concerned, joined the worldwide consumer revolution. (Zukin, quoting Easterling 2010: 81)

SHOPPING MALLS, FOOD AND CIVIC ENGAGEMENT

Unlike the public streets of traditional food space explored in the previous chapter, these food shopping forms were always privately owned, and thus self-evidently not part of the public realm. Yet their proponents viewed them or, at least marketed them, as centres for civic life. Cohen (1996: 1056) draws our attention to the fact that developers were not just motivated by the enormous profits to be made in developing suburban shopping malls, but saw their endeavours as a way of rationalizing consumption and supporting community building 'no less significant than the way highways were improving transportation or tract developments were delivering mass housing'. Mall promoter Victor Gruen, for instance, had in the 1950s argued that these interiorized spaces were analogous to the shopping streets of the traditional city, with 'stores lining both sides of an open-air pedestrian walkway that was landscaped and equipped with benches' (Cohen 1996: 1056). At Southdale, Gruen's two-level design cut down on long walks between shops and helped give 'downtown like' levels of bustle (Harris 1987: 322). Gruen (1964) argued that such spaces could give their users the same level of participation in community life as the Greek agora, medieval market place or the town square.

Yet these parallels were tenuous ones. Claims for malls and supermarkets to be viewed as the new high streets of urban life, simply reflecting consumer demands unmediated by external structural forces, were always hard to sustain. The mall presented a bland and 'lopsided' view of urban culture: 'even the largest malls are almost the opposite of downtown areas because they are self-contained and because they impose a uniformity of taste and interests' (Jackson 1985: 260). Just as suburbs themselves were managed to exclude those considered undesirable in some way, 'in promoting an idealized downtown, shopping centers like Garden State Plaza and Bergen Mall tried to filter out not only the inefficiencies and inconveniences of the city but also the undesirable people who lived there' (Cohen 1996: 1060). With 'public' circulation spaces designed to discourage sitting, strolling or other activities that could be undertaken without consuming, such malls did not address the social messiness of traditional streets. Instead, they re-created simulacra of these in internal, privately owned and controlled spaces that were physically enclosed and unresponsive to location (Knox 1992; Parham 1992, 1995, 1996; Crawford 2002). In fact, as argued in the early 1990s,

> Regional shopping centres are designed to avoid a convivial sense of urbanity. A common planning assumption, and much of the reality, is that most people drive to huge regional centres set in a sea of car-park; to consume in indoor, privatised shopping malls which are dead out of hours. ... It is easy to find things to criticise about 'lock up' malls. Urban designers have concentrated on their problems in achieving human scale, their lack of truly public space and the poor relationship of an alienated interior to the exterior urban world. Such shopping centres essentially lack conviviality, and this is at once a result of, and at the same time a motivation for, the gastronomic meanness expressed in food court design. (Parham 1992: npr)

In the latter part of the twentieth century, malls became extremely artful in presenting private space as public. Strategies to deal with mall decline in the face of other food-retailing formats included the 'stealth' mall, which submerged or hid its stereotypical features beneath a public veneer (Crawford 2002: 28). In reality, each food space remained 'simply a privatized, commercial space with a central management structure that can allow the public to use it – on their terms' (Scharoun 2012: 89). Thus, to eat at the mall meant being a consumer rather than a citizen (Barber 2002). As Voyce (2006) argues, the very notion of public space has been displaced in such shopping malls – replaced with narratives of property, order and ensuring safety that ensures an undisturbed capacity to consume. Work emanating from Australia has noted in comparing this author's notion of gastronomic quarters with food courts within shopping malls that

> [f]rom the environments examined in this paper it is evident that the gastronomic quarter is the food-centred environment that brings more vitality to urban spaces, and best connects the sale and consumption of food with the urban realm. The study has also illuminated that the presence of food brings diversity and activity to the spaces surrounding them, whether it be public space or controlled pseudo public space. The vitality that is present in shopping mall food courts is a contained vitality, wasted inside a controlled consumer environment. The exuberance brought upon by the presence of food, and people gathering together in the food court is an opportunity lost in the public realm. Gastronomic quarters bring this activity to a public setting and allow the liveliness and vigorous nature of the sale and consumption of food to become synonymous with the public spaces surrounding the gastronomic quarter. (Mand and Cilliers undated)

Further exploration of the issues of scale for food court settings and designs has revealed that shopping malls have been critiqued as both hugely over scaled and architecturally problematic:

> The scale of these suburban agglomerations dwarfs even the most ambitious projects of the Roman empire. These scatterations of unassuming generic structures of various sizes, lavishly sprinkled amidst a sea of parking and oversized commercial signs desperately crying for attention, can stretch for hundreds of acres. The nondescript repetitive architecture of the buildings and the vast expanses of asphalted surfaces which separate them negate the idea of spatial definitions and boundaries which makes it very difficult to comprehend their actual size. (Stanilov 2004: 179)

Post-war suburban food shopping models were built on certain assumptions about food consumption that remained in place through changes in retail formats. These resilient assumptions included that consumers would drive to do their shopping, would prefer to do all their food shopping in one store rather than at a series of separate shopping spaces and would do this weekly rather than daily, as had been the previous urban pattern. Seasonal and regional variations and ways of marking food's role in the pattern of life were replaced by weekly brand-based offers and markdowns. Food was processed and packaged and the geographical origin of food was replaced with company labelling that promised uniform quality everywhere. As Zukin (2004: 80) notes, 'discount stores founded in the 1960s – Wal-Mart, K-Mart and Target – ushered in a new order of shopping' in which their sheer size and locations outside cities 'raised shopping to a direct experience of mass consumer society'. These alterations were not just occurring on the consumption side. Behind the low-wage shop floor, where major supermarkets increasingly used universal product codes and scanner technology, they could start to rely on complex just-in-time logistics and large-scale global distribution systems, and these changes in turn reshaped the suburban food landscape spatially and economically. The balance of power shifted away from food wholesalers towards retailers. 'Advances in technology and scale allowed stores to grow bigger, and market integration (both horizontal and vertical) gave retailers (whose parent corporations are now often larger than their wholesale suppliers) increasing control over both wholesale and retail prices' (Eisenhauer 2001: 127).

This again became a model for outer suburban, large-scale retailing, attracting shoppers in cars from very big spatial catchment areas. Growing out of the shopping centre, the so-called big-box mall has been described as 'a product of stagecraft, proscenia for the enactment of consumption-oriented lifestyles' (Knox, in Whitehand and Larkham 1992: 208). Such food space was further supersized in the 1970s with the advent of the super-regional mall (Jackson 1985: 260). A development like Bluewater shopping centre outside London, although not actually built until the latter part of the 1990s, owes its shaping to these models. By the late 1980s, Wal-Mart was rolling out the first supercentres: huge combined grocery and discount stores (Ellickson 2011: 12). As Seiders et al. (2000: 190) point out in their study of the effect of supercentres on four retail markets in the United States, their entrance has taken significant market share from smaller food competitors. Since 1989, over 700 of these one-stop shopping formats have been established in both the United States and in places as diverse as Mexico, Argentina, Brazil, China, Germany and Indonesia, and it was expected that supercentres would be the fastest growing sector of retailing in the early 2000s (Seiders et al. 2000: 181). Food has been crucial to the success of these gigantic stores: a foundation for their ability to undercut competitors and drive down prices paid to their own suppliers.

> Supercenters have a different objective for their grocery operations than traditional supermarkets, whose ultimate objective is the profitability of their food operations. For Walmart and Target supercenters, the goal is overall store sales and profitability; food is often used to raise customer traffic and the spillover of shoppers to the general merchandise side of the store, increasing the sales of higher profit margin categories. Food more than general merchandise is particularly useful for driving up the frequency of shopper visits. Therefore, supercenters are willing to accept lower margins on food sales than traditional supermarkets, according to food-retailing executives. (Senauer and Seltzer 2010: 1)

TRADITIONAL FOOD SHOPS AND SPACES AND THE TRANSFORMATION OF 'BIG BOX' STORES

The dominance of big-box suburban food consumption forms in the latter part of the twentieth century is argued to be responsible for disastrous effects on 'mom and pop' food shops (Haltiwanger et al. 2010). It also adversely affected shopping in traditional high streets and main streets (Bromley and Thomas 2002; Oram et al. 2002; Oram et al. 2003). While that finding has been challenged in the United Kingdom and such supermarkets have been seen to bring certain benefits (Wrigley et al. 2009; Di 2007), research from Denmark tends to support the contention of adverse effects on small shops, showing both a dramatic decline in the number of specialty food stores and a significant decline in their market share in relation to supermarkets against which they are increasingly competing (Hansen 2003). More broadly, suburban shifts in the spatiality and economics of food shopping have been both clearly tied to notions of modernity and understood by traditional grocers and other food stores to be an existential threat to their livelihoods.

In fact, the predominant trend in food retailing over the latter part of the twentieth century (and into the twenty-first) has been the decline, and in some cases disappearance, of the food trade of traditional high streets (and smaller suburban centres), lost in the wake of supermarket and mall development (Jackson 1996). In the United Kingdom, for instance, national government moved belatedly to halt the spread of out-of-town shopping malls in the mid-1990s, with the then environment secretary, John Gummer, intending to use his 'planning powers to support local efforts to safeguard the vitality of towns and the economic viability of their retail centres in particular' (*The Guardian*, Thursday, 23 February 1995). However, the damage to many independent food retailers was already terminal, as Conisbee and Murphy (2004: 2) describe in relation to high street changes experienced in the 1990s:

> over the past two decades, the construction of large out-of-town shopping centres and waves of high-street bank-branch closures have driven many people away from town-centre shopping, resulting in the loss of thousands of independent traders ... we are losing our high streets as we have known them. Unable to compete on an unfair playing field, local retail is going up against the wall.

More recently, some researchers have discerned trends towards adaptation and resilience of high streets, in the United Kingdom at least, where both corporate and small food shops are seen to operate side by side (Wrigley and Dolega 2011). More broadly, the conventional argument in support of the global adoption of a suburban supermarket-based food consumption model suggests that these trends are a straightforward reflection of consumer demands for 'the convenience of longer store hours, shops closer to where they live, and easier access by automobile' (Bruegmann 2005: 201). In this view, the 'result is a proliferation of large supermarkets, shopping centers, discount centers, and big-box retail outlets like Wal-Mart or Target stores' (Bruegmann 2005: 201). Yet this simple cause and effect argument elides some complicated spatial, cultural, environmental and economic relationships and does not seem to adequately reflect the evidence from the field, including supermarket transformations evident in places such as Australia (Pritchard 2000). While some recent US-based research has concluded that shopping closer to home will not currently undercut a car-dependent shopping model, it equally suggests that offering more food shopping options within a smaller radius of home will make neighbourhoods more livable (Handy and Clifton 2001).

In the late 1990s and early 2000s, big-box retail outlets were themselves under attack by so-called category killer large stores like T. K. Maxx (Stone 1997). As noted above, rather than being an outstanding success in terms of consumer preferences, a significant number of shopping malls were subject to processes of retrofitting because they had become abandoned or underused grey fields (Dunham-Jones and Williamson 2009: 114–18). Moreover, Jackson (1996) points out in relation to American experience, and Humphery (1998) in regard to Australia, that, by the 1980s, the suburban mall had already passed its peak as a favoured food-retailing form, although in some other regions including Asia, malls are continuing to be built with increasing coverage in places including India (Srivastava 2008; Meyer-Ohle 2009). In an Indian context, in fact, it is argued that these malls are 'social fortresses' which allow the middle classes to distance and 'spatially purify' themselves from the needs of the poor (Voyce 2007). In a Western context, just as malls appeared to be completely ascendant as spatial forms, notions of modernity in food consumption upon which they were based themselves became outmoded (Humphery 1998). Malls increasingly began to be judged as

> so homogenized and predictable that they have lost much of their entertainment value. Revealingly, the number of centers under construction nationwide has been declining since 1988. Older shopping centers, in particular, have often closed or been razed. Some, like the 2.2 million square-foot Roosevelt Field complex on Long Island, with parking for 9,000 cars, have had a complete makeover in order to keep up with current trends. Smaller indoor malls, lacking the advertising budgets of larger operations, have encountered cycles of decline once associated with inner cities. The interiors of those structures have become ghost towns, with white butcher paper over the windows and specialty retail space perennially unleased. (Jackson 1996: 1120)

More recently, big-box supermarket spaces have been inserted into older, more traditionally shaped urban space, blighting surrounding urban fabric by destroying the fine grain of urban blocks and undermining the physical enclosure of space that creates outdoor rooms. Schuetz (2013) shows how this is playing out in the United States as large-scale food retailers have started to employ strategies of placing big-box stores into downtown areas (but also points out that this may help overcome the inner urban problem of food deserts in low-income areas). As design guidance notes, 'Large stores and other large "big-box" units that are often stand-alone, with exposed "dead" frontages, create particular problems for active and attractive streets' (Davies 2001: 43) and proposals for repairing such spaces by 'wrapping and capping' these overly large boxes are now encapsulated in some proposals for design that repairs the damage they have caused to urban fabric (Davies 2001: 43; Kelbaugh 1997, 2002; Dunham-Jones and Williamson 2009; Tachieva 2010). Senauer and Seltzer (2010: 2) show how this retailing strategy has played out economically and spatially, with both further economic monopolization within the supermarket industry and the emergence of more 'urban' stores:

> The growth of supercenters has brought a wave of consolidation to the retail food industry. Many small, frequently family-owned operators of one to several stores have lost the most sales to the new formats and gone out of business or been taken over by larger chains. This trend is likely to continue into the foreseeable

future. In addition, large traditional supermarket chains, such as Kroger and Safeway, may increasingly suffer from the intensity of the competitive rivalry from these new formats. Interestingly, Wal-Mart has announced that much of its future expansion will focus on smaller urban stores, rather than big-box, suburban supercenters.

These shifting locational decisions are being made by dominant retail players, who, having upsized catchments for suburban food retailing now have the opportunity to also establish themselves in finer-grained, smaller-scale food markets where their activities previously undercut competition from other stores. Transnational retailers have been forced to employ 'strategic localization' methods in any case at national level to deal with differing consumption cultures (Coe and Lee 2006), but this trend also reflects a scale shift within urban space. In Australia, for example, smaller 'Express'-type food stores began to be developed in the late 1990s by major food retailers on inner urban rather than suburban sites. One cited example has been 'the development of *Let's Eat*. A further variation of the standard suburban supermarket that opened at Melbourne's Prahan Market in October 1998. Management described this venture as a 'theatre of cooking combining fresh food, restaurant, take-away and cooking school within one site' (Pritchard 2000: 208). A focus on such store types has been argued as a possible way forward for Wal-Mart in Europe (Fernie and Arnold 2002). It is reflected in the following developments in its food-retailing model:

> In October 1998, Wal-Mart announced the opening of its first two 'neighborhood' stores, which primarily will offer food and a small amount of non-food merchandise. Wal-Mart's new format is intended to compete directly with the traditional neighborhood supermarket. The new stores will be located either between two supercenters in larger cities, or in towns that are too small to support even a single supercenter. (Seiders et al. 2000: 191)

A related recent development has been the super-regional mall located within an urban area, such as at White City in west London, and at Stratford in east London, the latter developed in the run-up to the London Olympic Games and benefitting from enormous public investment in transport infrastructure and brownfield reclamation. It is relevant to consider these design forms in a chapter on the suburban food landscape because they constitute the imposition of a very large-scale suburban food-retailing model within a denser, more urban setting, as was previously the case for suburban-style supermarkets introduced at more modest scale into urban locations, noted above and discussed in Parham (2012: 52, 265).

Finally, it is worth remembering in this section that despite its seeming ubiquity as a food shopping format in the latter part of the twentieth century, some food buyers have not been able to access supermarket food. In the last fifty years, supermarkets chose to avoid inner-city areas 'for suburban and exurban locations, which offered more land for parking, easier loading and unloading by trucks, convenient access to highways and arterials, and a development context for much larger stores' (Pothukuchi 2005: 232). As supermarkets constituted themselves as authorities in the foodscape and drove off competition from individual food shops, some urban areas were unable to secure or retain a supermarket presence in their area (Dixon 2007). Eisenhauer (2001: 126) demonstrates in her study of supermarket 'redlining' practices in the United States (techniques previously reported by Bennett 1992) that these stores not only disinvested in urban locations, but also avoided locating new stores in certain lower-income urban areas. This spatial limiting of urban dwellers access to good-quality, affordable food more readily available in suburban supermarkets meant that 'poor urban health is as much linked to twentieth century urban history as it is to individual, behavioral causes' (Eisenhauer 2001: 126). In Chapter 8, the food poverty and obesogenic environments associated with these locational strategies are further explored.

SUBURBAN SPACE SHAPING, FOOD COURTS AND FAST FOOD

The shopping mall food court is a foodscape closely tied to suburban gastronomy, although not confined to it, as the discussion in Chapter 4 demonstrates. In its earlier incarnations the food court was usually located in buildings without a connection to exterior space, while making reference to seasonal and regional links and food traditions from other cultures. It arose within an overall regional mall design and management that preordained the mix of food offered and generally proposed interior design with fixed furniture that disallowed personalization of the space by its users. In the 1980s, the most significant change to food court design was reported to be the move from four to two person table settings, it was claimed so that people did not have to sit next to someone they did not know (pers.comm). Design interest appeared to be largely instrumental in nature such as in mitigating noise levels (Navarro and Pimentel 2007). Such dining experiences might be thought to constitute the antithesis of the festive dining of Italian towns described by Field (1990) where dinners occur in enclosing though not closed public spaces that have grown out of the food production they celebrate and the civic engagement of the citizens that they reaffirm (Parham 1990).

Food court design at the super-regional mall level has become much more artful as such spaces themselves have attempted to come to grips with the issues outlined above. Bluewater, mentioned earlier in the chapter, has '330 stores, over 55 bars, restaurants and cafes along with 13,000 free car parking spaces attracting 28.1 million visitors per annum' (The Bluewater story). Its recently refurbished Wintergarden Food Court, with '(access from Rose Gallery) houses the family dining area with something to suit all tastes' and encompasses a series of chain restaurants and fast-food outlets, which enhance 'the guest experience'. The apparently more upmarket West Village dining and retail area within the mall is due to be expanded to add further restaurant space, and it is notable that food businesses across the huge 'food offer' appear to be almost entirely chain operators rather than single businesses. In addition to paying attention to design, underlining that this is in fact private not public space, in 2005, Bluewater's management introduced a 'code of guest conduct' which outlined that 'all groups of more than five without the intention to shop may be asked to leave the centre' and forbade 'unsociable behaviour that is detrimental to the Bluewater environment' (Bluewater website). Pets, 'shopping trolleys on the malls' and 'the wearing of any item of clothing which restricts the view of one's head/face (e.g. hoods) with the exception of religious headwear' are similarly not permitted, nor are unauthorized 'leafleting, canvassing or the conducting of third party interviews'.

At Stratford, discussed in the last section, the template of the super-regional mall has been inserted into a large brownfield site in east London where it is overlaid with an array of up-to-the-minute food spaces including high-end food halls modelled on the food available at hybrid markets like Borough (Parham 2012). Its food spaces include 'The Great Eastern Market', 'World Food Court', 'Chestnut Plaza', 'The Street' and 'Four Dials', configured as upmarket supermarkets, cafés and restaurants on internal and external 'streets', and interiorized food courts featuring state-of-the-art interior designs (although in Finkelstein's terms perhaps representing a series of parodic restaurants). Despite certain of its areas being named as 'streets' and 'plazas', the entire development is actually private space, and design in relation to these foodscapes is critical to the way each domain appears both modish and public: being artfully shaped to draw in customers and increase dwell times.

Although such super-regional malls demonstrate how food courts in these contexts have gone upmarket in food and design terms, more broadly, much of the literature about fast food is situated within discussions of food poverty, food deserts and obesogenic environments. Of course, this is not only a first-world issue: rising levels of obesity in developed countries are also associated with the prevalence of fast-food restaurants (Jeffery et al. 2006). However, it is worth noting here that the suburban arrangements for food retailing described above gave rise to particular kinds of eating spaces whose architecture is congruent with this low-density, car-oriented spatial form (Liebs 1995: 193). Much suburban dining outside the home has been found in fast-food chains which commercialize conviviality and also lighten the second shift for women (Parham 1992). In an American context, Leibs (1995) considers antecedents reaching back to the chuck wagon of the Old West, tea rooms on major tourist routes, family restaurant chains and food stands, while Kittel and Snow (1990) similarly trace a visual history of the diner as it developed from the mobile lunch wagon to a car-based quick-service restaurant form.

The growth in the 1930s of Howard Johnson's franchised chain of roadside restaurants is an instructive example of the way food, architecture and cars came together as Americans took to the road in increasing numbers despite Depression conditions. In Howard Johnson's, food was mostly prepared in a central commissary to speed up service while design features including easy-to-see bright-orange tiled roofs were employed to catch the attention of passing motorists (Levenstein 1993). Johnson 'tapped into the era's yearning for roots, tradition and community by making his restaurants New England "Colonial" style with white clapboard exteriors, homey lamps glowing in fake dormer windows, and roofs topped by prim cupolas and weathervanes. Inside, knotty pine paneling and ruffled curtains filled out a design one architectural historian has called "a beacon of traditional values"'

(Levenstein 1993: 48). As roadside dining space grew up alongside America's burgeoning highway system in the mid-twentieth century's automobile age, petrol stations also played an increasing role in convenience food vending (Jakle and Sculle 2002). In the United Kingdom and elsewhere, food space associated with highway service areas became a byword for bad food (although accounts of Italian autogrills are an exception), yet these road-related food areas may still aim to invoke cosy domestic imagery, with 'road pantry' shops at service stations likened in their branding to kitchen-based storage spaces.

The rise of fast food in suburbia has also been associated with the culinary transition discussed in Chapter 1, whereby food deskilling, the rise of pre-prepared, processed and ready-made foods, the relationship to food poverty and insecurity and the growth of obesogenic environments have all been theoretically interconnected. Despite its suburban connections to car culture, much of the literature on fast food, especially in the United States, relates it locationally to urban areas and poverty-stricken communities cut off from supermarket access. Originally associated with highways and ex-urban locations, fast food's theorized shift to urban areas has been related to the downturn connected to the 1973 oil crisis. Its rise is also connected to the redlining practices noted earlier in the chapter. Thus, 'while the growth of fast food in poor urban neighborhoods has increased steadily, supermarkets stocking fresh, high-quality food have simultaneously relocated to the more spacious and affluent suburbs' (Freeman 2007: 2226). In an analysis that situates fast food as a form of food oppression, supported by cruel treatment of animals, unsustainably monocultural agriculture and implicated in unparalleled levels of obesity and other ill-effects, fast food is described as

> highly processed and prepared using standardized ingredients and production techniques. Much fast food is deep-fried in partially hydrogenated oils (or trans fats), which lead to high cholesterol rates and heart attacks. Combined with starchy vegetables and sugary drinks, these foods have a high glycemic load, a factor that contributes to obesity and diabetes. Fast food also contains a large amount of chemical additives and often lacks accurate nutrition labeling. (Freeman 2007: 2225)

Yet fast-food outlets' historical association with suburban development has not disappeared entirely either in the United States or elsewhere. Like regional supercentres and their food courts, fast-food outlets share a standardizing, speed-driven approach. Many fast-food chains have achieved global reach, and a critique of their food's connection to obesity, nutritional deficits and poverty, problematic working practices and economics, sustainability and design has come from a variety of directions (Jeffery et al. 2006; Freeman 2007). Unlike shopping-centre-based food courts, fast-food outlets tend to deny even the need for some intensity of human activity as a context for their operation (Parham 1992) although accounts here demonstrate how standardization and speed have been subverted in certain circumstances. Suburban fast-food chains rely on car access so can be arbitrarily located, having lost the walking scale for food accessibility that earlier residential areas enjoyed. The food itself is problematized. Continual rejigging gives a new appearance to a narrow range of raw materials produced by agri-business (Schlosser 2002). Moreover, in design terms, fast-food outlets fail to achieve arrangements that relate to public space enclosure, as discussed in Chapters 3 and 4. It seems reasonable to posit that ubiquitous, uniform fast food does little to contribute to a distinctive sense of place where people can celebrate being connected as citizens through their foodways.

Here, again, there are ironies. In an exception to the car-based nature of much fast-food consumption, relatively recent work by Austin et al. (2005) in the Chicago region of the United States demonstrates that fast-food outlets cluster within walking distance around schools: 'we estimate that there are 3 to 4 times as many fast-food restaurants within 1.5 km from schools than would be expected if the restaurants were located around the city in a way unrelated to schools' (Austin et al. 2005: 1579). Given this clustering, the researchers argue that 'the concentration of fast-food restaurants around schools within a short walking distance for students is an important public health concern in that it represents a deleterious influence in the food environment that may undermine public health efforts to improve nutritional behaviors in young people' (Austin et al. 2005: 1579). These findings have been echoed more recently in the work by Davis and Carpenter (2009) and Howard et al. (2011) who found convenience store proximity connected to high levels of obesity among students.

SUBURBAN FOOD SPACE IN REVIEW

Suburban spatiality has undoubtedly had a profound effect on food's relationship with urban space, particularly in its latter twentieth-century manifestations. Suburban food space has traced a path from early, well thought-through Garden-City proposals for local food spaces and systems, to the rise of the supermarket and shopping centre in post-war America and elsewhere that radically reshaped both food and place. Responding to structural changes including mass car ownership, suburbia largely created a physical and social landscape for food that was at a much coarser grain than that of the gastronomic townscape that went before. Suburban shaping imposed a scale and design for consumption that mostly undermined the possibility of creating public outdoor rooms for food, while mimicking the more inclusive spatiality of the traditional city in some of its artful food space designs. Suburban design for food set the template for what was to come in the sprawling megalopolitan region that is explored in Chapter 8. Before that, however, we turn our attention first to the scale of productive space in the city in Chapter 6 and its peripheries in Chapter 7.

CHAPTER SIX

Convivial Green Space

INTRODUCTION

Green space in and around cities and towns has customarily been an important scale for growing food over millennia (Morris 1994; Rebora 2001; Gaynor 2006). At the urban scale such agriculture has been defined as 'the production for domestic consumption or sale of food grains, tree crops, fresh horticultural produce, fish and animal products within an urban area' (Koc et al. 1999: 13). Yet in the twentieth century, that critical relationship at the urban scale was dismantled in many places, which begs the question, to what degree should food production in a city take precedence over other meanings and uses of green space at domestic (or civic) scale? This is not a new question. There has often been a kind of moral judgement made upon those who chose to use open space for pleasure rather than production. Princely gardens of Hellenic times expressed the rulers' political power through the symbolism of control over nature. The early Romans disapproved of the decadence of the expression, not its intent, managing to emphasize the simplicity and frugality of vegetable gardening which they enjoyed, while maintaining the apparent austerity of their philosophical convictions on plain living.

In recent years, a reappraisal of the importance of urban food growing has begun to reshape urban food space (broadly) driven in the global north by sustainability concerns, while in the global south issues of food resilience have been to the fore. Ideas about embodiment, visceral geographies and political ecology noted in the Introduction have been theoretically important in framing the way geographers and sociologists see food's relationships to urban green space. For designers, by contrast, it is the spatial qualities that these parts of city form can encompass that are of particular interest. This chapter focuses in particular on some of the design and planning elements of the urban food-growing trajectory, and in exploring transformations in such growing practices, a central assumption is that conviviality can be expressed through these spaces and opportunities. It has been suggested that food-growing spaces' conviviality encompasses the non-human as well as the relationship of people to one another and

> this is not the conviviality of Ivan Illich's 'tools for conviviality' (1979) nor of Zygmunt Bauman's modus conviviendi (2003), who restrict their concerns to how people accommodate one another in the everyday business of living together. It is rather party to the kind of conviviality gathering force in the name of 'posthumanism'. (Hinchliffe and Whatmore 2006: 125)

While accepting that the notion of conviviality can be broadened out from its human-centred roots, its expression still has design dimensions for urban green space. These are considered later in a chapter which explores the evolution of food growing in cities with reference to a range of examples from urban agricultural practice, and from perspectives that include environmental history, urban sustainability, planning and design. Relationships with the agricultural countryside and peri-urban agriculture are touched on, insofar as the emphasis is on design interplay arising from servicing urban food needs. However, this chapter pays attention mainly to growing space *in* cities whereas city peripheries are the theme of Chapter 7. While much of the focus is on the interplay of cities' design with food growing, the chapter also considers green spaces designed for food consumption and pleasure, in both historical and contemporary situations.

HAVE WE ALWAYS GROWN FOOD IN CITIES?

Urban food growing has an extremely long history, with examples from a great diversity of cities (Mougeot 1994). As noted by Morris (1994), with the rise of farming, settled communities evolved in Neolithic times in places including Mesopotamia, Egypt and India, although Jane Jacobs (1972) has famously turned on its head the agriculture first-cities afterwards thesis, arguing that agriculture and animal husbandry arose in cities first. Her 'New Obsidian' of pre-agricultural hunters, a city described as located in the Anatolian plateau of Turkey, derived its wealth from trade in obsidian, with food bartering forming a crucial aspect of its economy only later (Jacobs 1972). Morris (1994: 304), however, demonstrates that Jacob's argument does not hold more broadly in the light of archaeological and other evidence. Rather, between 3,500 and 3,000 BC, some Neolithic villages began to transform into cities on alluvial plains, and one of the requirements for this 'urban revolution' was the production and storage of surplus food (Morris 1994: 5). Cities of the Indus River civilization, such as Harappa and Mohenjo-Daro 'were once specialized agrourban centres' (Mougeot 1994: 2) and the development of these early cities 'seems to have paralleled that in Mesopotamia, with Neolithic farming communities establishing villages on the higher plains away from the actual river courses during the fifth millennium BC, before becoming sufficiently well organized socially and technically to take on the challenge of farming the flood-plains' (Morris 1994: 14). In the city of Mohenjo-Daro, archaeological evidence shows substantial granaries, specialist shops and houses focused on courtyards (Morris 1994: 17), which could have included food-producing gardens. The presence of animal husbandry in urban areas is similarly evidenced in places like Ninevah in Mesopotamia, which had large open spaces for cattle grazing (Kostof 1992).

In traditional city form the need to ensure food supplies meant that perishable crops, in particular, were best located close to where they would be consumed. Growing food for personal and commercial reasons meant that productive land uses including urban gardens, allotments, orchards, market gardens, vineyards and bee-keeping spaces surrounded cities and were interwoven with them (see Figure 6.1). The cities of fifth-century BC Greece, for example, remained closely tied and dependent on agricultural hinterlands, while a plan of Rome shows that within the Aurelian wall (built in 272–280 BC) gardens covered ninety-eight of the available 1,386 hectares (Kostof 1991: 41). It can be reasonably speculated that some of these gardens were used for food growing to complement the small shops, warehouses and wholesale food markets that supplied daily food needs (Kostof 1991: 47). As Kostof (1992: 166) further points out, open land within the city limits was not at all exceptional in antiquity and the Middle Ages

> since the activities of the countryside did not stop at the city gates, fruit and vegetable gardens at the backs of houses spread patches of green through the urban form. Commons where town cattle could graze were often not much beyond the built up area, and doubled as recreation grounds.

By the eleventh century, population and food production were both growing, the latter allowed by the introduction of the three-field system, mills and other agricultural technologies (Hollister 1974: 150). These changes, in turn, saw towns revive: 'the European economy in the High Middle Ages remained fundamentally agrarian, but the towns were the economic and cultural catalysts' (Hollister 1974: 147) and were ready markets for local agricultural products. In medieval English cities, customary rights grew up in common law that stopped attempted enclosures of common pasture lands (Hollister 1974: 147).

FIGURE 6.1: Traditional productive urban space, Transylvania.
Photo: Susan Parham.

> In the fourteenth century the English town was still a rural and agricultural community, as well as a centre of industry and commerce ... outside lay the 'town field' enclosed by hedges, where each citizen-farmer cultivated his own strips of corn land; and each grazed his cattle or sheep on the common pastures of the town which usually lay along the river side. (Trevelyan 2007: 28)

Urban food production took on specific spatial forms, including the domestic kitchen garden, discussed in Chapter 2. Houses were interspersed with 'gardens, orchards, paddocks and farm-yards' (Trevelyan 2007: 28), and because agricultural practices were somewhat improved through feudalism, a food surplus became 'available to towns with their steadily growing populations of non-agricultural specialists' (Morris 1994: 69). In Italian city-states of the same period, meanwhile, the interdependence of town and country was equally well understood as not merely the consequence of landowning by citizens.

> The essential function of the great majority of towns was as the principal market centre for local commodities. Most towns were probably dependent on their own rural territories for grain, wine, meat, cheese, vegetables and fruit, a majority even for their hides and wool, a great many too for their oil and fish. (Waley 1969: 35)

With colonialism, and the opening up of new urban spaces, urban agriculture in the New World often reflected broader movements of economic change and immigration and took particular forms of immigrant gardening, such as those of America's mining communities around Lake Superior (documented by Alanen 1990: 161) and the Dutch 'truck farmers' who settled the Calumet region on the southern edge of Chicago from the 1840s. These urban edge farmers worked the rich soils as market gardeners, selling at both wholesale markets and taking their excess produce directly to Chicago's burgeoning industrial areas (Zandstra 2004: 120):

> with little money and no refrigeration, food needs were a daily project. Corner stores and street paddlers supplied fresh foods as well as other needs. Another strong tradition in those neighbourhoods was storing food for the long winter months. Canning, preserving, drying, pickling, 'krauting', or storing summer's abundance was an annual ritual. Dutch gardeners were a natural compliment to this industrial working population. Produce was often customised to meet the various ethnic requirements.

In the twentieth century, lucrative perishable vegetable crops, including turnip greens, mustard greens, collards and spinach greens, were grown to reflect the food preferences of black and white immigrants from the south, displaced by the Depression from their share cropping origins (Zandstra 2004: 120). Later, with an influx of Mexican families from Texas (originally to work seasonally on Dutch market gardens as labourers, later to stay as urban industrial workers), crops including 'cilantro, tomatoes, plus many varieties of peppers, such as jalapeños and poblanos, became staples in Dutch farmers' fields' (Zandstra 2004: 126).

In nineteenth-century California and Australia meanwhile, Chinese market gardeners supplied fresh fruit and vegetables in a wide range of urban and rural contexts (Chan 1996; McGowan 2005). These gardeners were associated with the influx of miners to various gold rushes: in Queensland, for example, arriving to work the goldfields at Palmer's River in the north of that state (Jack et al. 1984: 51). Here, Chinese peasants from Hong Kong and Canton arrived by steamer to the remote Cape York region via Cooktown to both pan for gold and to lease space to garden in a location with an acute need for such food services (Jack et al. 1984: 51). All the gardeners on Cape York's Palmer Field were Chinese, counting some sixty-eight by 1883, but by 1900 few remained, with records showing only one gardener of Chinese origin, Ah Toy, gardening here until the 1930s (Jack et al. 1984: 51). Archaeological evidence indicates fruit trees including custard apples, orange, mandarin and mangoes as well as rough leaf pineapple were grown, and traces of vegetable irrigation ditches remain extant (Jack et al. 1984: 51). Now understood as pioneers, such market gardeners employed innovative growing techniques and were particularly skilled at conserving water in arid locations (McGowan 2005: npr). Their under-researched history of urban agriculture is now being reassessed, demonstrating that

> Chinese market gardeners were ubiquitous in colonial Australia and for several decades after Federation. As early as 1863 one observer said of Victoria that there was 'scarcely a town but is now well supplied with all kinds of household vegetables by these celestial gardeners'. Many other contemporary observers made similar observations. In the Northern Territory in the nineteenth century the Chinese had total dominance of all agricultural and market gardening activity. (McGowan 2005: npr)

URBAN AGRICULTURE AS A RESILIENCE STRATEGY

Those writing about urban agriculture in non-Western urban settings have shown urban food growing to be a practice relatively unbroken over time, despite the imposition of unsympathetic colonial and technocratic regimes at the broad scale, and more individual scale problems of theft and predation (Freeman 1991: 98). Food security is at the heart of this activity (Pottier 1999). Understandably, with the extremely rapid urbanization experienced in developing-world cities since the 1960s, and the significant implications for food security of these burgeoning populations, the focus of much writing on urban agriculture in developing cities is its contemporary practices and future possibilities rather than historic examples. What accounts from Africa, Asia and South America tend to show, though, is how central food growing is to urban quality of life, and how these practices shape the spatiality of places as well as being shaped by them (Ikpe 1994; Binns and Lynch 1998).

Migrants to urban areas often cannot find jobs and undertake urban agriculture as a stopgap survival strategy 'not only to deal with food insecurity and poverty, but also to organise with fellow citizens and improve the quality of life in their communities' (Redwood 2009: 4). Cities as diverse as Nairobi (Freeman 1991), Kampala and Shanghai have famously been virtually self-sufficient in food produced from localized agriculture, and in certain places in the first-world urban and peri-urban agriculture still contributes an important source of food (see Figure 6.2). In Moscow, for example, around 80 per cent of the population was reported in the late 1990s to be involved in some level of food growing (Brown and Jameton 2000). Evidence from Australia in the early 2000s demonstrated that between every second and third suburban household grew some of its own food (Gaynor 2006). In relation to more organized urban farming,

> an estimated 33% (696,000) of the 2 million farms in the United States are located within metropolitan areas. These farms produce 35% of all crops and livestock sales. The United Nations document on urban agriculture reported that 25% of urban households in the United States are involved in gardening, including food gardens and landscaping. (Brown and Jameton 2000: 20)

In developing countries and continents, where food production is a crucial form of food support for many urban dwellers, such self-sufficiency is again on the urban agenda following the failure of the more technocratic approaches of the Green Revolution of the 1940s–70s (Freeman 1991; Redwood 2009). There is considerable literature on urban agriculture found within wider work on developing nations and food resilience, and this shows, as noted above, that much urban food growing is undertaken in the face of official discrimination, barriers or prohibitions

FIGURE 6.2: Urban agriculture project in Glasgow.
Photo: Susan Parham.

(Mougeot 1994: 7). Yet informal urban agriculture is increasingly evidenced in places as diverse as Accra (Obosu-Mensah 1999) and Latin and South American cities (Redwood 2009). In fact, in both developing and Western urban settings, urban agriculture has grown up alongside urbanization and is increasingly understood to make an exceptionally valuable contribution to individual and societal urban resilience (Midmore and Jansen 2003; Mougeot 2005).

Sometimes, the focus on urban food growing may be sanctioned and supported by government as it is driven by the exigencies of a particular crisis. In Havana's city and suburban areas, for example, the state of emergency brought about by an American trade embargo and the collapse of the Soviet bloc, which had previously offered support, led to an urban agricultural programme 'characterized by self-reliance, low input use, and organic farming practices' (Funes et al. 2002: ix). Cuba was forced to re-embrace and adapt traditional forms of urban agriculture, not only because of pressing external conditions, but also because of internal drivers such as the low quality of vegetables in urban markets and under-exploited spatial potential for production in urban areas (Altieri et al. 1999; Funes et al. 2002: 220). Such urban agriculture had to be responsive to a heterogeneity of urban and suburban conditions, have low impact to avoid the effects of toxic pesticides in proximity to people and be efficient in the use of water, maintaining soil fertility and cultivation of crops and animals, especially perishable foodstuffs (Altieri et al. 1999; Funes et al. 2002: 220). In Havana, the employment of a large number of people in urban agriculture (many of whom had emanated from rural areas) was one of its biggest social impacts, and driving these changes was the potential for increased income generated by selling produce from urban gardens; this attracted the attention not only of workers, but also of professionals from diverse backgrounds (Altieri et al. 1999; Funes et al. 2002: 222).

Individual food growers have also shown themselves highly reluctant to waste potential food-growing land, whatever position is taken by the state more formally. 'If one looks carefully, few spaces in a major city are unused. Valuable vacant land rarely sits idle and is often taken over – either formally, or informally – and made productive' (Redwood 2009: 1). In Quito, Ecuador, for example, around 35 per cent of urban land is vacant and some is used for food growing (Redwood 2009: 6). In Peru, urban agriculture is an effective livelihood strategy for the inhabitants of Lima's Carapongo neighbourhood where land reform in a peripheral area under threat from urban sprawl has allowed small farmers to work food plots, which were formerly the property of large haciendas (Villavicencio 2009). Experience from Africa, meanwhile, demonstrates that there is a positive correlation between urban agriculture and avoiding malnutrition (Maxwell et al. 1998). It also shows the critical role of women cultivators on vacant lots and in public land such as roadside verges, parks, drainage 'way-leaves' and institutional land: 'women are major producers of food in large cities now, as they have always been in rural areas' (Freeman 1991: 79). This can be critical for food security. Thus, in a city like Dar es Salaam in the United Republic of Tanzania, 'the proportion of families farming grew from 18 percent in 1967 to 67 percent in 1991 in response to food shortages, inflation and increased rural-to-urban migration (UNDP, 1996)' (Marsh 1998: 6).

URBAN FOOD GROWING – SUSTAINABLE URBANISM IMPLICATIONS

Despite these examples, a rapid decline in urban food growing appears to have been the prevalent mode in first-world cities' spatial, economic and cultural relationship with food production. The loss of market gardens, from examples such as their astonishing peak of urban productivity in the mid-to-late-nineteenth-century Marais in Paris, and the sharp decline in vegetable growing in the Ile de France region around Paris over the post-war period can be seen as symptomatic of broader trends in both attitudes to, and practices of, urban food growing (Stanhill 1977; Cockrall-King 2012: 84). The conventional argument about the loss of productive food space in cities over the twentieth century has ascribed transformations in the role and function of designed and informal urban green spaces as broadly an erosion in the perceived need to locate such food space close by living areas. It has also referred to structural economic changes impinging on traditional spatial practices including urban food growing, with these associated with the problem of time poverty under late capitalism. While this issue of compression of time available for food has been considered more in relation to food consumption, findings about the use of ready or partially cooked meals seem conceptually relevant here. Thus, while critically interrogating the term 'convenience food', Warde (1999: 521) notes that

> The appeal of convenience increasingly involves appeal to a new way of conceptualising the manipulation and use of time. It speaks to the problem of living in a social world where people in response to the feeling that they have insufficient time, set about trying to include more activities into the same amount of time, by arranging or rearranging of their sequence.

In some places, too, the narrative of food productivity decline is challenged, and although this relates to private gardens, the subject of Chapter 2, it is considered here because it also connects to broader perspectives on why food productivity has diminished. In her study of the Australian 'backyard', Gaynor (2006: 2), for example, argues that for private gardens in suburban locations, any characterization as a 'simple transition from production to consumption – from vegetable patch to swimming pool – is somewhat flawed', because in Australia at least, a considerable number of households continue to produce a proportion of their own food. So in relation to private green space, Gaynor's analysis contests the view that the loss of urban food productivity has been typified as a move from productive working-class back gardens towards private open space that is largely symbolic in its planting schemes (reflecting more middle-class conspicuous consumption of space). Such an analysis also challenges the theory that this alienation of traditional growers has been followed by a re-appropriation by middle-class households of vegetable growing for lifestyle and health reasons. Gaynor's alternative reading suggests, rather, that across the twentieth century, middle-class and 'respectable' working-class households in Australia have consistently grown food in their back gardens. 'For those people, home food production has long been a source of food valued for its freshness, purity and health giving qualities' (Gaynor 2006: 3).

Other theorists, too, have noted the material pleasures to be had through gardening and food growing. Perspectives include actor network theory which tends to rather problematize food-growing relationships, with the garden seen as 'an ephemeral and precarious outcome, whose achievement both symbolically and materially is constructed and negotiated through the interaction of different actors' (Power 2005; Hitchings 2003: 102). More applied approaches have tended to view gardening as both a private pleasure and an activity important to urban sustainability (Marsh 1998). In fact, sustainability analyses of city green space specifically for food-growing places the decline of urban public and private sites for food as a significant problem for, and in, many modern cities (Hough 1984). Food growing, and green space more generally, in most Western cities has become primarily symbolic in nature over the course of the twentieth century. The role of such space in material culture is increasingly distanced from food except at times understood to be of national emergency, when as many productive areas as possible have been given over to food growing. Wartime examples include Victory Gardens (see Figure 6.3) in the United States and Canada (Johnson 2009) and allotments in the United Kingdom (Crouch and Ward 1997), France (Jones 1997: 58) and the United States through which 40 per cent of fresh vegetables consumed by Americans in wartime were supplied by twenty million small gardens (Lawson, in Nordahl 2009: 17).

More broadly, urban sustainability theorists argue that our view of cities and their surrounding regions has become divorced from nature and thus from cities understood as appropriate sites for food growing (Hough 1984). Urban expansion

FIGURE 6.3: Second World War Victory Gardens poster, USA.

Photo: War Food Administration, Agriculture Department, Washington D.C.

has built over the open spaces necessary for private vegetable gardens; public green space is unproductive and energy profligate; and agricultural production technologies used for urban fringe production are wasteful, polluting and unsustainable (Hough 1984). In contrast to the notion of conviviality discussed at the beginning of the chapter, attitudes that underlie unsustainable approaches to urban development and management reflect the way in which a dominant view of landscape has become detached from notions of ecology (Moran 2006). Cities have increasingly been developed as 'pedigree landscapes' stressing horticultural management, rather than relying on earlier ecological traditions of the cultural vernacular expressed in location-specific built forms and open spaces (Moran 2006).

In design terms, the resulting 'fixed mould of aesthetic convention and maintenance procedures' has produced in many places a sterile food-free environment evident in a variety of open spaces: turfed parks, hard-paved streets, vacant lands, industrial zones, areas for waste disposal and other miscellaneous spaces (Hough 1984: 247). Sustainable diversity is now more likely to reside in the forgotten or abandoned sites spared the attentions of landscape gardening and scientific horticulture. Open space predominantly mirrors the industrialized countryside in which big agriculture is dominant, with its associated biodiversity, ethical and food quality concerns. Thus agricultural industrialization has urban space consequences: 'a lover of the countryside in the future will be more likely to find it in the old city and its fringe than in the land beyond' (Simmonds 1993: 101). Urban space may offer more fine-grained opportunities than the countryside for an ecologically based conviviality centred on food.

It is this perception in part that underpins the increasing interest by urbanists and sustainability theorists in the city as design and planning context for, and driver of, more sustainable agriculture. The analysis of urban food growing from different disciplines is increasingly interconnected because the effects are so clearly intertwined, and because urban food-growing promises a range of benefits (Garnett 1996; Howe and Wheeler 1999). Urbanists, for instance, argue that our current agricultural systems are unsustainable for urban dwellers for some of the reasons touched on above (Farr 2008: 179). Just as the loss of food-growing space in cities has negatively affected us (as is explored in depth in Chapter 8), urban growth in its sprawling incarnation is implicated in the loss of agricultural land on the urban fringe, which in turn has distanced those in cities from the immediacy of food production. As Barton et al. (2003: 30) explain, food is a key resource within an approach in which the 'basic principle of sustainable development is that buildings and settlements should use resources at sustainable rates and avoid polluting their own or the global backyard'. The social, economic and sustainability implications of shifts away from urban and peri-urban food growing are sharply illustrated in the American context. In the United States, there were more than six million farms in 1940, but by the turn of the century this had dropped to under two million (Nordahl 2009: 3):

> And so the agricultural paradigm had shifted. The pervasive ideology of the mid-twentieth century became that food production was no longer suitable in and around our cities, as it had been for centuries. Growing fruit and vegetables was no

longer work of community-minded individuals and families on small local farms, but endeavours better suited to corporate-owned, factory-like 'agribusiness' in more distant parts of the country. (Nordahl 2009: 3)

The sustainability case for urban food growing is also increasingly cogent. Not only can urban agriculture reduce the production of embodied energy currently associated with conventional agriculture, it has benefits in relation to reduction in greenhouse gas emissions and waste production, improvements in air quality and biodiversity and numerous social and economic advantages in relation to seasonality and local consumption (Viljoen et al. 2005). The need to focus on these sustainability benefits has been emphazised by climate change and the increased challenges this brings in relation to food resilience and security (Dubbeling et al. 2011).

This sustainability analysis also has a health focus, as food-growing and healthy cities are increasingly seen as interdependent (Barton et al. 2003; Wakefield et al. 2007). Action on food growing that integrates health and planning approaches is advocated (Hoehner et al. 2003). Yet while theorists are more and more concerned with the characteristics of healthy cities, they may discuss food as an aspect of healthfulness in terms of social justice and inclusion, but are not necessarily foregrounding thinking about the spatiality of urban food growing and agriculture in their analysis (Corburn 2009). Rather, where there is a food-growing focus, issues of health may be perceived as about reducing risk in relation to hygiene practices in production (Redwood 2009: 239). A welcome exception is the increasing theoretical and applied interest in the relationship between neighbourhood food environments and (Cummins and Macintyre 2006; Stafford et al. 2007) obesogenic environments (Lake and Townshend 2006; Lake et al. 2010; Bagwell 2011; Ludwig et al. 2011) and food deserts (Wrigley et al. 2003; Smoyer-Tomic et al. 2006; White 2007; McClintock 2011; Thierolf 2012). While some have challenged the existence of this linkage, arguing there is no relationship between obesity and sprawl or contested aspects of the notion of obesogenic environments (Eid et al. 2008; Guthman 2013), work in this area is one of the developing intersection points between health-focused urban research and analysis of city design and planning in relation to food growing and consumption, and these perspectives are discussed in detail in Chapter 8.

CONVIVIAL GREEN SPATIALITY

It is worth noting that gastronomers have been arguing for many years that food security and conviviality for urban dwellers are not givens, but increasingly compromised by unsustainable food-growing approaches, among other aspects of food's interplay with urban design and place (Parham 1990, 1992, 1993). More recent research findings reinforce the point that even in countries which are currently the most food secure, the most excluded will be worst affected (Gorton et al. 2010). The combination of climate change and variability in supply

> are affecting the food system—from production, processing, and distribution through to consumption. Particularly vulnerable foods include fresh fruit and

vegetables, intensive livestock and dairy, and seafood. Potential health impacts include compromised nutritional status because of modified food availability and affordability and higher incidences of foodborne diseases. (Edwards et al. 2011: 100S)

As this writer noted in 1993, 'the adverse signs are all around for those who wish to look: loss of market gardens, the decline of small centres on the urban fringe, increases in the long distance shipping of vegetables to urban markets ... soil erosion, land degradation, and increases in runoff pollution from hard paved garden spaces'. Gaynor (2006: 10) notes this view in her environmental history of growing food in Australian cities, arguing that the 'gastronomic positions on sustainable urban development' that have emerged are 'broadly ... characterised by a tension between the rural and urban commons: a desire for increased production of fresh food at home vies with a longing for more densely populated, cosmopolitan cities which (at least in theory) support a greater variety of food outlets, and local, sociable dinner companions'. This perhaps sets up the convivial green space argument as a dichotomy between productivity and conviviality when it is, rather, an attempt to meld the design and planning of food-centred urban space in ways that mutually support sustainable urban food production and consumption across urban, suburban and post-urban contexts.

More broadly, and perhaps understandably, surveying contemporary cities that appear largely to have abandoned food production, recent writers on sustainable food planning and design, Viljoen and Wiskerke (2012: 19), have declared that urban food growing 'is, or at least until recently, was often not an issue on the urban planning, development and/or policy agenda', noting the way that food growing has not been regarded as an urban activity (Pothukuchi and Kaufman 1999, 2000). At a more applied level, in a developing-world urban context, it is also suggested that urban agriculture has been discouraged because of a particular paradigm operating in relation to urban space: 'in the eyes of planners, architects, politicians and developers, trained as many across the world are, in the arts of town planning by European colonial systems, farming in the city was considered a practice either to be discouraged or ignored' (Redwood 2009: 5).

Yet not all planners, urban design writers or urban food practitioners have been oblivious to the problematic decline of food growing in these varying urban contexts: some have offered proposals to respond to these transformations in urban space, with alternative, productive examples and ideas. The disconnect between food production and cities has been challenged practically and theoretically over the course of the past fifty years not only by the actions of individual households (Gaynor 2006), but also by the countercultural urban farming movements of the 1960s and 1970s (Olkowski and Olkowski 1975), through proposals for gastronomic design and planning strategies for cities and for more convivial green space (Parham 1990, 1992, 1993, 1996), in a revival of specific urban growing forms such as allotment holding (Crouch and Ward 1997) more recently in the growth of interest in so-called edible landscapes and edible cities (Esperdy 2002; Vitiello 2008; Paull 2011) and in relatively current calls for agricultural urbanism (Duany 2011; Broadway and Broadway 2011; Phillips 2013) that seek to re-engineer urban form to more productive ends.

The rebirth of urban food growing has been driven by a diversity of factors. Some of these are long term. Immigrant experiences and food practices feature prominently here while others have been propelled by requirements for subsistence or national emergencies, as in Cuba's case. More broadly, in the 1990s and 2000s, the focus has been on a holistic and critical look at the current food system, and has resulted in increasing attention to urban food resilience in the face of issues including food scares, unethical production, anxieties about food industry monopolization, rising food poverty and sharpening climate change effects. From a study of urban agriculture in three US cities suffering from post-industrial decline, Meenar et al. (2012: 1) note that urban agriculture can be part of a conscious urban design and planning strategy of 'ecological and cultural regeneration based on a notion of "post-growth"'. Practical actions are occurring in many places (see Figure 6.4), often within an urban greening paradigm (Pincetl and Gearin 2005: 377; Birch and Wachter 2008). City-level policies for urban agriculture are a growing trend, as in the American city of San Jose (Schultz and Sichley undated). As one online commentator (cities.mrc) argues, 'Cities around the world are emerging as key locales for growing food. A variety of approaches are being piloted to enhance health and well-being, encourage local economic growth and self-sufficiency, enrich social cohesion and community development, and diversify urban greening and resilience'.

In some urban areas, traditional, convivial connections to food growing have remained relatively strong. In France, for instance, in the late 1990s, over 20 per cent of fruits and vegetables consumed were thought to be grown in family plots, with just over half of households owning a garden and two-thirds of these gardens including some vegetables (Jones 1997: 58). The *potager* has been seen as embodying the French garden, and is now fashionable again, but was in any case

FIGURE 6.4: Community Garden, Adelaide.
Photo: Susan Parham.

a tradition maintained across France by every class over centuries: 'the very word "jardin" long meant, to the vast majority of the French population, a kitchen garden' (Jones 1997: 8). Although the decline of kitchen gardening in urban areas in France reflected a post-war urban boom which 'played down country roots in the euphoria of modern appliances and paid holidays', by the 1960s there was again a 'nostalgic return to the earth and a war between organic methods and chemicals in gardening and agriculture ... today ... every tiny hamlet in deepest France reveals small kitchen gardens, tucked into the most unlikely corners' (Jones 1997: 8). Space constraints for food growing in cities have been met with ingenious solutions in many places: 'as populations grow and land becomes a luxury in the urban environment, smart young gardeners have adapted the *potager* to the smallest city spaces. It appears as a table-top garden on a roof terrace or confined to an oil drum – a useful solution for a mobile population' (Abbott 2001: 8).

ALLOTMENTS – DECLINE AND REVIVAL

The history of decline and revitalization of allotments in the United Kingdom is an instructive example of the renewal of food-centred green space in cities. Allotments were an English invention, which can be traced back to cities most affected by the beginning of the industrial revolution in the late eighteenth century when small gardens on the edge of town could be cheaply rented by working-class people, and 'where they could be out in the fresh air with their families, and incidentally raise food to supplement their diets' (Kostof 1992: 57). The allotment as a land use form increasingly appeared in European cities by the end of the nineteenth century and remained popular as a warm-weather leisure retreat (Kostof 1992: 57). In the United Kingdom, allotments were still commonplace in cities in the first half of the twentieth century, and the Tudor Walters report of 1918 proposed using 'backlands' in the middle of large suburban housing blocks as allotment sites (Whitehand and Carr 2001: 47).

Although allotments persisted as a feature of urban landscapes in British cities, by the latter half of the century, they were increasingly understood as a leftover urban form, no longer of use and representing a spatial management problem for local authorities (Whitehand and Carr 2001: 47). Crouch and Ward (1997) reflect on the threats to allotments in the 1980s through local authority sales to speculative house builders, yet note that despite these depredations, urban allotments have a special position in British cites, as a form of both urban agriculture and open space (see Figure 6.5). For instance, local authorities have a statutory duty to provide land for cultivators and most allotment sites cannot be sold-off (despite huge commercial pressure on urban land) without central government consent (Crouch and Wiltshire 2005: 125). The difficulty until recent times has been a range of external forces impinging on the perceived necessity for keeping allotment space:

> The original logic for providing and protecting this land was rooted in the poverty of manual workers over a century ago, however, and while allotments still provide for subsistence needs in many deprived communities, here as

FIGURE 6.5: Contemporary allotment garden, London.
Photo: Susan Parham.

elsewhere the availability of cheap supermarket foods and the claims of work time and alternative leisure time pursuits have undermined the revealed demand for allotments. (Crouch and Wiltshire 2005: 125)

A recent change in attitudes to urban food growing has seen the rebirth of the allotment movement, and now, again, allotments are perceived as important to urban agricultural practice in the United Kingdom and elsewhere. As Buckingham (2005) notes, in the United Kingdom allotment holding has shifted from primarily an older male, working-class activity to one in which younger and more middle-class women now take a more substantial role. In various local authority areas in London, allotments have become so popular that they have long waiting lists, or those lists are so extensive they have been closed to those wanting to join in urban food growing. This shortage is in part because authorities sold-off many of the extant allotments before this cultural change occurred. In addition, some well-known gardeners have taken up the allotments cause, arguing that the narrow cost–benefit analysis of economists which says that private allotments, 'or as they are officially known *leisure gardens*', are expensive to run is wrong (Don 2009).

Blogs in broadsheet newspapers trace the allotment year and a number of cookbook writers, and restaurateurs similarly have documented their allotment adventures in London (Clark and Clark 2007; Tulloh 2011), while a food critic's allotment is the subject of a recent book on kitchen gardens (McCulloch 2011: 86). These situate contemporary allotment holding as an authentic gastronomic, often immigrant, practice borne out of cultural preference as much as necessity. Those with considerable cultural capital in Bourdieu's terms are now joining in and benefitting from their allotment food growing (and it could be argued appropriating the cultural capital of their allotment neighbours through their books and articles).

As one allotment gastronomer reflects, this food-growing practice allows both spatial and cultural access that would otherwise be unavailable:

> That first summer I was rewarded with a place where I felt far away from the city without ever leaving it ... I learned that our site is an inspiring mix of Turkish, Portuguese, West Indian and English gardeners, each with their own approach and their own vegetables. My plot may be sandwiched between a busy road, a sewage works and a dump, but its fertile soil is dark and alluvial with the texture of very rich chocolate cake. (Tilloh 2011: 14)

Another says something remarkably similar:

> The first person we met was our neighbour Hassan: kind Mr. Charisma, who was to become our friend and mentor. He introduced us to other people on the allotment – Cypriots, Kurds, and Turks. We soon realised that we were among special people who thought differently about growing and could teach us much about cooking too. Our eyes were opened to things such as frying green tomatoes, cooking artichoke leaves, braising wild poppy leaves, and much more. The Eastern Mediterranean was alive in Hackney Wick. (Clark and Clark 2007: introduction)

One of the insights this newly acquired practice revealed was that people who traditionally grow food on allotments also cook sustainably from scratch. 'Cooking from an allotment means I can't help but cook seasonally. Now I pick first and then decide what to make with it. The pressure of trying to plan what to cook is gone; the garden does it for me' (Tulloh 2011: 12). Yet all is not entirely Edenic in urbanism terms in these magical food spaces. Reflecting on the fragility of allotment lands in the face of competing pressures on urban space, Clark and Clark (2007: viii) point out that at the time of writing, their 'century old allotment' area in east London was removed to make way for part of the London Olympics site, and they struggled to obtain an alternative site the following year. These tropes of the allotment as both working class and immigrant cultural form, and more latterly, stylish foodie accoutrement, map across arguments in support of allotments. Allotments have three very positive roles that would place them in the 'convivial green space' frame and reflect particular positive urban design elements. These are their role as part of the urban green, their contribution to the urban landscape as sites for autonomous and creative activity and their work as spaces in which people reconnect with each other 'as communities of propinquity and interest' helping to build local identity (Crouch and Wiltshire 2005: 128). The vernacular landscape of the allotment, with its ramshackle sheds often of found materials, is still sometimes seen as problematic in urban amenity terms, while also now valorized as part of the urban world's landscape charm and sustainability. As one cookbook-writing allotment holder describes their allotment 'infrastructure',

> All that I needed was a shed. This my husband built for me out of timber scrupulously salvaged from skips. But by the time he'd bought the clear plastic corrugated sheeting that I'd decided I had to have for a lean-to greenhouse, it

ended up costing nearly as much as a ready-made shed. My sister decorated the door with a pattern of parrots and the children painted a scribble of owls and numbers on one side. (Tulloh 2011: 14)

THE EMERGENCE OF URBAN AGRICULTURE AS A MOVEMENT – URBANISM IMPLICATIONS

In recent years, renewed interest in urban agriculture in the first world has emerged from a variety of directions (see Figure 6.6). Among other justifications, urban food growing has become known as a healthful pursuit supporting individual well-being, a way of avoiding pesticide-ridden commercial crops, seen as reflecting broader concerns for the sustainability and resilience of cities and enhancing food security for individuals on low incomes (Hallberg 2009). Defined as occurring in three waves, the revival in the popularity of allotments is situated as responding to issues including about food safety, globalization, food miles, food quality, increased urbanization and climate change (Cockrall-King 2012: 73). Recent urban agriculture proposals are not simply alternatives to commercial fruit- and vegetable-growing operations in and around cities (and increasingly removed from them spatially) but often act as a practice-based critique of these mainstream spatial arrangements. As Flores (2006: 2) argues,

FIGURE 6.6: Skip garden, King's Cross, London.
Photo: Susan Parham.

'Whether you live in an apartment, in the suburbs, on a farm, or anywhere in between, growing food is the first step toward a healthier, more self-reliant, and ultimately more ecologically sane life'. Civic agriculture in these terms is defined as embodying

> a commitment to developing and strengthening an economically, environmentally, and socially sustainable system of agriculture and food production that relies on local resources and serves local markets and consumers. The imperative to earn a profit is filtered through a set of cooperative and mutually supporting social relations. Community problem solving rather than individual competition is the foundation of civic agriculture. (Lyon, quoted in Nordahl 2009: 10)

The 2000s have seen the rise of a vast range of urban food-growing projects, schemes, networks and strategies, with urban governments and city dwellers increasingly aware of food-growing issues and prepared to make changes in their own food production and consumption habits including adopting permaculture approaches (Cockrall-King 2012; Viljoen and Wiskerke 2012). As Nordahl (2009: ix) reports in the American context, with the economic downturn and unstable food supply as one of the effects of climate change on cities, public networks of food-growing opportunities on underutilized public (and private) land are increasingly seen as feasible. It has been realized that it is in urban rather than rural or urban fringe locations that the best opportunities may be found 'to realize the environmental, economic, and equitable benefits of a more local system of agriculture ... in and among the places we pass by daily on our way to work, home, school, commerce, and recreation' (Nordahl 2009: 8).

By the early 2000s, in Canada and the United States, 'seemingly everywhere, cities were forming food-policy councils; community gardens were multiplying; and municipal governments were voting on whether to allow households to keep a few urban chickens or a beehive; or to permit commercial farming to coexist with other commercial pursuits in their cities' (Cockrall-King 2012: 15; Lawson 2005). In the South Australian capital city of Adelaide, for example, proposals emerged as part of development of a long-term vision for the city to convert part of the parkland area encircling the central city into a city farm. Thus, in a process of engagement based on ideas for enlivening the central city and framed by its 5,000 postcode,

> 'Farming the City', or bringing sustainable agriculture, rooftop crops and hanging vertical gardens into the CBD and inner suburbs, has emerged as a key theme among the 1000 ideas already uploaded to social media sites. A range of uses for the 50ha city farm – mocked up during 5000+ forums to generate debate – includes wheat crops, citrus orchards or huge lettuce gardens. (Monfries 2012)

As this author argued previously (Parham 1993), opportunities for urban food production are available at a range of scales, from private gardens, through productive streets, to public and community gardens, allotments and farms. Recent work on rooftop and 'leftover space' food-growing opportunities offer a number of urban retrofitting possibilities (Engelhard 2010; Parham 2012). Nordahl (2009: 9) asks urban dwellers to give attention to the food-growing potential of plants in all of a city's public spaces: 'fruit trees and shrubs along streets and in medians; orchards in parks; herbs and vegetables in planters located on plazas and sidewalks in our commercial areas; and roof top agriculture, to name a few'. In part, the wellspring for such action is to bring people in touch with food; loss of connection being considered a hallmark of the modern food system. Reflecting on his own regional food background, Nordahl (2009: 11) notes a high level of gastronomic alienation from locally produced foods: 'many Iowans have never eaten tofu, tasted soy milk or ever heard of edamame-yet Iowa is the largest grower of soybeans in the country'.

URBAN AGRICULTURAL PROJECTS – COMMUNITY GARDENS, URBAN ORCHARDS AND BEYOND

As urban agriculture has gained momentum, a renewed interest in urban food growing has seen a focus on settings such as old industrial regions (Reid et al. 2012) and explored actions from front yard farms (discussed in Chapter 2) to edible estates (Haeg et al. 2008), community gardens (Irvine et al. 1999) and orchards, guerrilla gardening (Severson 1990; Tracey 2007), food networks and crop swaps (see Figure 6.7). Often, these have an overt focus on community support and resilience in the face of food poverty while wider benefits to well-being are also noted (Hynes and Howe 2002). They also have potentially dramatic spatial effects. Reflecting on her experience of community food growing in Eugene, Oregon, Flores (2006) suggests turning wasted lawn spaces, including parks and front gardens into urban food production sites. 'Food not Lawns started several more gardens and circulated seeds, plants, and information. We planted food all over town-vegetables and fruit trees in public parks, berries along the bike path, squash down by the river-anywhere that looked like it would get water and sunshine' (Flores 2006: 10).

FIGURE 6.7: Community Garden Scheme, North London.
Photo: Susan Parham.

Examples of fresh produce grown on public land include those at a significant scale such as in the shrinking city of post-industrial Detroit with its well-known urban farmers movement (Bonfiglio 2009; Giorda 2012). These also encompass a plethora of small-scale instances in the form of community gardens, vacant lot plantings and reuse of blighted or abandoned land (Wakefield et al. 2007). In Detroit, one among many examples is a series of vacant lots in a deprived area taken over by a non-profit organization for fruit and vegetable growing (Choo 2013). A diversity of such urban agriculture schemes are underway in cities including Los Angeles, Vancouver, Toronto, Milwaukee, Detroit, London, Paris and Havana (Cockrall-King 2012). These are sometimes a contested process (Staeheli et al. 2002; Smith and Kurtz 2003). At other times, as in London's, *Capital Growth*, a partnership between the public sector and community groups supports food-growing opportunities, training and education, with its 2013 campaign, 'Growing a Millions Meals for London' directed towards making London again a food-producing city (Capital Growth 2013). Capital Growth's website includes a 'capital growth space finder' through which people can find local food-growing spaces and opportunities:

> We have a large garden at Sotheby's Mews Centre which we would like to develop into a community allotment where members of the centre, older people living in Islington, can grow and harvest food which can be used by themselves but also by the centre. Part of the project will be a community kitchen where food can be cooked and served to members of the community. We will aim to set up a gardening club who will maintain and look after the raised beds. As far as possible we will use recycled materials. We will have the facilities to compost food and garden waste and harvesting rain water. As many members are BME we will encourage them to grow food from their own cultures. (Capital Growth Space Finder 2013)

At the other end of the urban spatial spectrum, in Todmorden, a village of some 17,000 inhabitants in west Yorkshire, a residents' group decided to start growing edible plants in a number of public locations around local streets and underused spaces on public land in an initiative that became known as *Incredible Edible Todmorden*. Reportedly tired of food plans, strategies and policies, in 2008, the group began a process of appropriating 'neglected verges, and spaces around halls, council buildings, the canal, and the local railway station, for food growing' (Paull 2011: 28). Paull (2011) explains that villagers took a while to catch on to the idea of 'open source food' in which it was permissible to pick food in public places planted by someone else. These 'public space food plantings' were described by one of the scheme's main proponents as '"propaganda gardens" – something to use as a tangible expression of a set of bigger ideas – including growing local, eating local and fresh, eating seasonal, and knowing the provenance of food' (Warhurst, quoted in Paull 2011).

The development of community gardening projects has sometimes been highly contested, as in the example of New York where a neoliberal agenda on the part of city government saw attempts to sell off community garden space to speculative property developers to realize the property market value of the land (Smith and Kurtz 2003). In Toronto, meanwhile, community gardens have been described as providing opportunities by which (to paraphrase) individuals and communities actively make places, construct culture and in so doing assert their food citizenship (Baker 2004: 305; Palassio and Wilcox 2009). Toronto's 110 community gardens are seen to be part of a food security movement in the city which challenges the corporate food system, allows participants to actively shape their communities and engage cross-culturally (Baker 2004: 305). Such gardening offers the chance 'for people to dirty their hands, grow their own food, work with their neighbours, and generally transform themselves from consumers of food into "soil citizens"' (Baker 2004: 305, quoting Esteva and Prakash 1998; DeLind 2002). While it is acknowledged that people grow food in these gardens to save money, to exercise and to pursue their cultural food preferences, it is also suggested that these food-centred activities contribute to a broader, explicit strategy by the community garden movement to increase urban food resilience. Of particular interest in relation to the spatiality of the community gardens is their location in often-leftover spaces that reflects both the cultural marginalization of their proponents and the messy nature of cities' spatial structure:

> Hidden away in corners of public parks, on apartment-building properties, in backyards, on rooftops, and behind churches are Toronto's 110 community gardens, many of which reflect the city's increasing ethnocultural diversity in the faces of the gardeners and the varieties of plants they grow. Immigrant gardeners bring local knowledge from around the world and adapt it to urban gardening spaces in the city of Toronto. As in the example of New Orleans's Vietnamese market gardens (Airriess and Clawson 1994), many of Toronto's community-garden plots reflect the landscape memories of their gardeners. (Baker 2004)

Urban agricultural interests have also turned to productive street trees and urban orchards, and while the latter are being notably revived on the urban fringe (as discussed in Chapter 7), some more city-based examples also exist or have been proposed.

> We should plant orchards of fruit trees on common land and in parks and gardens everywhere. The ubiquitous Queensland Box could be replaced by quince, pear, apple, or plum trees in our streets. Orchards give the land an almost magical quality. They also let people who are out of touch remind themselves of the seasons, harvest, and provide a source of fresh food. (Parham 1992: 4)

The Headingley Community Orchard, for example, has been developed by residents and workers from the Headingley area of Leeds in the United Kingdom, in the form of a 'collective dispersed orchard' which is located on underused land owned by others but available for use (Tornaghi 2012: 349). This 'dispersed' orchard is described as an example of urban agriculture which offers the opportunity for 'new ways of sharing spaces and experiencing conviviality in public spaces' (Tornaghi 2012: 349), while the example of the Union Street urban orchard, meanwhile offers a 'worked' case study of urban food production as an interim or temporary use carved out of what might seem on first acquaintance a very unlikely site in an inner post-industrial area of London. Conceived as part of an architecture festival, over four months in 2010, a vacant site in the Bankside area of Southwark was transformed from its derelict state to become a temporary orchard planted with eighty-five fruit trees housed in remodelled pallets (Bost 2010). The Bankside Open Spaces Trust were approached to take a role in creating the temporary orchard space that was designed with elements including a scrumping shed, seed shed, table tennis skip, 'living ark' and 'nest', and the range of fruit trees included apple, pear, quince, medlar, plum, greengage, peach, cherry and apricot. It was noted that

> the tight timescale, the lack of money and the temporary nature of the Orchard raised concerns, but as that long winter melted away so did our concerns. The central idea of transforming a desolate interim car park into a stage for a life affirming interaction with art and nature became a daunting but irresistible prospect. (Bost 2010: 48)

Over the course of the few months of its brief life, a series of events and parties were run in the urban orchard, including a harvest celebration, and the site attracted over 10,000 visitors. At the end of the project the trees were transplanted to a number of locations in the local borough including housing estates, sheltered housing and existing community gardens (Beirne 2011: 73). As Bost (2010: 50) points out, the temporary nature of the orchard meant that its appeal was somewhat limited to the 'sophisticated cognoscenti' who immediately connected to it, rather than permeating out to the less hip. But, equally, its evanescent nature meant 'the Orchard did not have to suffer the indignity of a slow decline because it disappeared so quickly' (Bost 2010: 50).

The art of foraging for food from publicly accessible sources in an urban context has also gained popularity in recent years and is linked to urban orchards in the sense that some foragers are also active in planting productive plants suitable for foraging visits (Pollan 2006). This urban gleaning is the continuation of a long tradition of gathering leftover food from fields and forests that would otherwise go unused, and in an urban context has had something of a revival. Reflecting on his own foraging experiences, Pollan (2006) situates this as both a personal gastronomic pleasure and a precursor to a better understanding of where food comes from, while Fowler (2011: 11) argues that urban foraging allows her to pick foods such as fruits that are 'far more beneficial to me and my surroundings than anything I buy out of a packet'. Nordahl (2009: 76) similarly suggests that foraging for urban food is not only better for increasing food system understanding but also offers other very real benefits including enhancing environmental responsibility by locally sourcing some foods that would otherwise require shipping in and providing extra calories for snacks and even (potentially) whole meals.

In the previously mentioned city of Adelaide, olive groves planted in the parklands ringing the central city from the nineteenth century onwards have remained a strong feature of the parkland landscape even after significant olive oil production from them ceased in the twentieth century. Post war, Italian and Greek migrants to the city began to harvest the olives, and, in recent times, a new set of middle-class urban gleaners have begun to do the same. Nordahl (2009: 76) suggests that in his area of Berkeley he has seen 'oranges, lemons, cherries, persimmon, fennel, the occasional tomato vine, as well as rosemary, thyme, sage, and other herbs – all occupying space between the sidewalk and the street' and references the Fallen Fruit foraging group in Los Angeles (Nordahl 2009: 77) that collects publicly accessible fruit and now plans 'fruit parks', that is, community orchards. Today's urban foraging practices, both individual and more organized, reflect a range of motivations: from ameliorating food poverty and hunger to exploring sustainability possibilities and reflecting gastronomic preferences of some restaurateurs and others. It may well be that such foraging can be criticized because it fits a hipster habitus, but this should not bar us from being interested in design which supports foraging opportunities, and other forms of convivial green space, and is hence considered in the next section.

WHAT ARE THE DESIGN REQUIREMENTS FOR URBAN FOOD GROWING?

Theorists from urban biodiversity, sustainable cities and urban design perspectives are increasingly tying together notions of food-centred productivity, with reshaping and retrofitting urban built form (Dunham-Jones and Williamson 2009; Tachieva 2010). Drawing on pre-urban patterns from European experience, Vall-Casas et al. (2011) argue that the rediscovery of rural grids and watercourses can form the physical design and landscape basis for remaking suburbia as more sustainable urbanism, including for urban agriculture. Notwithstanding the critiques of radical political ecology which problematizes biodiversity corridor and framework approaches within its Marxist analysis of urban landscapes, it appears that when successfully designed, urban food-growing spaces can contribute to making convivial outdoor rooms on which public green space is focused. The idea of design for 'nourishable' places (Mouzon 2008) has emerged, while constructs such as the urban bioregion seek to interconnect cities and their landscape contexts and opportunities in much more resilient ways, as explored in Chapter 9. Issues of urban food security are sharpening interest in the possibilities for the design of convivial green space. Among conceptual approaches is that of 'continuous productive urban landscapes', the 'CPULs' which seek to insert an unbroken chain of open spaces across cities, linking urban and rural space and providing opportunities for urban food growing from inner city to fringe locations (Viljoen et al. 2005: 11).

As noted in Chapter 2, Alexander et al. (1977: 792) offer notably holistic proposals for remaking city form in ways sympathetic to food growing. In *A Pattern Language* (Alexander et al. 1977: 302), it was argued that (to paraphrase) urban green space provides a kind of flip side to lively, engaged streets; forming the quiet backs, still water and cool greens that urban dwellers turn to when the pace and connection of civic life threatens to overwhelm their sense of self. Just as city dwellers need promenades to which the café is the proper adjunct, they also require quiet places where 'the mood is slow and reflective', where only natural sounds can be heard, where there are pools and streams nearby, but where walking is protected by a comforting wall at one side (Alexander et al. 1977: 302). Within this wider set of spaces, a series of patterns for urban gardening are laid out that echo in spatial terms the kitchen gardens of traditional cities and include terraces and bunds on which to plant vegetables and orchards, wild gardens, garden walls, trellised walks, greenhouses, garden seats and vegetable gardens (Alexander et al. 1977: 302). The designers are very conscious that these productive spaces can (and should) be very beautiful and robust over the long term, as in the example of the *limonaie* of Lake Garda described by D. H. Lawrence which form exquisite 'terraced labyrinths', some of which have borne fruit for 150 years (Rudofsky 1964: npr).

Farr (2008: 179), meanwhile, proposes that in terms of urban food production, 'planners and architects have the opportunity to bring back what years of irresponsible practices have taken away' and documents a number of built and unbuilt examples for well-designed space for growing. This design schema integrates food production and food access into urban and neighbourhood plans and includes both individual domestic and community scale interventions (Farr 2008: 179). These encompass rooftop and household gardens, greenhouses, community gardens, orchards, aquaculture, edible landscapes and farms, and include opportunities for reusing compostable material to make soil (Farr 2008: 180). Duany (2011: 8), meanwhile, suggests employing a holistic design and economic model of 'agrarian urbanism' in which 'society is involved with food in all its aspects: organizing, growing, processing, distributing, cooking and eating it'. In his model, 'the physical pattern of the settlement supports the workings of an intentional agrarian society' and includes saving existing farmland, cultivating land within existing cities and suburbs and allowing urban working farms. Rather than situating this as some kind of nostalgic return to traditional (often backbreaking) agricultural labour, Duany (2011: 8) argues for design and process that is a pragmatic response to present difficult urban conditions. Thus, agrarian urbanism would learn from successful placemaking examples including Garden Cities, and employ modern management practices to be 'profitable, popular and reproducible' (Duany 2011: 9).

Allotments, discussed earlier in the chapter in relation to their historic evolution, have also interested contemporary urban designers who can have useful roles in helping find ways to accommodate the 'living landscape and culture of the allotment' within previously mentioned continuous productive urban landscapes (Crouch and Wiltshire 2005: 130). As a result of European cities' history of allotment garden provision, many cities have extensive allotment garden holdings, with cities including Copenhagen, Freiburg, Paris and Vitoria-Gasteiz rich in this regard (Beatley 2012: 22). Yet, although allotments are more accessible and human-scaled than the industrialized countryside, designers have been somewhat ambivalent about them, as these apparently unkempt spaces may undercut notions of order and clean lines in urban development (Crouch and Wiltshire 2005: 127). In fact, the 'unhappy marriage' of allotments and imposed design has been 'a persistent feature of manuals and government-inspired advice since the 1930s, and even before that Garden City design included neatly arranged allotment plots, although on the traditional layout.

But the gardeners persist in softening the edges, disrupting the lines, etching their culture into the landscape' (Beatley 2004: 129).

Allotments in contemporary cities may still send 'mixed messages' to urban designers, seen by some as simply leftovers from wartime austerity, and viewed as 'semi-derelict eyesores' while in some instances attracting lengthy waiting lists of would-be holders (Barton et al. 2003: 160). In the United Kingdom, this contemporary design friction between allotments as 'chaotic but well tended urban landscapes melding rural and urban' and normative preferences for a more apparently orderly landscape reminds allotment experts of previous attempts to give these spaces a design makeover (Crouch and Wiltshire 2005: 130). Related to the Thorpe Report of the 1970s, ill-fated proposals to turn allotments into 'leisure gardens', with European-inspired design models including spiral-shaped spaces, turned out to be prohibitively expensive to impose and were resisted by allotment holders (Crouch and Wiltshire 2005: 130).

Of course, there are legitimate design and equity problems of accessibility, as early adopters can keep allotments for generations, restricting access to garden space to a privileged few and in a de facto way enclosing public land into semi-private spaces from which most are excluded (Crouch and Wiltshire 2005: 127). Despite such failures, successful allotment design offers a unique opportunity to contribute to a productive and aesthetically satisfying urban landscape. In fact, 'many organizations now see the allotments renaissance as representing an important tool for delivering sustainability to neighbourhoods', especially in their greening, health, social inclusion and food-growing opportunities for low income and elderly people (Barton et al. 2003: 160). On this basis, Barton et al. (2003: 31) suggest that food growing should be part of the urban designers' checklist of good urban form, with designers asking 'is the potential for households to grow their own food at or very close to their dwelling being maintained, even in this era of low participation in vegetable gardening?' Good design can thus be seen as critical to balancing private and public accessibility and use. Designers' skills could make allotments more open and more widely valued green space, 'through designs which encourage and enhance the gaze, stimulating the viewer to ponder the merits of buying a fork and joining in, while protecting crops and property from misadventure with softened but appropriate security, and integrate the loose-fit character of new allotments into wider design schemes' (Crouch and Wiltshire 2005: 130). In addition to the allotments themselves, a well-designed allotment area could include shelter belts, a sustainable urban drainage scheme, a wildlife garden and balancing pond, community composting, a community garden, a community orchard and pedestrian and cyclist cut-throughs (Barton et al. 2003: 160).

Finally, in this section on productive design opportunities, like allotments, the previously mentioned community orchards are a design area relevant to both urban and rural settings. In some cases they have replaced allotment areas where use has declined. In others, vacant industrial land in urban areas has been reused in a temporary way, as in the Union Street example referred to earlier in the chapter. Urban orchards have been seen to both improve the environment and the biodiversity of places including housing estates, industrial estates, hospitals and schools (Barton et al. 2003: 161). Echoing the allotments discussion above, in relation to the design of orchards, it is argued that 'the presence of orchards adds an experience that has all but vanished from cities the experience of growth, harvest, local sources of fresh food, walking down a city street, pulling an apple out of a tree and biting into it' (Alexander et al. 1977: 795).

GREEN SPACE AND FOOD CONSUMPTION

While this chapter has so far focused on green spaces in cities primarily in relation to food production, open spaces in cities have long been the sites for outdoor eating, as touched on in Chapters 4 and 7. Eating outdoors is well recognized as a feature of farm working, and has given rise to particular meal structures and settings in that context (Burnett 2003: 21; Fenton 2003: 39). Similarly, in cities, from necessity or choice, eating outside has long been a feature of both working lives and leisure activities. Burnett (2003: 26) notes, in fact, that the urban working class of nineteenth-century English towns often lacked adequate kitchens at home, worked long hours and were forced by their circumstances to buy minimal rations to eat on the street and doorsteps from both street sellers and a plethora of small shops. Food available on the street in nineteenth-century London included hot eels and pea soup, pickled whelks, fried fish, sheep's trotters, baked potatoes, ham sandwiches, pies, boiled puddings, cakes, tarts, gingerbread, muffins, crumpets, ice-cream and coffee (Burnett 2003: 26).

Traditionally, more conviviality in food's interplay with open space has been available to those further up the social hierarchy who have been able to enjoy al fresco meals in parks, pleasure gardens and tea gardens, with notable examples, as discussed in relation to 'Les Wauxhalls' in Chapter 4. In London, for example, the fashionable Ranelagh Gardens offered 'lawns and walks, flowers and shrub berries, fountains and statuary, dancing platforms and stages for musical and theatrical performances' and such riverbank-based or semi-suburban green spaces provided more affluent urban dwellers with airy locations to which they could bring their own food and drink or buy refreshments which were served in booths (Burnett 2003: 26). Others spent evenings at Hampstead Heath, although its current reputation would suggest today that sex rather than food in the open air is the main attraction for some. Picnics were also a feature of French social life outdoors for all social classes, and while generally held in a rustic or rural setting, meals out of doors became a feature of aristocratic town life and revolutionary street banquets (Csergo 2003: 145). At an aristocratic level, the *fête champêtre* was a common 'outdoor' dining form in the reigns of Elizabeth I and Louis XIV, for which often extremely ornate pavilions were constructed in the open air to house banquets. An example of such a fête in the garden of the Earl of Derby at The Oaks, in June 1774, is included in *The Works in Architecture of Robert and James Adam* (Vol.111, Plate 22) and can be seen at the Sir John Soane Museum. Here diners are shown at an enormous table in the grandest of (temporary) ballrooms. By Victorian times, large-scale picnics such as that detailed by Mrs Beeton (1861) were mammoth affairs requiring substantial planning.

In continental Europe, a long tradition of beer gardens and outdoor café threads through urban life and in contemporary practice of the use of urban street space for food, dining outdoors is sometimes situated by place users as denoting a 'European' sensibility (Parham 2012). Fieldwork in London suggests that sitting outside to eat and drink is both highly valued as a food practice (Parham 2012) and for designers and place marketers widely understood as a marker of urban vibrancy. Work from the United States suggests that urban parks are key sites for some Hispanic communities to socialize around food (Rishbeth 2001). While tailgating parties outside football stadiums have become ubiquitous. Yet within the context of arguments about development density and place design, café culture and outdoor dining have sometimes been argued to be citified imports redolent of the urban *flâneur* rather than reflecting more home-grown suburban mores centred on the private realm of home and family.

Despite such cultural anxieties about urbanity, outdoor restaurant dining is ubiquitous across cities, and sometimes ingenious space capturing, heating and cooling arrangements are employed by restaurateurs to make this possible in what would otherwise seem unpromising urban conditions. Such dining is at once a prime example of commercial conviviality and an area of spatial contestation reflecting changing attitudes, including in relation to tobacco smoking. Research from South Australia (Miller et al. 2002), where bans on smoking in both indoor and outdoor food settings have been in place for some time, suggest that this has increased the enjoyment of such spaces for a substantially higher percentage of diners than prior to the ban coming in. More recent research from another Australian state also found majority support for smoke-free outdoor dining areas (Walsh et al. 2008). Arguments over smoking in food space suggest changing, and more informed, perspectives on what constitutes convivial green space.

CONVIVIAL GREEN SPACE AS PUBLIC POLICY

Design and planning for urban food growing is increasingly recognized as a public infrastructure and policy matter that fits within both wider urban sustainability initiatives and those more recently focused on food resilience and conviviality (Neuner et al. 2011). It was reported as early as the beginning of the 1990s that aspects such as water infrastructure which would support urban food growing were increasingly subject to both guidance and practical retrofitting interventions. 'Governments are looking at more environmentally conscious approaches to waste water and sewerage – ponding and woodlotting, retaining grey and stormwater on site to water gardens is quite likely to become ubiquitous. There are plans afoot to dig up the concrete around our creeks and create corridors for wildlife out of ugly drains' (Parham 1993).

In more recent days, existing food-growing spaces and further opportunities for them have started to be reflected in strategies, plans, policies and design guides from national, regional, city and local governments, as well as design experts (Cooper-Marcus and Francis 1997). The planned poly-nucleated city of Almere in the Netherlands, for example, has been argued to be based on principles of sustainable development. Its designers have proposed an entirely new growing zone adjoining the town, with the intention of the Agromere proposal to reintegrate food growing into Dutch city life (Jansma and Visser 2011). There are also a considerable range of food-growing policy initiatives in the United States, including the City of Seattle Municipal Plan of 2005, which 'recommends an increase in number of gardens city-wide as well as a target of one community garden for each 2,500 households located within designated villages throughout the city' (Neuner et al. 2011). The Central Bedfordshire Design Guide (2010: 55) from the United Kingdom proposes allotments as part of a town's 'green infrastructure' and sees these as an important potential element of any public realm design. The London Food Strategy (2006: 72) notes a need to 'Expand individual & community growing in response to demand (e.g. allotments, community gardens, parks & open spaces, school grounds, etc.)' while in Melbourne, Australia, there has been an attempt to develop a policy for Food Sensitive Planning and Urban Design (FSPUD) that encompasses a range of urban food-growing opportunities within a sustainable city paradigm (Donovan et al. 2011). At an 'in principle' level, Nasr and Komisar (2012) also propose ways to reintegrate food and agriculture into urban planning and design practice.

The benefits of explicitly directed policy and projects do not just accrue in first-world settings. The Food and Nutrition Bulletin (1985) shows how a 'garden city in the slums' in Lima is ameliorating food poverty, while Marsh (1998: 5) notes how, in very poor areas in developing countries, urban food gardens can have significant benefits in relation to supporting food security, as well as reflecting available resources, food preferences and cultural traditions. While opponents tend to argue instead for vitamin supplements and targeted subsidies, often their analysis of cost–benefit is extremely narrow, tending to focus on one element such as vitamin deficiency. A wider view finds that even in very marginal situations, homestead gardening and other more community-based urban food-growing opportunities contribute strongly to well-being (Marsh 1998: 5). Once again, design sensitivity is critical, as there have been situations in which, where government is involved,

> 'improved' gardens are planned and developed for which the effort and costs for the household often outweigh the benefits, leading to eventual abandonment of the gardens after the project subsidies terminate. Were the improved gardens to build on the characteristics and objectives of traditional gardens in the region, many resource constraint problems could be anticipated and avoided. (Marsh 1998: 5)

CONVIVIAL GREEN SPACE IN REVIEW

Growing food in cities is an ancient practice, as is eating food outdoors. The close interplay between outside living space, and food sources and practices, traditionally reflected resource constraints and led to growing and eating spaces and approaches that varied by place and time but always acknowledged that local supply of food mattered. Food growing and eating were not just borne of necessity, though, but in many cases sources of pleasure and enhancement of the convivial green space of the town. In Western cities, by the mid-to-late twentieth century, however, these realities no longer seemed to apply on the production side while spaces for outdoor food consumption became more prominent. Growing food close to living space no longer appeared necessary for health, sustainability or economic necessity; thus growing spaces in cities dwindled and interest in food production likewise declined. Exterior food spaces at the same time were increasingly seen as markers of urban vitality, although were contested in terms of civilized behaviour, including in relation to damaging practices like smoking.

By contrast, in the developing world, the absolute requirement for local food sources never went away, and the role of urban agriculture in supporting resilience, mitigating food poverty and dealing with climate change has been increasingly recognized despite technocratic arguments associated with notions of modernity that discount such place-based necessities. In the West, along with a rump of those who had continued to grow food in urban areas through allotments, community gardens and urban farms, the radical social movements of the 1960s and 1970s ushered in new perspectives on food and revived not only community food-growing spaces and practices but also awareness of the design basis and requirements for these to succeed; a cultural change that has again been revived and extended in the early twenty-first century. As an understanding of the fit between urban agriculture, sustainability and food resilience has grown, the burgeoning urban agriculture movement has taken off. In cities currently, there is a rich array of urban food-growing and foraging activities which have reacquainted urban dwellers with the vital design and practice interconnections between cities as places for living and convivial green space for food. Similarly, on the consumption side, outdoor dining spaces and opportunities offer a diverse set of designed settings in the public domain that support convivial urbanism. In this chapter, a fundamental conclusion is that at the scale of the convivial green, such urbanism offers substantial design opportunities for interweaving food production and consumption space. In the next chapter, we will expand to the scale of the urban periphery where contestation between living and food growing is perhaps at its sharpest.

PART THREE

Food Space and Urbanism on the Edge

CHAPTER SEVEN

The Productive Periphery

INTRODUCTION

Just as the last chapter explored food's design interplay with green space in cities, this chapter moves the physical scale of the analysis outwards by considering the urbanism of the urban edge in food terms. The chapter looks at the way in which some cities and towns have maintained and strengthened the gastronomic landscape of their urban peripheries, and in so doing contemplates the complex, interrelated elements that support positive peri-urban food design and planning. It contrasts success with other less desirable experience of edge food space, investigating whether an aspect of declining conviviality and sustainability is an urban failure to achieve a close-knit physical, social and economic relationship to the surrounding productive land (Hough 1984). The chapter draws on a range of theoretical sources and reflects on fieldwork undertaken in Italy and France on productive peri-urban areas that contribute to a rich gastronomic landscape.

The nature of the urban edge has long interested scholars, but in the 1970s, the food writer Jane Grigson (1978: 14) drew gastronomic attention to a process occurring in many places in which the most valuable, productive and beautiful rural land around cities was being replaced by urban development. As Grigson (1978: 14) observed, 'In my more optimistic moments, I see every town ringed again with small gardens, nurseries, allotments, greenhouses, orchards, as it was in the past, an assertion of delight and human scale'. Evidence from a number of regions and city fringes suggests that a process of alienation is accelerating: destroying or fragmenting the operation of small market gardens, orchards and viticultural areas, by making peri-urban land more desirable for both formal and informal settlements of housing, as well as large-scale retailing, distribution and customer fulfilment centres, than for food and wine production (Parham 1990, 1993b; Deelstra and Girardet 2000; Aguilar et al. 2003; Couch et al. 2007; Leontidou et al. 2007; Huang et al. 2009).

The interaction between urban development and food on the periphery is important because the nature of its agricultural production, food distribution, retailing, consumption and waste arrangements over the long term represent critical gastronomic resources for cities and citizens (Parham 1992, 1993b). Yet transformations of peri-urban space do not reflect a straightforward causative relationship between urban expansion and the decline of food space. Urban edge food resilience is the result of a complex interplay between critical shifts in the nature of urban expansion and also of changes that are internal to the evolution of productive landscapes and consumption spaces. While much peri-urban food practice can clearly be seen to operate within the modern food system, as discussed in the Introduction to this book, there are urban edge focused food-policy makers and producers, retailers, restaurateurs and consumers who attempt to maintain more place-based food strategies and practices. This chapter explores the design and urbanism issues this struggle raises.

HISTORIC FOOD PRACTICE AT THE URBAN EDGE

The symbiotic relationship between the city and its surrounding countryside has historically been a critical one in shaping urban growth and development. Poverty has driven the rural poor to cities or towns, while wealth creation in towns has allowed urban dwellers to purchase country houses and land. Rural wealth has acted as the foundation for the acquisition and expression of urban power. The spatial relationships created by this interplay have given rise to an extraordinarily diverse range of landscape circumstances at the urban edge, but a near constant has been the presence of food growing and other food-related land uses. 'City' and 'wall' are interchangeable terms in some languages, with the circumspection of the urban edge a principal characteristic of city form (Kostof 1992: 11). While city walls themselves often became unnecessary defensive mechanisms, food has always been integral to land use at, and just beyond, such city edges. Thus, traditional rites to mark city boundaries have often been associated with ploughing and crops, while the presence of pastureland around the city is equally ancient. Suburban areas that combined houses, farms and cattle stalls, surrounded by fields and gardens, were

FIGURE 7.1: Peripheral food-growing space, Transylvanian village.
Photo: Susan Parham.

found outside the walls of Middle Eastern towns as early as the third millennium BC (Kostof 1992: 47). This extramural zone has historically attracted food-related uses that need cheap space: for example, Roman cities demonstrated 'attenuated roadside suburbs' with this apparent in places as far afield as London and Timgad in Algeria (Kostof 1992: 48). Food goods being brought into the city from this kind of peri-urban zone would pass a customs boundary where tolls were extracted, and these areas developed their own landscapes of food magazines, wine depots and animal pens (Kostof 1992: 13).

In the much more recent past, English common lands around cities similarly acted to both limit a city's growth and to ensure that the urban centre stayed within a workable distance of the countryside (Kostof 1992: 55). The English green belts associated with nineteenth-century urbanization were an example of this spatial interplay: both an urgent response to the sprawl made possible by the advent of mass transportation and an opportunity for agriculturally productive space to ring proposed new towns. For Ebenezer Howard (1902: 9), whose Garden City ideas about food and peri-urban productivity were discussed in Chapter 5, the green belt around the City was for farms, market gardens, allotments and animal pasturing. As noted there, local produce would be sold to the townspeople, and farm rents provide revenue to the City, while the town's food refuse would be used as compost for soil, thus maintaining a productive spatial, environmental and economic relationship with the town it served.

Traditional forms of urban edge food cultivation have tended to continue in a relatively unbroken line in places where spatial relationships between city and surrounding countryside have been less disrupted by urbanization. This was, for example, traditionally the situation in Vietnam, where the major urban markets of Hanoi and Ho Chi Minh City have been served by peri-urban gardens supplying a substantial proportion of their vegetables (Marsh 1998: 9). On the fringes of Chinese cities this intensive food production has been similarly long term. 'In Shanghai, for example, about 3.6 million farmers supply 40 per cent of wheat, 90 per cent of eggs, and 100 per cent of milk needed by the city. ... In Beijing, between 1970 and 1990, about 70 per cent of the nonstaple food in the city-region was produced in open fields and plastic-covered greenhouses in the suburbs' (Laquian 2005: 317). In Hong Kong, meanwhile, traditional fish farming on the marshes abutting the city not only played a role in the environmental protection of wetland habitats but also formed part of a gastronomic patrimony now in decline (Cheung 2007: 37). As is further explored in Chapter 8, with urbanization in Asia (and elsewhere) now being transformed by the development of megacities, there have been attendant shifts away from peri-urban, family-run smallholdings and towards vertically integrated larger-scale food businesses (Laquian 2005).

Sometimes historic growing practices have been maintained to some degree in contemporary edge space (see Figure 7.1). Farming in Rosario, Argentina, encompasses peri-urban *quintals*: family-run vegetable farms that produce food for the families themselves, as well as for the adjoining urban market (Prospersi 2009). This mainly horticultural family farming is both a substantial and historic sector in the Rosario region, with considerable influence on settlement patterns, and a 'significant number of people, including farm workers and their families, live on the quintals' (Prospersi 2009: 167). These urban fringe farming activities can also be practices tied to migration as was briefly touched on in Chapter 6. Recent work on peri-urban allotments in London and Lisbon, located on vacant 'leftover' land and in informal settlements, suggests that traditional gardening practices of migrant communities act as both methods for increasing biodiversity on the fringe and are experienced as emancipatory by their users (Cabannes and Raposo 2013).

Certain edge-of-town food-growing forms have remained despite urban change around them. The green zones around French towns can be situated spatially and culturally somewhere between the big city allotment (discussed in the last chapter) and the rural family's home garden (Jones 1997: 65). Around the town of Cajarc in the Lot, for instance, these productive food-growing zones are physically separate from individual plots in home gardens, but are owned by those who work them and are noted as places of extreme productivity. 'The smallest market communities in the most isolated provinces still display such strips and stripes, usually sited along the river or stream on which the town was built. Such gardens provide a kind of missing link between country and city kitchen gardens, largely unrecognized by garden historians much taken today with urban allotments' (Jones 1997: 65).

The *maraîchers* of Picardy, meanwhile, offer another French example of traditional if unusual peri-urban food production, although one that is no longer practiced at scale (Larkcom 1996). Not unlike the swampy *chinampas* plots of Teotihuacan civilization in Mexico (Mougeot 1994: 2), here intensive 'floating' vegetable gardens known as *hortillonages* take up over 300 hectares of garden plots on the eastern edge of Amiens, in a space that is broadly the shape of a triangle and covered by a network of canals (Larkcom 1996). A landscape originally produced by humans cutting peat left deep gorges in the Somme and Avre rivers. These became bogs, and, eventually, a web of canals enclosing land which proved ideal for cultivation, as well as providing a kind of safety valve that filters water and stops floods in Amiens (Jones 1997). As Jones (1997: 90) reports,

> Vegetables were already grown here in the time of Julius Caesar to provide food for Roman soldiers, who called the gardeners hortulani (a derivative of Horus meaning 'garden'). Today's market gardeners still go by the name of hortillons (Jones 1997: 86). ... Until the advent of refrigerated transport in the 1920s, market gardeners regularly packed their barges with fresh vegetables for the water market of the Oxtail canal next to the cathedral.

Today the *hortillons* require constant upkeep, and by 1997 fewer than ten market gardeners remained, so that weekend vegetable gardeners and volunteers have been needed to keep the area functioning as a productive landscape. Yet, paradoxically, while the expansion of Amiens eroded the gardens, the *hortillons'* area is now understood much more widely as of unique gastronomic and biodiversity value (Jones 1997: 86). The vegetable gardens were proposed for inclusion as a UNESCO site of significance and have since been listed as part of the national inventory of French sites (with world listing expected).

EDGE SPACE FOR PLEASURE

It is worth noting too, that edge space is traditionally shaped for pleasure as well as production. Again from France, one example is the historic form of the *guinguette* which became particularly important in the eighteenth century as a kind of tavern on the outer edges of the Parisian suburbs, serving alcohol and food and allowing outdoor space for singing and dancing to its largely urban working class and artisan clientele (Brennan 1984, 2005). Crucially, these convivial spaces lay beyond the urban tax barrier and could thus serve much cheaper alcohol than their urban counterparts. As the Marquis de Mirabeau commented, 'The entire population of Paris leaves the city on holidays [to go to guinguettes]. ... Half of the people come back drunk, gorged with adulturated wine, paralyzed for three days ...' (Brennan 1984: 154). However the chance to get drunk cheaply was not their only attraction. The *guinguettes*' location and use as a kind of modern prefiguration of the Sunday leisure time visit to the countryside made them the convivial goal of elaborate excursions where 'Men and women spent a day or an afternoon and evening promenading through the faubourgs and beyond, stopping at various guinguettes and walking on. The drinking and walking were conflated, in many people's testimony, into one recreation, an activity that was clearly distinct from work and the work day' (Brennan 1984: 159).

The pleasure gardens of suburban London that grew out of suburban fairs were a somewhat more polite peri-urban spatial form than the *guinguette*, but shared qualities including the chance to escape unpleasantly humid urban summer conditions. They offered a kind of inclusive conviviality that was notably welcoming to different ranks and to women (Conlin 2008: 25). 'In a society defined by rank, pleasure gardens were exciting, if occasionally unsettling, places to be' (Conlin 2008: 25). London's Vauxhall pleasure garden was such a success, in fact, that it spawned a number of European imitators in the years between 1764 and the French Revolution, widely referred to as 'les Wauxhalls' (Conlin 2008: 25). 'Promoters presented Wauxhalls as a means of encouraging those who would never think of entering a guinguette to experiment with a new type of entertainment space in which middling and noble ranks could encounter each other in pleasant surroundings. They would, it was hoped, also serve to nudge an overly stratified French society in an "egalitarian" or "patriotic" direction' (Conlin 2008: 25). The guinguette itself remained a strong cultural symbol of food- and drink-based conviviality, if not in fact such a common spatial form in twentieth-century France. Duvivier's 1936 film *La Belle Equipe* made at the time of the Popular Front tells the story of a group of friends who develop a *guinguette*; with the *guinguette* idealized as an Edenic pastoral land use form which expresses working-class solidarity and allows its occupants to escape urban capitalist exploitation (Ousselin 2006: 957):

> In what becomes an onscreen social experiment that mirrors Frontist ideals, the five friends form a workers' cooperative, buying a dilapidated house in what was then the Parisian countryside. They use their skills to renovate the house and its surroundings into a popular seasonal restaurant, or guinguette, hoping to provide a convivial gathering place for their former neighbors and extended community.

EXPLORING THE CONTEMPORARY URBAN EDGE THEORETICALLY IN FOOD TERMS

The production, processing, exchange and consumption of food on the urban edge are not just of historic interest. This is still a critical food space, albeit one that is increasingly hard to capture theoretically given the complex interweaving of town and country (Buxton and Choy 2007). There is a need to understand this space not just as a landscape interplay or hybrid space between country and city but as a distinctive place in its own right that is also contested ground (Boume et al. 2003; Simon et al. 2006; Qviström 2007). Although tending to be approached in rather functionalist terms, the relationship between the aesthetics and physical structure of the fringe are also being reconsidered and attempts made to document and analyse the mixture of land uses found there in ways that rethink their ecology (Tjallingii 2000; Gallent and Andersson 2007). There is equally a need to distinguish between the rural-urban and the urban-rural frontier, avoiding any over-simplistic analysis of sprawl that will 'reify and obscure, rather than illuminate, the complexity of economic and sociospatial forces shaping the edge' (Audirac 1999: 7). Drawing on his work in Alnwick, in the United Kingdom, the great morphologist M. R. G. Conzen (1960: 58) described the peri-urban as 'a belt-like distribution of land use units which for one reason or another seek peripheral locations ... it presents a distinctive group, including certain industries, institutions, community services, small houses, and further out isolated larger houses and open spaces. These are all typographically associated with town fringes'. Food's role in this complex space has also been made more complicated by the place-specific transformations in the nature of food production on urban edges over time, as in the area around Mexico City over the nineteenth and twentieth centuries as traced by Losada et al. (1998).

In the latter part of the twentieth century, the changing spatial, social and economic nature of such city edge urbanization meant new ways of trying to conceptualize this space, some of which directly referenced the landscape as food space: 'edge city, megalopolis, technoburbs, flexspace, peperoni-pizza cities, a city of realms, superburbia, disurb, perimeter cities, outer cities, technopolis, heteropolis, exopolis, and perimetropolitan bow waves' (Oatley 1997, in Adell 1999: 6). Theorists of urban structure in the United States, for instance, noted that a kind of 'exurbia' was developing between the city and its rural hinterlands, which they depicted as a new form of urban development outside traditional city boundaries but within commuting range (Nelson 1992; Davis et al. 1994; Carruthers and Vias 2005). This is very much a contested ground in land-use terms, including in relation to food (Nelson 1999). Houston (2005, 2009), meanwhile, draws on Mckenzie's 1996 definition of this area as 'generally understood to comprise the zone of transition between the edge of the newest suburbs and the outer limits of the commuter belt'. Buxton and Choy (2007) likewise give prominence to the heterogeneous nature of the peri-urban, its spatial dynamism and its 'jumble' of different kinds of land uses blurred into an unstable relation with one another (Audirac 1999: 13; Lapping and Furuseth 1999). This is a problem that is seen to be especially serious, given its sheer scale, around megacities like Beijing (Zhao 2010).

Various theorists have attempted to define food's role in contemporary peri-urban spatiality within this zone, including by functional typology, in terms of food security and in relation to connections between urban and rural food systems, such as work on alternative food networks and urban foodsheds (Cofie et al. 2003; Paül and McKenzie 2013). Building on Hedden's (1929) notion of the foodshed, developed largely in relation to examining New York's food supply, Getz (1991) has argued for an urban foodshed responding conceptually to transformations in productive space wrought by suburbanization. Defined as 'the geographic area from which a population derives its food supply' (Peters et al. 2009: 2), the urban foodshed can act as both a conceptual and methodological unit of analysis for understanding not only the way that food growing around an urban area is spatially organized but also how it can be better aligned to the needs for food resilience and conviviality (Kloppenburg et al. 1996: 33; Peters et al. 2005). 'In the foodshed, efforts would be made to increase the level of local and intra-regional food production, processing, and distribution and so to retain economic value and jobs' (Kloppenburg et al. 1996: 38). The foodshed approach to understanding peri-urban food space is very much about place and is grounded in an attempt to bound and respatialize the food relationships between country and city (Feagan 2007: 5). Examples, such as around Chinese cities, demonstrate how municipalities are spatially 'over sized' to allow for a local foodshed which enables them to be nearly sufficient in perishable crops (Mougeot 1994: 4). The foodshed can be further broken down to consider the role of the peri-urban in flows for particular food items, as work on the 'milkshed' around Addis Ababa demonstrates (Tegegne et al. 2006: 461). The foodshed equally has strong conceptual connections to protected designations of origin, patrimony and of 'terroir' which are dealt with in depth in Chapter 9.

A more design-based approach is that of the transect, in which peri-urban areas are conceptualized as part of a complex spatial design configuration of conditions that range from city to country, urban and semi-urban, through semi-rural to rural and suggest particular forms of urbanity with intensity generally decreasing with distance from the city centre (Duany 2002; Talen 2002; Dunham-Jones 2009: 37). Duany (2002: 253) provides a short history of the transect, defining it at as 'a natural law', that is, 'a principle derived from the observation of nature by right reason and thus ethically binding in human society', which is discernible from ancient settlement patterns onwards. From its antecedents in Patrick Geddes work, through Ian McHarg's (1969) *Design with Nature* to Alexander et al. (1977) in *A Pattern Language*, the transect has now emerged in a refined conceptual form by way of New Urbanist theory and practice (Alexander et al. 1977: 254). The transect

> has heretofore been understood as an ordering system deploying a geographic gradient to arrange the sequence of natural habitats. This conception proved to be extensible to the human habitat, as every component of urbanism also finds a place within a continuous rural-to-urban gradient. ... Beyond being a system of classification, the transect has the potential to become an instrument of design. The correlation of the various specialized components by a common

rural-to-urban continuum provides the basis for a new system of zoning, one that creates complex, contextually resonant natural and human environments. (Duany 2002: 255)

In food terms the transect as an organizing principle offers two rural zones (reserves and preserves) and a number of urban ones all of which offer space for various food elements suitable to their place in the continuum. For regional and peri-urban food space, the transect has particular integrative capacities that cut across different scales. Thus, the 'transect is a fractal that allows design to integrate across scales: from regional tiers, to community codes, to architectural standards. At the largest scale, that of the region, the transect geographically allocates urban-to-rural tiers to accept varying degrees of environmental protection and development' (Duany 2002: 258).

Others, meanwhile, have situated peri-urban space as neither rural or urban but rather suburban; as in Stilgoe's (1988: 9) burgeoning 'borderlands' of the United States, evolving as a zone between rural and urban, with food uses such as market gardening interspersed with residential development (Stilgoe 1988: 74–5). Bunker (2002: 61) points out that the rural-urban fringe around Australian cities has a 'complex and dynamic character' typified by strong outward urban growth, with its inner edge seen as 'the expanding suburban edge'. In food terms, the outer reaches of this edge space constitute 'a periphery further removed, where urban influences are apparent in rural production and pursuits, and where many people reside in towns, small settlements or areas of country living but are closely tied to the parent metropolis one way or another' (Bunker 2002: 61). What is clear is that there is often no longer a firm spatial division on the urban periphery that sharply separates farming from urban dwellers. In Jutland, Denmark, for example, five different kinds of urbanization have been identified in this confused area, with complex outcomes that are both spatial and socio-economic (Madsen et al. 2010).

It is clear that food's role in the spatiality of peri-urban areas can vary enormously, and there has been an attempt to capture these complex flows and interconnections within work on the 'peri-urban interface' (Adell 1999; Allen 2003). As foreshadowed in the chapter's introduction, edge-of-town locations around Western cities have often compromised a predominantly food-focused landscape in the twentieth century, as part of modernism's spatial project, and some urban hinterlands have acquired complex land use mixes in which food is just one of many elements. Theorists exploring peri-urban space in central Italy, for instance, have developed a typology to make sense of the intricate and unique ways that mixtures of urban, industrial and rural landscapes have grown up around Tuscan cities and towns: these are peri-urban spaces in which agriculture is found in both residential and industrialized countryside (Leonardi 1994).

Around cities in developing countries, too, peri-urban areas are now often suffering strains induced by massive urbanization, while retaining a critical role in food security. On the peri-urban edge of Hubli-Dharwad in south-west India, for example, increasing connectivity to its two urban centres, rapid industrialization, speculative urban development of its landscape and movement of population towards urban areas are all drawing people away from traditional farming pursuits and reshaping the food-growing mix in ways that penalize poor farmers (Brook and Dávila 2000). In Central Africa, meanwhile, city hinterlands are the 'peripheries of the peripheries', places where food-related survival strategies for dealing with post-colonial crises are practiced by the urban poor in the interstices between the urban and the rural (Trefon 2009: 15). Similarly, in areas around sub-Saharan cities, research findings emphasize the important role played by peri-urban agricultural activity in avoiding hunger at the household scale (Cofie et al. 2003).

CONTEMPORARY PRACTICE IN URBAN EDGE FOOD SPACE

Given its centrality to how peri-urban space functions, the interconnectivity between edge food space and urban systems and pursuits is a theme requiring some further interrogation (Figure 7.2). It is important to note as a context for this examination that food growing has not disappeared from the peri-urban zone even around Western cities, although rurality is being reconfigured and reconstituted (Murdoch and Marsden 1994). Australia's peri-urban fringes similarly show these areas account for up to a quarter of all agricultural output in dollar terms, with Sydney's market gardens contributing around 40 per cent of food consumed locally (and this figure increases to 90 per cent for mushrooms, lettuce and bok choy), while peri-urban rural council areas around Melbourne make up over 10 per cent of total farm business turnover in the state of Victoria. However, broadly, in Western countries, trends to urban commuting from the urban periphery, the location of business on the edge and the settlement of retirees and others in the rural hinterland have changed the framework of peri-urban settlement and the map of land uses around cities. A considerable grey area of land uses has grown up of semi-urban–semi-rural development, including small-scale hobby farms run by those deriving income from primarily urban sources. In this peri-urban patchwork a range of competing interests are at work, leaving food space vulnerable. Brueckner (2000), for example, notes that the costs of commuting are not fully reflected in this settlement pattern and as Bourne et al. (2003: 253) say in relation to Toronto's experience of this transformation,

> The peri-urban zone is also contested ground. It is the interface, the transitional setting, in which processes of urban growth and development intersect with the pressures for rural preservation. It is, for example, the location at which the varied demands of urban dwellers for new housing and living space, of builders, investors and property-owners for land and speculative profits, of employers for more efficient production space, and of almost everyone for accessibility to recreation and breathing space, come into conflict with the desires that rural residents place on that landscape.

Farming viability has traditionally been measured by the size of operation, but attempts to define farming space have become increasingly inadequate to the complex ways that space is now used in peri-urban areas, income generated on the urban periphery

FIGURE 7.2: Edge-of-town vegetable growing, Lincolnshire.
Photo: Matthew Hardy.

and its effects on the landscape properly costed. On the urban edge, as elsewhere in the farming sector, the cost-price squeeze, in which the costs of farming have risen sharply but prices have only moderately increased, has resulted in increasing productivity linked to primary industry restructuring and diversification (Bunker and Houston 1992; Deslauriers et al. 1992; Napton 1992). In particular, the size of farms has grown, and the rural workforce has shrunk, while part-time and hobby farmers have entered the industry in considerable numbers (Bunker and Houston 1992; Deslauriers et al. 1992; Napton 1992). Interesting research results from the peri-urban area around Copenhagen suggest that traditional landscapes have proved resilient, remaining despite socio-economic changes including the loss of farmers and increasing non-farm activities, although rapid land use change is expected in the future (Busck et al. 2006).

Positively for gastronomy on the urban periphery, in some cases, stringent economic conditions are responsible for increasing farm product diversification as farmers have sought to find new products and new destinations to augment or replace declining prices for traditional commodities on domestic and export markets. In certain regions of Europe, the Americas, Australia, New Zealand and South Africa, wine for the export market is one of the most obvious examples of this diversification and niche marketing activity in recent times, while other products found in various of these places include non-traditional fruits, vegetables and livestock, flowers, seeds, nuts, herbs and spices, native flora and fauna products, aquaculture and organically grown crops. Many of these products are best suited not to broad-acre farming but to horticulture and intensive agriculture to be found close to major cities (Leonardi 1994; Parham 1995). Urban edge food areas may thus be best situated to develop new products and markets as they possess unique land resources and climatic conditions that give them considerable flexibility for the development of a wide range of traditional and emerging rural products. Seen in combination with their locational advantages and the pre-existence of an agricultural industry and infrastructure, some of these areas take on a strategic significance in the pursuit of the new market directions that are emerging. (Bunker and Houston 1992)

Peri-urban agriculture in fact demonstrates an exceptional, and possibly undercounted, contribution to national economic health (Houston 2005). Thus, just as cities have proved to be incubators for innovation in the global economy (Parham and Konvitz 1996), city fringes can be the location for food's cutting edge. Ironically, this may be associated with the way that an increasing number of farms on the urban periphery in both Western and developing peri-urban areas may derive a substantial part of their income from sources away from direct food production (Tacoli 1998: 158; Lanjouw and Lanjouw 2001; Bah et al. 2003). These may not operate as traditional farms in an economic sense at all, even if they conform to spatial definitions intended to reflect economic viability. Farms studied in places as diverse as those around Florence, Montpellier and Adelaide are the sites for multiple functions and activities, combining, for example, bed and breakfast, family retreat and some traditional farming activities (Parham 1995). Such findings are part of so-called pluriactivity affecting farming practice and can be situated as an element of a broad 'post-productivist transition' towards non-farm diversification but based to a considerable extent on food services (Reardon et al. 2007: 3). Such changes have been noted in specific places, such as around Copenhagen and more broadly across European hinterlands (Præstholm and Kristensen 2007: 13; Briquel and Collicard 2005). Of course, the complex interplay between rural and urban activities has been a long-term feature of developing-world cities. Evidence from Central African experience, for example, shows members of urban households shuttling back and forth to peri-urban areas to engage in both subsistence and commercial food-related activities including producing food, raising livestock and collecting wood for cooking fuel (Trefon 2009: 17).

WHY URBAN EDGE FOOD SPACE MATTERS: SUSTAINABILITY ARGUMENTS

The countryside around cities has continued to be seen as possessing desirable qualities for the good life, with many wanting to work in the city but to live in pastoral surroundings, and the scale of the sustainability issues this raises is rapidly growing. Both these incursions and the 'dislocation' of the city from natural systems is an issue in the peri-urban region and wider regional context. With huge middle classes emerging in both India and China, for example, 'peri-urban housing estates are occupying farmland and increasing food prices for rich and poor alike' in places including the Pearl River Delta and around Shanghai and Beijing (Douglas 2006: 18). Yet these urban hinterlands may no longer be perceived as necessary for sustaining a city's food supply (Dixon et al. 2009). There has been a 'divorce of everyday life in urban areas from biological reality'. This is allied to business, big farming and governmental preferences for functional foods, whereby increasing 'efficiencies' are tied to technologically driven rather than place-based approaches (Morgan et al. 2006). Agricultural land on city edges may remain the location for a limited supply of high value-adding foods that cannot be produced industrially. However, it is the capacity to generate income rather than a concern for feeding local populations, conserving traditional landscapes or a perceived connection to the health of the ecosystem of the city's bioregion that leads to such land use retention.

Despite the positioning of peri-urban agriculture as increasingly irrelevant to urban food supply, various sustainability arguments challenge such technocratic public policy and economic development paradigms (Ilbery et al. 1997). Although food miles debates are complex, in as early as 1995, there were forecasts that long-range transport of foodstuffs was increasingly environmentally and economically unsustainable (noted in Parham 1995), a view reinforced by more recent evidence (SDC 2008, 2011). Although recent work from Coley, Howard and Winter (2011) suggests that because of the differences in emission caused by the mode of transport being as influential as sheer distance, food miles cannot be used as a reliable indicator of sustainable consumption, many cities are now reaching an environmental crisis point in their attempts to reconcile the desire of increasingly large numbers of urbanites for living 'in the country' and being close to the city, while also maintaining a sustainable peri-urban food system. Over the later part of the twentieth century the loss of the gastronomic and spatial landscape of small-scale scattered market gardens at city edges is one obvious example of a transformation to increasingly large-scale, distant, food production and distribution that is reshaping peri-urban areas and challenging urban sustainability.

Sustainability theorists argue that the effect of urban incursion into the city's regional hinterland is a long-standing environmental problem (Hough 1984, 1990). This reflects a lack of focus on the peri-urban as a unique and complex interface in environmental planning terms (Allen 2003). Problematic effects of urban growth can be recognized in specific peri-urban regions, such as that around Mexico City where urban farmers found innovative, ecologically positive ways to deal with the environmental degradation brought about by massive urbanization (Losada et al. 1998). However, the point remains that if nature is understood as separate and different from cities then urban edges will not be recognized as integral to natural ecosystems. This in turn keys into a 'perverse energy system' in which, to paraphrase Hough (1990: 126), resources are taken from the country, through agriculture occurring at huge environmental cost, exploited for city needs and then expelled as waste into a hinterland constituting a polluted sink for urban excess. Thus, cities

> act as vast nutrient and energy sinks for the surrounding countryside. The nutrients and minerals are quite literally stripped from the hinterland and are freighted en masse to the cities or overseas for export. The nutrients, once accumulated in the urban areas, become an enormous problem in the form of sewage and other pollutants, which cannot be easily dealt with given current funds and technology.
> (Flannery 1994: 400)

The experience of China's recent massive urbanization is instructive in this regard, with peri-urban areas particularly prone to the effects of urban and industrial pollution (Chen 2007: 9). Although not limited to China's experience of rapid urbanization, this (as elsewhere) could be argued to demonstrate the food-related consequences of viewing the city and its peri-urban setting as mutually exclusive and unrelated zones, while spatially confusing their boundaries. It would appear to perpetuate an apparently cost-free urban expansion in which there is no clear nexus between the use of land and other resources and their finite nature, or between polluting practices and damaging effects (Hough 1990). Attitudes to land management in twentieth-century cities, it is argued, are tied to a horticultural aesthetic rather than an understanding of landscape ecology, so that an ethic of constant growth fails to respect the constraints and opportunities imposed by climate, topography, agricultural soils and water supply (Hough 1990). Moreover, as the natural landscape becomes more industrialized, the edges of cities may now have a unique role in promoting biodiversity that such development strategies undercut (Simmonds 1993: 101).

Notions such as the ecological footprint, ecosystem services and the urban metabolism have been developed to help conceptualize and offer applied tools to better understand and measure how far into its own region (and beyond) a city absorbs food and other resources and creates carbon and other negative outputs (Rees 1992; Wackernagel and Rees 1996; Girardet 1999). Within an ecosystem services approach, food for consumption is not only one of the provisioning services provided to the city by its hinterland and further afield, but also has a role in cultural, regulatory and supporting services. Ecological footprints are continuing to grow. 'Urban dwellers depend on the productive and assimilative capacities of ecosystems well beyond their city boundaries – "ecological footprints" tens to hundreds of times the area occupied by a city – to produce the flows of energy, material goods, and nonmaterial services (including waste absorption) that sustain human well-being and quality of life' (Grimm et al. 2008: 759). And the expansion of urban ecological footprints has particular implications for the peri-urban food interface:

> in terms of both increasing pressures on its carrying capacity and missing production opportunities, for instance when food is imported from distant regions rather than supplied from the city's hinterland ... through the expansion of the urban ecological footprint, the supportive reciprocal relations between cities and their hinterlands tend to break down, promoting unsustainable patterns of natural resource use and the transference of environmental problems to distant regions. (Allen 2003: 140)

While Allen (2003) suggests that even a global food footprint may be manageable 'if side-effects or risks are contained within acceptable limits and thresholds', it follows that the way city edges are spatially dealt with in food terms (the city's 'environmental shadow') can be critical to rethinking unsustainable approaches that undermine gastronomic health. With a productive symbiosis between cities and urban edge agriculture, it seems possible that the urban periphery could, in the longer term, provide an urbanism-informed basis for an improved gastronomic and environmental relationship with both the broader rural and urban landscape.

URBAN EDGES – CONTEMPORARY FOOD TRANSFORMATIONS

Interest in issues of the growth of cities into their surrounding agricultural regions in Western countries has waxed and waned. Prior to our current emphasis on problematizing sprawl (Rome 2001; Frumkin et al. 2004; Couch et al. 2007) and improving food resilience in the face of climate change (Aguilar et al. 2003; Richardson and Chang-Hee 2004; Paül and Tonts 2005; Torres et al. 2007; Chen 2007; Grimm et al. 2008), the last major focus on the perceived problem for agriculture posed by urban expansion was in the 1970s in response to rapid urban population increases. While economic recession in the 1980s tended to overshadow concern for the health of the city's countryside (Bryant and Johnston 1992), in the twenty-first century, the problem for food of the transforming nature of the city's countryside is again on the agenda, as pressure intensifies on edge space for urban development of various kinds, and farming itself transforms. In a developing-world context, for example, evidence from Dar es Salaam suggests that an ongoing process of urbanization leads to a continual outward movement of the peri-urban edge (Briggs and Mwamfupe 2000: 804). As noted earlier, food growing can be seen as both a survival strategy and a temporary land use while opportunities are sought to realize value through more urban development, which will attract higher land prices:

> Many bought land and moved to the edge of the urban area from the city itself, primarily in response to a need to produce their own food to survive the growing economic crisis of the early and mid 1980s. The land vendors, mainly indigenous Zaramo people, frequently then moved further out to buy land at a lower price, to farm in the immediate term, but with the longer-term hope that land prices will rise as Dar es Salaam expands spatially, and then the whole process will be repeated. The advantage of these areas for incomers was that they offered people opportunities to engage in farming, and hence secure a household food supply, whilst also engaging in non-farm jobs in the city. (Briggs and Mwamfupe 2000: 804)

Around Western cities, too, ex-urban settlement of peri-urban space has accelerated as incomes have risen and high levels of car ownership have made travel easier. For many it is now possible to live in a semi-rural environment and commute to the city to work. As Laquian (2005: 198) demonstrates, the establishment of urban nodes within peri-urban spaces in Asia has in certain places been accomplished without destroying surrounding agricultural land. 'Urban agriculture has flourished around Asian cities and there are a number of successful efforts to use the waste of a city as an input to production' (Laquian 2005: 198). Yet the bulk of this settlement is not in established rural townships or newly developed compact nodes, which would fit with transect-based design approaches described above, but on land outside developed urban space, and in the form of self-contained, single-use developments (Southworth and Owens 1993). These fragmented clusters of residential development can be very disruptive to peri-urban landscapes in rapidly expanding metropolitan regions (Buxton and Choy 2007). Generally, land holdings are considerably smaller than that required to run a viable farm, ranging from less than a hectare to ten hectares, or larger in a few cases. This ad hoc process of settlement increases land values and undermines not only the agricultural but also the landscape qualities of the environment, intensifying various forms of environmental risk, including, in the Australian context, water quality and bushfires (Bunker and Houston 1992). Such development also imposes infrastructure requirements, the costs of which must be spread across the community including the agricultural sector, thus imposing extra strain on farm viability, although as Bryant and Johnston (1992) suggest, understandably rural populations usually share ex-urban residents' desire for better services.

This process of urban transformations on the periphery is highly related to a 'presumption of primacy' for urban development and results in an effect, first identified in the United States, which has been defined as the 'impermanence syndrome' that sees farmland as simply 'suburbs in waiting' (Bunker and Holloway 2001: 13; Cook and Harder 2013). This, in turn, has been 'traced back to the simple failure of planning policy effectively to insulate rural activities from urbanization pressures, especially rising land values' (Bunker and Houston 1992: 24):

> The unspoken yet powerful assumption that urban development *will* occur, emphasises the farmer's sense of impermanence and leads to a progression of self-reinforcing changes in the way that farmers invest in, manage, and use their land. These include reversion to low input farming systems, perhaps to the extent that land becomes idle or is used just for agistment; cost-cutting on management of land, farm operations and capital items; ultimately sale of land for hobby farms or residential purposes. (Bunker and Houston 1992: 24)

Added to the problem of the presumption of primacy is the widely held belief among farmers that they possess development rights to sell their land for urban prices, either to fund farm acquisitions or as a form of superannuation (Bunker and Houston 1992: 24). The spatial implications can be profound because monetary aspirations are likely to extend across a much larger area than will ever experience conversion from rural to other land uses (Bryant and Johnston 1992). At the same time, farmers in areas of limited or marginal agricultural value, trapped on farms which have become unviable, or without a next generation wishing to farm, will be particularly attracted to realizing on their land asset in this way, often to the detriment of the overall landscape (Bunker and Houston 1992). Reviewing land price expectations in peri-urban farming areas in the United States, Plantinga et al. (2002) argue that the situation may warrant governmental intervention to avoid the loss of such agricultural land.

Apprehension about the loss of farmland occurs most acutely in areas near cities as these tend to encompass the largest proportion of high-quality agricultural land. Many New World and developing countries have a coincidence of urbanized areas and good farmland, and population is often the densest in those areas where economic return per agricultural hectare is the highest. Thus, rates of conversion of land from agricultural to urban use have often been cited as a measure of the fragility of an agricultural region. Canada, for example, converted nearly 150,000 hectares of rural land to urban use between the mid-1970s and the mid-1980s (and the majority had a high agricultural rating). In Asia, similarly, significant amounts of agricultural land have been converted to urban development.

It is clear that the cost-price squeeze described above is a major factor in alienating farmland at the edges of cities. Farmers experience pressure to expand in order to increase productivity and thus deal with the twin difficulties of rising costs and lower prices for commodities (Bunker and Houston 1992). Land values, however, tend to rise as a result of urban expansion and thus constrain commercial farming operations' ability to grow. Essentially, where farmers cannot afford to pay urban land values to produce agricultural products such products are priced out of competition. Land costs are reflected in final product costs and can be critical in determining farming viability (Bunker and Houston 1992).

While the edges of many cities are still superficially rural, as ex-urban dwellers move in, conflict over amenity issues such as crop spraying, noise and smells from agricultural production, and traffic generated by the movement of produce can emerge as an additional source of constraint on or even cause abandonment of food production. Such land use conflict appears to be widespread in a diversity of peri-urban conditions. In the northern Swiss context, von der Dunk (2011: 149) has developed a typology of different kinds of such conflicts, including noise pollution, visual blight, health hazards, nature conservation, preservation of the past and changes to the neighbourhood, which is intended to tease out the complex interplay of issues and separate out 'real' problems from the contingent. Anxiety over this threat to food production has seen 'right to farm' legislation enacted in the United States, although with some unintended consequences including being used to limit urban agriculture in US cities (Hamilton 1998; Norris et al. 2011).

RECONFIGURING FOOD'S RELATIONSHIP TO URBANISM ON THE EDGE?

From a variety of perspectives – environmental, social justice, gastronomic and economic – a convergence of view is emerging that peri-urban food planning, design and management needs reconfiguration. Farming and other food-related land uses on the edge of towns and cities require not only a supportive investment climate but also protection through metropolitan planning, design and resource management policy, if the urbanism of the edge is to remain productive. Part of this is an increasing awareness of landscape needs, which is finding its way into the spatial planning of regions. Tools such as multifunctional landscape plans might offer assistance here (Selman 2002). Already, the advent of strategic and regional area planning, and the associated research and information gathering techniques, has greatly improved the sensitivity of approaches to peripheral development in places like Canada and Australia. Techniques for strategic planning, including land capability assessment, urban design studies and environmental auditing, are all now employed by governments as essential components of regional growth planning. In a situation of competing claims and conflicting expectations, the need for constructive approaches to food space is particularly critical on the urban edge for gastronomic reasons, and also because of the connection between a high level of economic opportunities, intense pressure for urban expansion and their importance to bioregional sustainability.

A question remains whether there is a right balance to be achieved. Is it possible to ensure a productive diversity of land uses, encompassing farms, houses, business, shops and services as a sound basis for food health? Certainly, there has been well-established knowledge since the 1990s of the need for gastronomically informed production on which sustainable food product exports and job growth can be based; underscored by supranational work about a number of countries, cities and regions (Parham 1995). In Australia, for instance, a series of Federal Government studies around that time included work produced by the Taskforce on Regional Development (1994) which identified viticulture and wine making as major economic strengths of the Adelaide region and noted the rich potential of wine and food production in developing niche export markets, in terms of tourism, and in providing major scope for job generation (Parham 1995). In Italy, regional governments have similarly developed a strong focus on rural niche markets for high-quality food goods produced in large part around their cities, as a basis for their economies. This is a focus which is further explored in Chapter 9 in the discussion of the critical food region. The main point in the context of this chapter is that many rural-urban fringe areas are well placed to exploit these food strengths but continue to need to look more closely at the food dimensions of the urbanism issues facing them and the opportunities for better spatially configured food relationships. Further attention to food and refinement of policy responses is required in order to take most advantage of the special economic, cultural and landscape possibilities presented by peripheral areas.

If the urban edge is clearly a food landscape of critical importance in urbanism terms, how to best deal with it spatially has elicited a diversity of responses. Comparisons between sprawling and more 'managed' growth approaches demonstrate slower agricultural land conversion rates and reduced costs when the latter are employed (Burchell and Mukherji 2003). In certain places (most famously Portland, Oregon) urban growth boundaries have been used as regulatory devices to draw a line around existing built-up areas altogether (Nelson and Moore 1993), although these appear to have diverted rather than strictly limited ex-urban growth (Jun 2004), and their effect on surrounding rural landscapes has been under-researched (Harvey and Works 2002). Greenbelts, too, offer a form of growth boundary traditionally employed in the United Kingdom and elsewhere to protect agricultural fringes from development, and are a critical component of Garden Cities discussed in Chapters 5 and 6. Other techniques have included development charges and proposals for hypothecating the costs of growth to those who choose to settle on the fringes of cities (Clinch and O'Neill 2010). This would contrast to the more normative use of fiscal, taxation and planning policy instruments to prop up or cross-subsidize expansionary development choices.

So far, the gastronomic costs and benefits, measured in implications for conviviality and sustainability, have rarely been factored into discussion of peripheral food production and other foodscapes. At the same time, techniques developed in recent times to try to better fit together urban peripheral land and the uses to which it is put have a (largely) unacknowledged food component. Landscape audits, for example, can define the best use for land within broader criteria that properly encompass food's role (Aalders and Morrice 2012). Meanwhile, environmental costing and ecosystems' service analyses also offer opportunities to value food's role on the edge (Williams et al. 2010). Such instruments emphasize that adequate multidimensional approaches are required for food-conscious decision making about productive space at the edge of cities if its urbanism is to be fully aligned to its food possibilities.

THE URBAN EDGE AS A GASTRONOMIC TOURISM LANDSCAPE – REAL AND IMAGINED

Many people are now so alienated by the ugly cities created through industrialization and the post-industrial spatial practices of modernism that although they may be active in producing these spaces they are eager to avoid them while travelling or when envisaging an alternative future for themselves and others. They seek out places where integrated food and landscape qualities have been maintained. Desired gastronomic landscapes are very often found around the vernacular cities that reached their zenith, or at least their physical fabric setting, prior to the industrial revolution (Morris 1994). In Europe, Asia, Africa, the Middle East, the Americas and elsewhere,

FIGURE 7.3: Farmers' market, Norway.
Photo: Susan Parham.

such locations share certain food space characteristics which are perceived as expressing both beauty and authenticity: reflecting a complex, sophisticated food culture with a strong relationship to location and the seasons. In spatial terms this may be given physical expression in urban edge space through, for example, small-scale mixed farming and viticulture, traditional farm buildings and the presence of vernacular food shops, markets and restaurants that use local produce (see Figure 7.3).

While these landscapes are discussed in more depth in Chapter 9's consideration of the region in food terms, it is worth noting here that food-centred or gastronomic tourism situated as a growing subset of cultural tourism can be typified as of visitors primarily interested in a peri-urban region for its diversity of good local food and wine products and the landscapes that support them (Parham 1995, 1996; Bessière 1998; Richards 2002; Hjalager 2002; Hjalager and Richards 2004; Santich 2004; Kivela and Crotts 2006). Defined as tourism where 'an opportunity for memorable food and drink experiences contribute significantly to travel motivation and behavior' (Harrington and Ottenbacher 2010, quoting Wolf 2006), such tourism plays an increasingly important role in peri-urban areas (Boniface 2003). It forms a kind of synchronicity with edge space agriculture and other food-related land uses which challenge industrialized food approaches (Boniface 2003). Ecotourism and agritourism, often closely related to attractive viticultural and horticulture landscapes, fit well within value-adding niche marketing of local food products that stress a close and direct relationship to the place in which they are produced (Boyne et al. 2002, 2003; Frochot 2003). These attract a visitor catchment that can be wide ranging, as demonstrated in South Africa (Rand, Heath and Alberts 2003) and elsewhere (Hall 2003).

Perhaps rather pragmatically, tourism authorities have been advised to learn how to 'gastrospeak' to increase their gastronomic tourism potential and to support local foods to enhance the visitor's perception of authenticity (Fox 2007; Sims 2009). Similarly, in this respect, the hard copy and more recent rise of online food, wine and travel guides, magazines and blogs, which have augmented regional tourism brochures, offer a somewhat paradoxical backdrop. These are able, if well deployed, to provide valued information, reviews and opinion about dining, accommodation and distinctive products of particular regions, which are supportive to artisan and small-scale producers and restaurateurs (Okumus et al. 2007). Yet such materials may also be situated as reflecting elite preoccupations, and for the individual, offering tools for developing a particular habitus that demonstrates their wealth of cultural capital. Food magazines, newspaper lifestyle pages and foodie blogs might be understood as primers for those wishing to position themselves in a certain way, teaching their readers how to become more sophisticated and thus higher status consumers of dining and travel. They can certainly support to a substantial degree a display of conspicuous tastefulness (Hall and Sharples 2003) as part of individual habitus formation.

Yet how many can actually take part in these experiences? While many may desire to, say, take a cooking class with Giulliano Hazan near Verona, visit the kitchen at Mugaritz outside San Sebastian, view the courtyard herb garden of Blue Hill's Stone Barns Center for Food and Agriculture in New York's Hudson River Valley or dine at the Napa Valley's French Laundry, far fewer will have the resources to do so. Not only may peri-urban representations thus remain in the realm of aspirational food fantasy, but also the viewer can see these places and spaces depicted in apparently pristine circumstances, with any uncomfortable or unsightly features, context and details erased. It could be argued they need not engage with actual threats to fragile gastronomic resources at real urban edges and ignore real food production, distribution and consumption arrangements. Accounts of that gorgeous little *trattoria* in the hills outside Florence, or photographs of the flawless Tuscan food landscape do not reveal the ugly rural factories, autoroutes or strip shopping and sprawl housing developments that actually mar sections of the countryside at urban edges here and elsewhere. At this level such contextual artefacts might suggest a level of distortion and commodification of food experience and may be uncivil in Finkelstein's (1989) terms.

Equally, cookbooks that instruct buyers about peri-urban and regional food and provide insights into the ways of life of local producers who practise gastronomic approaches to production, emphasize the acquisition and performance of conspicuous competence rather than gaining structural insights into the nature of peri-urban edge as food space. At the same time, it seems reasonable to posit that the extraordinary popularity of such locationally grounded cookbooks and other food-related material both reflects and itself enlarges the sizeable appetite for information about food with strong connections to place.

Despite this growing virtual life of the peri-urban in the gastronomic imagination, in reality, such information sources also reflect the interests and purchasing power of increasing numbers of well-informed local, national and international visitors who are primarily motivated by the opportunities to experience peri-urban landscapes, enjoy high-quality local restaurants, taste regional wines and purchase products from wineries, mills, farm shops, nurseries, apiaries and markets. The growth of somewhat anarchic markets at edge space locations could be an indicator of an economically subversive gastronomic approach insofar as these bypass the vertically integrative economic arrangements of conglomerate food suppliers, wholesalers and retailers. Food purchased here is also likely to be fresher, cheaper and economically more supportive of small-scale growers.

The interests and tastes of such visitors also point to the increasing importance of forms of high-quality food and accommodation which are quite unlike the traditional notions of luxury tied to large-scale, all-inclusive resorts. Although luxury is evident in many peri-urban food places, this may be in the form of an adjunct to a good restaurant, an upmarket guest house or traditional farming building ensemble, an idiosyncratic bed and breakfast, or a historic cottage, but will stress excellent local food, wine, personal service and (often very) comfortable surroundings, in buildings of architectural or heritage interest. Especially in emerging gastro-tourism regions, as, for example, the Maramures area of Romania, accommodation may be extremely modest and small scale (Regoli et al. 2011). The gastronomic tourist cuts across several categories developed by tourism authorities to define the visitor population into clear market areas (and therefore predictable market niches for operators and regulatory schema for land use planners). For these reasons it will no doubt continue to be difficult to adequately quantify the environmental impact and economic importance of food tourists to the health and viability of the urban periphery but more attention should be paid to such growing travel forms as part of the approaches to edge space urbanism. The strength of some cities in pursuing an alternative vision, spatial form and cultural relationships based on food production and consumption on their urban edges is encouraging. These places show that urban expansion need not necessarily undercut food production, which reflects its location, and that travel can be organized around gastronomic health in both city and its hinterlands.

EXPLORING PERI-URBAN FOOD SPACE – AN ITALIAN CASE STUDY

Results from primary research into the food space of two urban peripheries in Italy and France demonstrate how certain regions have dealt relatively effectively with the relationship between pressures for urban expansion, high-quality peri-urban agricultural production, food tourism and the landscape and sustainability of urban edge terrain (Parham 1995). For reasons of brevity, the case study work from Tuscany is the focus here, although detailed primary research was also undertaken in Languedoc Roussillon around Montpellier, with similar findings (Parham 1995). Despite Tuscany being overworked, both as a destination and as a vision of a desirable food landscape, its experience offers useful insights. These range beyond its obvious tourist appeal as a series of discrete historic towns set in an often beautiful rural landscape of traditional small-scale artisan agricultural production. Tuscany is also a real place subject to real threats and opportunities to food space on the peri-urban edge in which an aesthetically pleasing landscape, as seen by tourists, may not be quite the same as a practical, productive one experienced and required by farmers and others (Galli et al. 2010). In a wider farming context in which dominant neoliberal regimes have defined as superfluous the very existence of upland farming, even Tuscan farmers are now experiencing problems associated with deregulation (Shucksmith and Rønningen 2011; Orsini 2011). Claudia Roden has nicely summarized aspects of the changing nature of agriculture and viticulture in Tuscany's urban peripheries from a food writer's point of view (Roden 1989: 92–8) while from a more academic perspective Leonardi's (1990) typology of land uses demonstrates a complex pattern of development emerging across the Tuscan landscape since the 1950s, including an industrialized countryside, industrialized districts, tertiary areas and more residential countryside.

Recent research findings from the peri-urban area around Pisa (Orsini 2013) replicate earlier case study research results (Parham 1995) which indicate that urban 'hobby' farmers continue to help revitalize aspects of peri-urban food production, protect desirable landscapes and support a robust gastronomic tourism economy. Such farming practices on Tuscany's peri-urban edges offer a useful example because they demonstrate how agricultural producers (new and traditional) undertaking gastronomically informed food production have been able to shape the outcomes of a visitor influx to their own advantage and to the benefit of the landscape. A region of this kind is special because it has to some degree resisted or renegotiated the forces that make food and travel experiences more likely to be manufactured and appropriated out of relationship with context. Tuscany self-evidently is a place of the most robust sense of place – spatial, gastronomic and cultural – which has used tourism to reinforce a strong export base of high-quality agricultural products. The income derived from success in these (and a number of other) niche markets supports a favoured cultural base as well as the gastronomic landscape of the region's peri-urban areas. Tuscany has also had high levels of urbanization of its countryside, demonstrating that the melding of rural and urban may be possible in ways that relatively successfully interconnect a food-centred spatiality on the urban edge.

Research findings demonstrate that new entrants to farming have brought to bear both investment capacity derived from incomes as urban professionals and highly developed gastronomic sensibilities to peri-urban farming practice (Parham 1995; Orsini 2013). These incomers tend to view productive edge upland landscapes as part of the region's cultural heritage, not simply as units of privately owned productive space to be exploited (Orsini 2013). Meanwhile, their food production and processing activities have helped revitalize the economic capacity of city peripheries as places of agricultural innovation, while maintaining or renewing natural environments and food spaces such as hillside terraces and abandoned wine caves. New farm-related activity and new non-farm land uses connected to food around Florence (Parham 1995) and Pisa (Orsini 2013) have exploited the associated benefits of agricultural production and *agritourismo* in proximity to urban space and melded rural and urban derived funding in a way that gastronomically supports the regional economy, culture and environment. These transformations have shown that urban and rural land uses need not undermine one another, but can produce a symbiotic relationship at city peripheries that plays a part in landscape revitalization and supports conviviality. For instance, Tuscan research shows a high level of concentration on producing and marketing (at home and beyond the region) of significantly value-adding rather than high-volume food produce, supported by very geographically and producer- specific labelling (Parham 1995). This approach has been to the peri-urban area's advantage in terms of attracting travellers interested in trying and buying local food products of high quality.

In the Tuscan case, strong links built between agriculture, viticulture and *agritourismo* also protect landscape quality, in turn making its peri-urban areas more attractive to travel in, as well as sites in which to invest and live. Strong links have been developed between built and cultural heritage; between food production and gastronomic tourism; and from artisan producer to traditional central city food market places (Parham 1995). There has been considerable flexibility about mixing farm and non-farm uses on rural holdings close to urban space, and rural policy makers have seen 'boutique' food producers as important to innovation and product development (Parham 1995). People live on and around existing farms near the city rather than replacing them with urban residential developments (Parham 1995). In fact, support for small-scale production, considered marginal in a mainstream farm economics sense, has been especially important because of its flow on effects in tourism, landscape protection, social cohesiveness and regional ecology. This has been recognized by regional government that appears to have a highly developed gastronomic consciousness, which shows in policy terms; those interviewed during the research process clearly saw food and travel as central to public life as well as private pleasure (Parham 1995). Tuscan experience, then, reinforces subtle interconnections between regional, environmental, social and economic health and the critical part that peri-urban agriculture can play in such food-centred, convivial urbanism.

DESIGN RESPONSES TO THE PERI-URBAN AS FOOD SPACE

Food space design has important but perhaps under-explored implications for both conviviality and sustainability on the urban edge as it deals with an essential design paradox: how to give people the access they desire to both a wild and productive countryside without continuously sprawling into that space, and thereby destroying valuable food landscapes and built forms (see Figure 7.4). Whitehand (1998) traces back to the German geographer Herbert Louis the morphological design concept of an urban fringe belt with food-related land uses, while more recently, various design solutions have been proposed to contribute to improved spatial, social and environmental outcomes on this urban edge. Murrain (1993: 85), for example, points out that designers can assist in avoiding the retreat from interactive urbanism and instead pursue design solutions for urban expansion. These would emphasize fine-grained, mixed-use town forms that are walkable and transit oriented, and through their compact approaches slow down incursions into agricultural space (Murrain 1993: 85).

FIGURE 7.4: Productive landscape design, southern France.
Photo: Susan Parham.

Designers informed by landscape ecology have been alert to the importance of design's role in connecting and shaping the urban edge in biodiversity terms with green belts, green fingers, wedges and corridors (Spirn 1984; Beatley 2000; Barton et al. 2003; Girling and Kellett 2005; Hester 2006; Pickett and Cadenasso 2008). This supports the 'biophilic city' configured for biodiversity while in certain places supporting 'food webs' and reducing ecological footprints (Beatley 2010; Ignatieva et al. 2011: 17). Cities including Helsinki and Copenhagen have instigated substantial, long-term, formal 'green fingers' plans which create a green backbone to structure urban form (Bruel 2012: 85). These offer opportunities for productive food space, and in Helsinki's case, provide linked greenways (Beatley 2012: 117). The city of Vitoria-Gasteiz in the Basque Country, meanwhile, identifies its peri-urban area as part of a highly valued ecological and productive landscape requiring protection and focuses on city-level planning and design support for a compact city within a peri-urban agricultural and natural mountain system (Bruel 2012: 157). These European green cities may offer a model of good design and planning practice, with obvious lessons in relation to 'physical layout and architectural design', especially in the light of food insecurity among other climate-change-related trends (Beatley 2012: 216). Thus,

> [h]aving compact, mixed use, transit-oriented urban scales, amenities within short distances of residences, interesting walking environments, and investments in the public realm leads to lifestyles that are better for human and environmental well being. ... European cities have also invested much in green infrastructure-parks, natural systems, urban agriculture and greenery ... and the transformative benefits are equally evident. (Beatley 2012: 216)

In historical practice, Ebenezer Howard's design for a food production zone at the urban edge of the Garden City was discussed earlier in the chapter. However, it was with the advance of modernist approaches to city shaping that these food-related elements in urban edge space development were largely lost. Frank Lloyd Wright (1935) famously suggested dispensing altogether with cities as previously configured, and instead developing a semi-rural settlement form that he called Broadacre City. This was to be composed of square miles of blocks in which houses would be set one deep around the edges, along roads providing fast 'city' access. Food was important to this vision. Fruit and nut trees would be planted here and there, eventually providing both profitable crops and 'giving character, privacy and comfort to the whole city' (Wright 1935: 348). These super block ideas influenced New Town and broader peri-urban design for urban expansion but without any agrarian component remaining. As Fishman (1998: npr) notes, 'Wright came dangerously close to being correct here, as Broadacre City type development really did fill whole regions in the United States after 1945'.

More recently, other designers have proposed a city form of urban fingers advancing into rural areas and a rural form of country fingers extending right to the city's heart (Alexander et al. 1977). Such an entwined landscape relationship also presupposes the protection of the most fertile valley land for agriculture, with hilly areas given over to buildings (Alexander et al. 1977). The continuous productive urban landscapes discussed in the previous chapter, meanwhile, show peri-urban food design elements at the outer reaches of linked greenways between these townscapes (Viljoen et al. 2005: 11). Both the growth of suburbia and, to some extent, the city planning of the colonial and post-colonial world of the Americas and Australia were a conscious rejection of this 'fingers' pattern; instead, those on the peri-urban edge sought to create a personalized countryside at the scale of each house block.

Given the ubiquity of such suburban spatiality, a strong theme in recent urban design for food on the edge is to conceive this as part of a broader urban sprawl retrofitting activity (Dunham-Jones and Williamson 2009). On the basis of the principles of urbanism, which are spatially organized in line with transect approaches, these include specific design proposals to deal with agricultural aspects (Duany 2011). Similarly, proponents of green urbanism seem well aware of the role food-centred design could play to 'shorten supply lines and reform urban metabolism', while recent proposals include the notion of eco-belts between cities (Beatley 2000; Beatley and Newman 2009; Beatley 2012: 22). Other contemporary design approaches, emerging from landscape architecture, meanwhile, include variants such as landscape urbanism (Waldheim 2006; Shane 2006; Steiner 2011) and ecological urbanism (Waldheim 2010). These might, by implication, be expected to support a gastronomically informed design approach that addresses problems sprawl causes for food production and conviviality on the urban edge. However, these approaches tend to rely on very low densities of development and, as a result, '[w]ell-intentioned landscape architects who wish to open up more habitat and breathe it into our suburbs could in turn increase the city's overall sprawl balloon and place even greater demands on both private and public transport systems' (Weller 2008: 260).

THE PRODUCTIVE PERIPHERY IN REVIEW

This chapter has looked in detail at food at the scale of the peri-urban edge, finding that the city and its hinterlands have always been strongly interconnected in food terms, both for production and pleasure. In certain places traditional food production has continued or been revived to considerable gastronomic and landscape benefit; however, the dominant trend has been towards food space decline on the edge. While capturing theoretically what exactly constitutes the productive periphery has proved difficult – spatially, economically and culturally – it does seem clear that the alienation of peri-urban food space as a gastronomic landscape became a marker of twentieth-century attitudes and practices with largely negative food effects. With a presumption of primacy for urban development, food space on the urban fringe suffered in many places, paradoxically at the same time as its crucial role in urban food resilience became increasingly evident.

As case study research shows, contemporary peri-urban farming and tourism practice centred on food can help maintain or reshape peripheral locations as gastronomic landscapes, increasing both their conviviality and sustainability. Sensitive planning, management and design are all critical to this process, and designers have conceived a variety of schema for supporting food-centred urbanism, with the most promising emerging from transect, sprawl repair and agricultural urbanism perspectives. In the next chapter attention turns to the landscapes that have often replaced these productive peripheries, as the food space implications of a post-suburban spatial scale are explored in depth.

CHAPTER EIGHT

The Megalopolitan Food Realm

INTRODUCTION

Cities' outward growth used to be conceived broadly as taking the form of suburban expansion giving way to the peripheries that were discussed in the last chapter. However, these spatial assumptions no longer hold. The rise of vast settled regions around cities has provoked a great deal of theoretical attention in geography and related disciplines, but research into such spaces' food implications has been somewhat circumscribed. Although Pillsbury (1998: 209) has identified 'cuisine regions' on the basis of particular megalopolitan conditions across the United States, there is perhaps an understandable emphasis on food poverty and obesity in the interrogation of post-urban and post-suburban sprawl: topics this chapter also explores. Just as Chapter 7 considered the scale of the urban edge predominantly in relation to the changing nature of food production and consumption, this chapter focuses on the implications of the larger 'megalopolitan' scale (Psomopoulos 1987: 41). It explores the spatiality of food at this scale in relation to production, distribution, retailing, consumption and waste. Applying a food-centred design analysis, the chapter delves into some of the new urban forms that develop through megalopolis and considers some of their food-related socio-spatial effects, including obesity, food deserts and obesogenic environments. Examples are drawn from a variety of locations, from apparently welcoming dystopia to an emphasis on more place-specific, vernacular and traditional design solutions. Insights into transforming food space include those from urban design and urbanism which focus on retrofitting sprawl.

THE DEVELOPING POST-URBAN CONTEXT FOR FOOD SPACE

To be better understood, food space transformations wrought by massive urbanization need to be situated in relation to large and arguably unsustainable levels of population growth forecast within the next fifty to one hundred years. These, in turn, are expected to result in the development of vast urbanized regions stretching across much of the globe (Kasarda and Dogan 1989, in Galantay 1987; Perlman 1993; Gottdiener 1994; Laquian 2005). By 2010, there were forecast to be more than 511 cities with over a million inhabitants (Gottdiener 1994). More recent figures suggest that by 2015, there will be twenty-three megacities with over ten million inhabitants. While 3.3 billion people lived in urban areas in 2009, an estimated growth in numbers will increase that to five billion by 2030. By 2025, we can expect to see around 155 giant urbanized regions along coastal edges and inland plains across the world. Each of these giant population 'centres' is likely to have more than four million people living within it, and altogether these spaces will contain a total population of 1.26 billion people (Gottdiener 1994). Other forecasts, meanwhile, have suggested there would be over 160 'megalopoli' created by the year 2000, of which 53 would be interconnected into still larger megalopolitan networks (Psomopoulos 1987). Towards the end of the first half of the twenty-first century, these megalopoli will gradually be replaced by the next higher order settlement, the urbanized region, with these settlements following the 'axes of urbanisation' along the world's main inland and coastal plains (Psomopoulos 1987). For example, UN figures from 2005 show enormous levels of urbanization expected between 1994 and 2025, especially in the global south and Asia (Laquian 2005). While the growth of the urbanized middle class will continue, it is likely that a considerable proportion of this development will be informal or slum like, housing the 'post-urban' poor (Davis 2005).

The growth of megacities thus created is slower in the West than in the developing world as there are twice as many people in the latter cities (Perlman 2005: 169). However, shifts in the location of population across nation states and regions, and changes in preferred household formation arrangements have significant urban effects everywhere (Gottdiener 1994; Davis 2005). The post–Second World War era has seen the rise of huge metropolitan regions outside traditional urban centres, and the twenty-first century will see a continuation of this trend worldwide (Perlman 2005: 169). These processes seem to conform to a continuum of urbanization in which centrifugal forces on the location of population are superseded by decentralization outside the city and shifts in industrialization to other regions (Hall, in Brotchie 1985). Population and economic activity in the United States, for instance, has shifted from inner to outer cities, from larger to smaller cities, and from the north-east of the country to the south-west, and this process of decentralization has been allied to a massive deindustrialization of the traditional industrial heartland (Hall, in Brotchie 1985). These changes have led to shrinking and bankrupt cities in the north-east, of which Detroit is perhaps the best-known example, with troubling and intriguing results in food space terms.

In the United Kingdom, a similar movement of people has occurred from north to south, while in Italy the trend is broadly migration from south to north. Australia's situation is symptomatic: population has migrated to the north-east and to the west (away from the south) with a rapid extension of conurbation development across its eastern seaboard, concentrated around Sydney and other urban centres (Essex and Brown 1997). These burgeoning urban regions were already forecast in the late twentieth century to have the capacity to eventually form linked settlements along Australia's Pacific coastline (The Resource Assessment Commission 1991) such as a linking of Sydney-Wollongong-Newcastle and Brisbane-Gold Coast-Sunshine Coast

(Burnley 2003: 276) while similar, although smaller, settled regions are emerging around the cities of Melbourne, Perth and Adelaide. More recent work reinforces the sense that conurbation areas around Sydney, for example, are turning into post-urban environments (Anderson, Kay 2005: npr), albeit ones that are thought ripe with the 'potentialities of new suburban forms, imaginaries, and governance structures'.

The speed of conurbation development has been slower in some countries and regions than in others, but at the same time the rise of the fully urbanized region (Gottdiener 1994) expresses predominant trends including inter-regional population shifts towards economically stronger areas, high levels of household formation out-pacing population growth especially in outer urban areas, the rise of single-person households, population ageing and high levels of urban edge housing development and employment activity. The growth of multi-centred metropolitan regions composed of a variety of different realms – residential, business, cultural, recreational, service and retail – that are occurring in Europe (Bontje 2004; Phelps et al. 2006; Catalán et al. 2008; Helbich and Leitner 2009), Asia (Laquian 2005) and elsewhere, has produced the kinds of food spaces that challenge notions of what is urban. The boundedness of Ebenezer Howard's Garden City again has a particular resonance as 'now our challenge is to escape from the low-density "anti-city" (to use Mumford's term) that has sprawled out over whole regions and has de-concentrated the central cities far more radically than the garden city activists ever envisioned' (Fishman 2002: 59).

The sheer number of neologisms employed to describe the new kind of city that began to emerge in the latter part of the twentieth century may be an indication of the difficulty in capturing theoretically or fully articulating the complexity of these emergent spaces. Patrick Geddes coined the term 'conurbation' to deal with early-twentieth-century urban sprawl, referring to the London polypus, while more recent notions include Gottmann's (1964) megalopolis, emerging on the eastern seaboard of the United States, the non-place urban realm (Webber 1964), the non-place (Augé 1995), the megacity (Perlman 1993; Marshall 2004; Zeiderman 2008), the megalump (Tibbalds 2012), the heterotopia (Foucault 1967), heteropolis (Jencks 1993), privatopia (McKenzie 1994), exurbia (Nelson 1992), post-suburbia (Kling et al. 1995; Teaford 1997; Phelps et al. 2006), in-between-city (Sieverts 1998, in Helbich and Leitner 2009), the fractured metropolis (Barnett 1995), the edge city (Garreau 1991), nerdistan, generica, mallcondoville and metrourbia (Chaves et al. 2011), post-urban region (Fishman 1987), post-Fordist city (Amin and Thrift 2002) and the airport-focused aerotropolis (Kasarda 2001). This bewildering array of descriptors has been developed to define salient aspects of these complex, transforming spaces of metropolitanization from a variety of perspectives. It is telling, though, that books dealing with the 'gastropolis' consider traditional cities rather than megalopolis (Hauck-Lawson and Deutsch 2009; Brooks et al. 2012).

In this splintered or scattered urbanism, traditional notions of concentric rings of urban development have given way to more friable places that can no longer be understood as bounded localities and in which everyday life now has different dynamics (Graham and Marvin 2001; Phelps et al. 2006; Fiedler and Addie 2008). The entire post-urban region has been conceptualized as a phenomenon primarily responding to communications and other high technologies, with the technoburb described as a 'peripheral zone perhaps as large as a county, that has emerged as a viable socioeconomic unit' (Fishman 1987: 184, 1996):

> Spread out along its highway growth corridors are shopping malls, industrial parks, campus like office complexes, hospitals, schools, and a full range of housing types. Its residents look to their immediate surroundings rather than to the city for their jobs and other needs, and its industries find not only the employees they need but also the specialised services. (Fishman 1987: 184)

As suburbs are replaced by a post-urban world that provides jobs, housing and food services to its residents, but without the presence of traditional urban forms, an underlying question remains as to whether the city as traditionally understood still exists (Fishman 1987; Habermas 1989). A kind of urbanism is developing that may render earlier classifications such as urban and rural or city and suburb into 'zombie' categories (Beck et al. 2003). By the late twentieth century, it was estimated that up to 86 per cent of the population of the United States were living in such post-urban regions (Gottdiener 1994). Many of these were defined as edge cities in which social life, including in relation to food, largely took place in privately owned spaces including indoor malls, business and office park atriums, gyms and airports (Garreau 1991). This has not just been an American phenomenon. In Europe, too, edge city inflected spatiality has emerged around a number of cities, including Paris, Berlin, Budapest and Madrid in a process known as 'euro-sprawl' (Hardy 2004: npr; Pumain 2004; Bontje and Burdack 2005). From Los Angeles, often cited as the paradigm for post-urban transformation, Edward Soja (1989: 208) reports on 'a sprawling and poly-nucleated decentralisation process' that has 'characterised the historical geography of the capitalist city since the nineteenth century', replacing older industrial and residential subregions.

For some, these new forms are understood as still solidly rooted in their suburban antecedents (Teaford 1997). Edge city style spatiality is traced back in conceptual terms to the Garden City and planned urban industrial dispersion (Hise 1997: 11). Yet surveying the nature of twenty-first-century space suggests a kind of 'ad hoc, laissez-faire urbanism. The result is a mix of stuff all over the place ... in other words, left to its own devices this is what a modern, car-oriented, telecommunications-rich society would naturally generate' (Marshall 2009: 243). Such conurbation development has been linked to a 'crisis of place', representing the end of any direct connection perceived between personal well-being and the physical environment: 'the inevitable and desirable expression of our new technologies and hyper-individualized culture' (Calthorpe 1993: 11). A kind of deterritorialized meta-city is thus posited, comprising those with little interest in the local (Graham and Marvin 2001). It follows that place-based interest in food might also suffer.

Central area decline in Western Europe, Australia and elsewhere in the West has followed a path similar to American experience. Spatial shifts in the pattern of urbanization include decline in (or disappearance of) manufacturing, rising levels of service employment, new industry based upon high technology production and massive post-urbanization (Phelps et al. 2006; Phelps and Wu 2011). A substantial exodus by those able to leave central cities has seen a 'movement of middle-and upper-income residents to outlying areas which has drained urban tax bases, weakened secondary labour markets, and led to increased isolation and segregation of low income minorities in the urban cores' (Kasarda and Dogan 1988: 8). From the latter part of the twentieth century, a fine-grained resorting process began with activities that could be decentralized being moved out of the city to cheaper locations (Hall 1993). Yet this created space that was 'building a massive future liability – an inefficient, inflexible urban form ill-suited to meeting future challenges such as rising oil prices, financial market instability, or the productivity and competitiveness of business' (Blais 2010: x). Despite the loss of primary focus on place, as households increasingly work, travel and pursue leisure options in a wide variety of locations, there is also thought to be a countervailing strengthening of emotional and environmental connection with the landscape. How food has both shaped and responded to this paradoxical transformation is explored in the next section.

EXPLORING THE POST-SUBURBAN WORLD AS FOOD SPACE – THE EMERGENCE OF PRIVATOPIA

Just as megalopolitan space created by massive urbanization and urban restructuring cannot be understood spatially as simply part of a city-suburb hierarchy, neither do its food expressions fit easily into traditional urban categories and food practices. Food is still grown, distributed, sold, consumed and forms waste, but now this often occurs in autonomous regions in which traditional city centres have not only lost their dominant food role, but may also have disappeared altogether. This fast-growing post-urban context offers an array of food spaces that reflect settlement forms revolving around (and as far a food is concerned often experienced in) gated communities, distribution and customer fulfilment centres including 'dark stores' (Benedictus 2014; Butler 2014), business and office parks, 'big box' food stores (Basker et al. 2012) and hypermarkets, fast-food outlets and chain restaurants, petrol station forecourt 'road pantries' (Parham 2005) and the food courts of outlet and megamalls.

The rise of 'privatopia', a neologism coined by McKenzie (1994), has been identified in the United States, where elements of Howard's Garden City vision have been transmogrified into an urban spatiality which at once plans for building entire communities yet turns its back on the public realm, including its traditional food spaces. By 1998, it was estimated that around sixteen million Americans were residents of master-planned and gated communities (Low 2003), including upmarket golf course – focused developments (Sorkin 1992). Gated communities have been theorized as forming a part of the so-called Disneylandization (Davis 1996) of post-urban space, and, in fact, Disney-owned theme park landscapes have helped shape the 'fun enclaves' of development of the new city of Marne-la-Vallée in France (Giroir 2007: 236). Prior to its disastrous subprime collapse, the United States' 1990s property boom left the country with 'some remarkable new landscapes. Private, master-planned communities have appeared – or at least begun to appear – around every large metropolitan area, creating a series of "artful fragments" that seem likely to prefigure the post-suburban form of the *fin de millennium* metropolis' (Knox 1992: 207). These forms of exclusion operated along a continuum rather than a sharp differentiation between open and closed: 'ranging from no-through streets, superblocks, environmental areas, gated communities and privately managed communities' (Charmes 2010: 357).

Both within and outside the United States, gated developments emerged in a wide variety of post- urban settings as a response by those with economic resources to a complex range of external conditions including fear of crime, and in pursuit of particular leisure interests (Blakeley and Snyder 1997; Glasze et al. 2002; Glasze et al. 2006; Macionis and Parrillo 2006). Gated communities are mushrooming in places as diverse as Canada (Grant 2005); Russia, in the *poselki* or planned developments around Moscow (Blinnikov et al. 2006); Istanbul (Geniş 2007); the Arabian Peninsula (Glasze and Alkhayyal 2002; Glasze 2006); Lebanon (Glasze 2003); around Chinese cities (Miao 2003; Pow Choon-Piew 2009; Wu 2010); in South Africa (Gnad 2002; Landman 2002; Hook and Vrdoljak 2002); Ghana (Grant 2005); India (Falzon 2004); Indonesia (Leisch 2002); Australia (Burke 2001; Rofe 2006; Costley 2006); the United Kingdom (Atkinson and Flint 2004); the Caribbean (Mycoo 2006); Central America (Suárez Carrasquillo 2011); and Latin America (Coy 2006), including Brazil, Chile and Argentina (Caldeira 2000; Salcedo and Torres 2004; Roitman 2005). In Shanghai's 'emerging consumption-scape, shopping malls, gated communities and skyscrapers have become a ubiquitous sight in the city' (Pow Choon-Piew 2009: 35). In China's experience of this phenomenon, the leisure-focused gated community reverses the paradigm in which leisure and tourism are by-products of urbanization processes, to one where they become 'producers of specialised urban development' (Giroir 2007: 236). While some theorists have decried the fragmentation of urban governance in these places as representing a form of 'medievalisation of the modern city, others see it as a shift back to something more natural, after a 20th century experiment in municipal socialism: yielding back to the market certain municipal management functions' (Le Goix and Webster 2008).

Food spaces associated with gated communities are thinly represented in the research literature but include onsite 'gourmet restaurants' and other restaurants and supermarkets (Blinnikov et al. 2006: 77; Giroir 2007). As Pow Choon-Piew (2009) notes, Bourdieu's notion of the habitus appears well suited to describing lifestyles which model distinction through luxurious food consumption within such developments, often in the context of great inequality in the surrounding society. For example, in the 'serene, environmental paradise' of the Hiranandani Gardens development, 18 kilometres north of downtown Mumbai, a gated complex of residential and related developments is conceived as 'a city within a city' offering residents a range of food services including restaurants, shopping malls and food courts (Falzon 2004: 153). The shopping centres at Hiranandani Gardens are 'paragons of global consumer culture ... well-stocked with various national and international brands' and the 'food courts' serve 'anything from Mexican tacos, Italian pasta, Chinese chopsuey, and our very own masala dosa' (Falzon 2004: 153). In this inward-facing enclave urbanism, the spatial design and regulatory focus is on strict control of food space from the level of the dwelling (vegetable gardens are discouraged) to that of the preferred (large) scale of food retail outlets (McKenzie 1994). As firsthand accounts from a gated community in Scottsdale in Arizona demonstrate,

> The houses all had matching furniture, specific themed rooms, a high-end entertainment center, vaulted ceilings, and a modern, if not restaurant grade, kitchen. These attributes signify the amenities people build into their own small-scale community, their own home, with vast amounts of disposable income. (Romig 2005: 75)

> We feel out of place here. We don't drink wine. We don't like going to fancy restaurants. We often go down past Shea (South Scottsdale) to eat. (Case 2003; cited in Romig 2005: 81)

It is notable, too, how the marketing literature for gated developments may appropriate the agrarian landscape and more food-centred ways of life of rural communities that the new development displaces, or invoke that of places entirely disconnected geographically but symbolizing supposedly appealing and authentic aspects of food consumption. Thus a residential scheme in post-urban Texas at once replaces rural land and food-growing activity (and historically an indigenous community) while employing an imagery of continuity to provide place identity:

> Where Las Colindas grows today, once grew the maize and squash of the Tejas Indians. Over the hills and prairies came buffalo herds, Spanish adventurers, wagon trains, cattle drives to Kansas, ranchers, farmers and the mustang descendants of Spanish horses gone wild. The preservation of the heritage of Texas is an integral part of the commitment which Las Colindas makes to the future. (Dillon: 8, cited by Ellin 1996: 87)

SUPERMARKETS, HYPERMARKETS, MALLS

Meanwhile, other food spaces, with their seeds in the suburban landscapes explored in Chapter 5, have come to be seen as representative of the post-urban. Emerging most strongly from the 1980s, very large supermarkets, superstores and hypermarkets became central features in the post-urban retailing environment in Europe and elsewhere. From an original 1963 hypermarket developed south of Paris, Carrefour became the world's second largest retailer after Wal-Mart, by the early twenty-first century expanding throughout Europe, as well as into South America and East Asia (Teaford 2011: 31). Evidence from Portugal offers an example of this trajectory and suggests that the advent of the hypermarket as a food-retailing form occurred there from the mid-1980s, as the country was integrated into the European Economic Community and then the European Union and opened up to foreign investment. These food retail shifts had significant effects on food consumption practices, as was discussed in Chapter 5. Shoppers deserted traditional food shops in droves, preferring the free parking, low prices and wide range of products on offer, especially for groceries, drinks and frozen foods (Farhangmehr et al. 2000). Large-scale superstores have shown a great deal of resilience despite the advent of more High Street–focused policies, and 'though food security and reductions in carbon footprints are official government policies, it is not easy to see these reflected in retail policy' (Elms et al. 2010: 825, based on Hallsworth et al. 2008). From eastern Europe too (Dries et al. 2004) the opening up to Western economic activity, the globalization of retail ownership by multinational companies and 'softer' planning regimes allowing more out-of-town developments saw the advent of a number of very large food stores including the hypermarket type, including on the edge of Prague (Spilková and Šefrna 2010). Yet, as experience from Germany and Korea of Wal-Mart's attempts to develop its presence there suggests, a straightforward process of market penetration is not always assured. The desire to impose standardization of the retailing model on diverse food cultures and spatial circumstances saw the firm failing to build stores in what were seen by host communities as convenient locations (that is, compact, urban ones), and this proved to be a significant area of difficulty in these settings, with implications for other places too. Thus,

> Wal-Mart has always tried to keep their format standard in their international operations. Cultures that are similar to the American culture or those that are highly influenced by the American culture have accepted the Wal-Mart formats as is. But, countries like Germany and South Korea have not accepted this format. To be successful in India, Wal-Mart will have to learn from their German and South Korean experiences, and make suitable changes to meet the need of the Indian consumer. (Halepete et al. 2008: 709)

Similar conclusions in relation to Korea also foreground very clearly the issues of imposition of the norm of a large spatial scale that requires driving to shop for food services meeting an embedded cultural preference for food shopping as a convivial, walking-centred undertaking:

> Wal-Mart seems to have missed fully understanding the local Korean retail culture which thrives as a festive and social setting. Korea's retail market is composed of thousands of small retailers that are typically dispersed in local neighborhoods and form both a marketplace and a social center. The successful domestic retail supercenters aimed to recreate the festive, noisy atmosphere of the outdoor markets within their stores. (Ramstad 2006b in Halepete et al. 2008: 705)

Turning to the food space of the megamall, this developed as another food-retailing form whose antecedents could be seen in the shopping centres of traditional suburbia

(Teaford 2011: 27), the 'archtypical consumption space' of late modernity Zukin (1998: 828) describes, which were considered in Chapter 5 (see also Figure 8.1). As the ownership structure of retailing became more concentrated in many countries, through neoliberal regimes which increased the rate of extraction of surplus value, income distribution altered towards a concentration of wealth, on the one hand, and creation of more low-income households, on the other (Crew 2000; Koc et al. 1999). The response documented in the United States was of regional megamalls and outlet malls developed as an attempt to keep mall customers shopping longer and buying more. In this way, malls became the new centres of metropolitan life for an increasing proportion of urban dwellers (Rowe 1991: 109; Crawford 1992). In fact, 'superregional malls at freeway interchanges … became catalysts for new suburban mini cities, attracting a constellation of typically urban functions' (Crawford 1992: 24–6). A variety of niche malls developed which 'eliminate social and public functions to allow more efficient shopping' (Crawford 1992: 24–6). Sometimes understood as predominantly a Western phenomenon, the trend has also been noted in places including India, where malls have become ubiquitous as middle-class customers move from traditional 'kirana' stores to mall-based food consumption (Goswami and Mishra 2009).

It is possible to argue that in megalopolis, an urban form has been created that starves its inhabitants of opportunities for sociability and conviviality while, at the same time, given its vast spatial extent, rendering more of them subject to this narrowing down effect (Parham 1990, 1992). One way that this has been conceptualized is as a broad process of McDonaldization in which 'the principles of the fast-food restaurant are coming to dominate more and more sectors of American society as well as the rest of the world' (Ritzer 1995: 1, 2008). Despite well-documented sustainability issues such as those caused by feedlot meat production

FIGURE 8.1: Super-regional mall 'café' foodspace, London.
Photo: Susan Parham.

(Schlosser 2002; Horrigan et al. 2002), fast-food chains now have a global reach, and their evolving food-retailing formats and selling techniques have been widely adopted by other food retailers and industries (Ritzer 2008: 3). Pilcher's (2008: 73) exploration of the 'McDonaldized taco' is an illuminating case study, in which the American Taco Bell chain eventually opened outlets in Mexico City, expanding into a food culture and place from which tacos had nominally originated.

Certain other spatial trends also seem to bear out the McDonaldization thesis in the geography of the post-urban environment. From the car-based streamlining of food consumption allowed by the fast-food drive-through window to the efficiency, calculability, predictability and control experienced at a fast-food outlet in a mall food court, the McDonald's model has become a dominant method for much out-of-home eating in megalopolis (Ritzer 2008: 15). This is a model in which the convenience of car-based food consumption is central (Jakle and Sculle 1999: 323). It points to the very close interplay between fast food and malls:

> Shopping malls and McDonaldized chains complement each other beautifully. The malls provide a predictable, uniform, and profitable venue for such chains. When a new mall is built, the chains line up to gain entry. For their part, most malls would not exist were it not for the chains. Simultaneous products of the fast moving automobile age, malls and chains feed off each other, furthering McDonaldization. (Ritzer 2008: 35)

For food within a mall setting, West Edmonton Mall, in Alberta, Canada, was seen to have altered the practices of everyday life (Shields 1989). It may have been the apogee of the form of the giant mall-based food space, with over 110 eating places; while on the edge of Madrid, 'Xanadu' was claimed to be Europe's largest shopping mall, with thirty restaurants (Teaford 2011: 30). As noted in Chapter 5, the megamall adapted and changed in conjunction with the rapidly shifting conditions of the post-urban realm; and food court enlargement and refurbishment has been a key strategy (Zukin 1998: 830). Unlike attempts to design and manage out urban teenagers from using the mall as a hangout space (Matthews et al. 2000), the ways mall-based food spaces are configured, and the dishes served, has sometimes had to respond to different cultural preferences despite the universalizing systemization of the McDonalds' consumption model.

Thus, while Asian consumers accepted the McDonalds system of queuing, self-service and self-seating, in Hong Kong, for example, the firm responded to persistent food preferences for rice-based dishes to ensure viability (Sun and Chen 2007: 287). A similar situation occurred in Japan where regional food variations tried out on local McDonalds consumers included Chinese fried rice known as 'McChao', a fried egg burger called 'tsukimi baga' (moon viewing burger) and a soy sauce flavoured chicken sandwich known as 'chicken-tatsuta' (Ohnuki-Tierney 1997: 163). Referencing work undertaken in Beijing, a perhaps greater area of rebellion was demonstrated by the way that users effectively slowed down the food service model of short 'dwell times' and made fast-food outlets in malls into spaces for socializing (Yan 2000). 'In many parts of East Asia, consumers have turned their local McDonald's into leisure centres, after-school clubs, and social and dating places. The meaning of "fast" has been subverted in these settings where it refers to the delivery of food, not to its consumption' (Sun and Chen 2007: 287).

In places like Saudi Arabia and India malls continue to sprout (in the former case in part because women have so few sanctioned spaces to be outside the home), but by the early 2000s, new construction of traditionally configured malls had ground to a halt in the United States. Those malls that were not regenerated into superregional centres were replaced by big-box stores, internet shopping and other retail formats with minimal public space for food consumption (Sharoun 2012: 109). In the United States, at least, greyfield or 'dead malls' became a feature of redundant post-urban space nationwide (Parlette and Cowen 2011; Benfield 2013). 'Mall-over' strategies to 'de' or 're' mall these spaces began to emerge (Crawford 2002: 29; Southworth 2005; Drummer 2012). As commercial property specialists note, distressed malls are being revived by reintroducing 'public' space, and focusing on food in a performative way:

> Even malls that continue to thrive are being redesigned as town squares – adding more entertainment and service elements. Simon Property is remodeling 15 to 20 malls a year, adding such amenities as electric-car charging stations and stadium-seating theaters.
>
> Malls today have to 'provide a unique set of shopping, dining and entertainment experiences', Simon's President and COO Richard Sokolov told the New York Times, including scheduling 20,000 events a year to draw traffic, such as cooking demonstrations. (Drummer 2012)

Two food-related consumption spaces of increasing importance have been implicated in the decline of regional malls: these are the hybrid mall and the big-box retail store (Sharoun 2012). Of course, big-box retail is not new but proliferated as an ultra-low-cost alternative that drove many malls to extinction. The so-called Wal-Mart Effect (Fishman 2006), which reshaped consumption and pushed other retailers out of business, became commonplace across post-urban America. Such food-retailing shifts are now evident elsewhere, as large players such as Wal-Mart and Carrefour have penetrated retail markets in countries including the United Kingdom and India (Halepete et al. 2008). In huge emerging markets such as China, foreign firms including Carrefour, Wal-Mart and Germany's Metro, as well as Asian and home-grown retailers are all operating hypermarkets (Wang and Guo 2007: 265). In fact, in China, since the 1980s, 'a variety of new consumption spaces have been created ... all new retail formats that had been created in the Western economies were introduced. These include supermarkets, hypermarkets, factory outlets, warehouse retail stores (including membership clubs), and shopping centers' (Wang and Guo 2007: 265).

While both Wal-Mart and Target (the United States' second largest retailer) offer small food areas in some of their stores, 'a community space akin to a mall atrium or food court is lacking in the store design layouts' in a pared-down, no-frills retailing template (Sharoun 2012: 132). Increasingly popular, meanwhile, are hybrid malls which mix big-box retailing with more traditional malls and so-called lifestyle centres. Offering both enclosed and uncovered spaces, these retail formats may encompass food shopping, leisure uses, entertainment and even spaces shaped like streets, but which are not in fact public space. Judged by some retail analysts as a failure to capture trade (other than in the big-box elements), the addition of lifestyle activities offers 'the look of Main Street' with 'street oriented retail' that could incorporate upmarket operators (DMR Architecture 2013; Sharoun 2012: 118). As examples from the south-east of the United States suggest, food is central to these shifts yet also showcases the loss of truly public space focused on food (a theme explored later in the chapter):

> Parktowne Village in Charlotte, NC is a good example of the benefits that can be created with a cross-shopping experience. A traditionally challenged site, the now successful Parktowne combines fountains, outdoor eating areas and open air seating to invite shoppers to spend more time and sample more retail offerings. (DmrArchitecture undated)

> Westfield Brandon, east of Tampa, FL is another example of the value of adding pedestrian friendly areas to enhance the shopping experience. Shoppers at traditional anchors like Sears and Dillards can stroll over and dine at California Pizza Kitchen or Bahama Breeze Island Grille, stretching both their legs and their shopping day. (DmrArchitecture undated)

FOOD, BUSINESS PARKS AND DISTRIBUTION SPACES

A focus solely on mall-based food consumption space would fail to appreciate the ways that the post-urban realm also encompasses food in employment space and food distribution and logistics settings. Just as the rise of the supermarket re-ordered food distribution in ways that increased its scale, as discussed in Chapter 5, the advent of big-box and super-regional-sized retail centres again precipitated another growth in the scale of these operations. With the arrival of online shopping, cybermalls and customer fulfilment hubs and spokes, food distribution chains were reshaped with radical effects on food's spatiality which emphasize 'crowding out' by larger retail chains of smaller food stores (Ng 2003; Pozzi 2011). Storper and Walker (1989) have shown how distribution produces places, and this occurs in nodes within a post-urban region (Amin and Thrift 1992). This new geography of warehousing including for food has been facilitated and encouraged by the 'increased importance of air and highway transportation accessibility in a more time-sensitive economy' as part of a fundamental reshaping of the connections between metropolitan, regional and national economies (Bowen 2008: 386). These changing flows of goods and logistic infrastructures were pioneered in the United States and the United Kingdom, but are now emerging in Europe and elsewhere (Hesse and Rodrigue 2004). Their resulting land requirements, for aspects including 'goods handling, inter-modal transhipment, outdoor storage and vehicle parking' (McKinnon 2009: 5300) are likely to significantly increase in future. Firms prefer edge space to more traditional locations as they perceive fewer constraints in terms of 'traffic jams, the rigidities of planning requirements, or the power of trades unions – factors that are more prevalent within urban regions than at their periphery' (Hesse and Rodrigue 2004: 171).

These distribution changes have also altered the food consumption landscape. The emergence of so-called dark stores or dotcom fulfilment centres which are packing and distribution spaces connected to the rise of internet shopping is one of the results. These dark stores are effectively large grocery warehouses, where food and other products are selected by pickers on the basis of online orders from customers (Benedictus 2014). Described as the 'supermarkets you never see', dark stores are a rapidly growing retail form, with a number of supermarket chains operating them now, and planning more in future, as online sales continue to grow enormously (Benedictus 2014). In the United Kingdom, for example, dark stores built or adapted from existing warehouses are mostly located in suburban and conurbation areas like the south-east of England, where they service densely populated areas in which internet shopping is a rapidly growing market. Their design varies. Some have been shaped around robotic systems, with the pickers mostly staying in one place, as robots move goods from tower stacks. Others 'look almost creepily similar to normal supermarkets. In Hanger Lane, west London, Waitrose operates a dark store in an old John Lewis carpet warehouse. Inside, professional pickers roll baskets around the aisles much like civilians, except they are wrapped up in coats and scarves against the refrigeration system' (Benedictus 2014: npr). In the United Kingdom, a number of large supermarket chains (Tesco, Asda and Waitrose) are expected to double the amount of space they dedicate to dark stores from a current total of 1.8 million square feet (Butler 2014). As Benedictus (2014) asks, 'in future, when people are doing all their boring and heavy shopping through dark stores, and all their interesting and urgent shopping through convenience stores and local shops, what is going to become of the big boxes?'

Punakivi et al. (2001: 746) notes from Finnish and other experience that the rise of electronic grocery shopping (EGS) has profound logistic implications. 'To become a viable option for consumers the EGS has to be supported by a completely new logistics structure in which the Internet is used to connect all parties in the supply chain to the same real time information. As a result, the supply chain, all the way from the supplier to the household, needs to be redesigned'. Early adopters have been suffering massive financial losses as they develop these new systems. In terms of the developing spatiality of online-related distribution, Tanskanen et al. (2002) have suggested there may be possibilities to develop clusters of delivery boxes in certain locations including at entrance points to dwellings, to allow goods to be left for consumers, rather than requiring them to be present to receive them. The emergence of 'click and collect' systems is one early example using the existing infrastructure of local shops. As a footnote to this section, an intriguing aspect of the burgeoning online food shopping world is the emergence of 'ethnic food portals' as researched by Fonseca (2009), in which great play is made of nostalgia for particular culturally embedded food products to increase their appeal to consumers using online methods. Other findings suggest a continued preference for buying such food products in more conventional ways, as in the example of Irish specialty products (Canavan et al. 2007).

Turning to food's interplay with employment sites, it is fair to say that the office, business, science and industrial park has become a ubiquitous feature of megalopolitan space. There has been a remarkable growth in food service and catering over the thirty years to the early 2000s (Lang and Heasman 2004: 167). As an increasingly extensive land use form, business parks are prime location for modelling these changes. Lang's (2003) and Lang and LeFurgy's (2006) concept of the edgeless city foregrounds the particular condition of business uses 'as a form of sprawling office development that does not have the density or cohesiveness of edge city' (Lang 2003: 1). Both in the United States, and in regions including Latin America, the United Kingdom, Asia and Australia, huge retail facilities and ex-urban office parks have flourished, with forms including science parks, innovation centres and technology parks (Massey and Wield 1992; Laquian 2005: 335). Marketing literature puts emphasis on proximity to the natural world of green spaces and wilderness areas, combined with a preference for architecture that signifies modernity, and there are proposals for eco-industrial parks which would further develop parks as ecosystems (Côté and Cohen-Rosenthal 1998). Paradoxically, these sites are often at once lavishly landscaped yet generally free of any food plants and likely to be most often enjoyed in an attenuated way from a groomed outside space, the window of a car or an office or laboratory building.

Food consumption in these post-urban developments, meanwhile, both occurs in and responds to spatial isolation from other facilities and the corporate nature of the park in which predictability and standardization is highly valued (in keeping with the McDonaldization thesis touched on earlier in the chapter). At the upper end of the business park model food 'outlets' on site may encompass canteens, restaurants and food courts: for others the choices may be limited to a chain café, a fast-food outlet, a foyer café bar or externally catered space offering coffee, snacks, sandwiches and pastries. Such food services generally appear to be provided by corporate food businesses rather than individual cafés, restaurants or other food shops as would be much more likely in an urban or even suburban setting. Thus, as an example, Spectrum Business Park in County Durham, in the United Kingdom, advertises 'convenient high quality food outlets including Café Thorntons, Starbucks, Subway and McDonald's' with a 'proposed Phase 2 development to include hotel, supermarket, cinema and restaurants (Spectrum Durham undated).

Similarly, in India, in close proximity to Bangalore's new international airport, the 110 acre Manyata Embassy Business Park offers a mix of office, food court and co-working space and 'is an integrated facility with commercial, residential, hospitality and retail components' that also includes 'a multi-vendor, multi-cuisine food court [and] 24X7 café' (embassyindia). Like the totalizing environment Zukin (2010) refers to in relation to the supermarket in its heyday, the food space of such developments seeks to present a complete, standardized format that offers businesses everything that they need. Also, like the supermarket environment, this comprehensiveness only succeeds in reflecting the way business parks themselves are devoid of convivial food space. It is worth noting that business parks in various places are also sometimes seeking to become sites for employment related to food technology and food product manufacturing, as part of regional economic development efforts (Roberts 2004).

FOOD AND THE CAR

The ubiquity of the car has played a critical role in supporting post-urban development and shaping its relationship to food (Frumkin 2002; Frumkin et al. 2004) in the context of a posited 'hyperautomobility' (Freund and Martin 2007). As discussed in Chapter 5, shopping centres grew from their antecedents in the 1920s and 1930s when they first began to be developed around the automobile, with car use emancipating shoppers from retail locations that were a function of existing transport arrangements (Harris 1987: 320). By the time the post-urban realm emerged, car-based living patterns were largely determining the spatiality of all kinds of peripheral spaces, including those relating to food, in both developed and developing countries. While cars have historically allowed for exceptional personal and business flexibility, their use imposes a high social, environmental and economic cost (Duany et al. 2000; Camagni et al. 2002; Giuliano and Narayan 2003; Muñiz and Galindo 2005; Newman and Kenworthy 2006; Zhao 2010). They have serious implications for food production, as shown in Chapter 7, and food distribution and consumption as explored here and in Chapter 5. While 'peak car' may have passed, in post-urban space the car retains its dominance (Newman and Kenworthy 2011).

That ascendency has been well represented in megalopolitan food space, with roadside fast-food restaurant chains presenting a kind of frontier for capitalism in making places for food consumption (Jakle and Sculle 1999: 327). Fast-food chains, whose advent was discussed in Chapter 5, have also consistently been the focus for protest over poor wages and conditions (Klein 2000; Royle and Towers 2002). The state in a de facto way has propped up a business model which may not provide a living wage on one side, or a sustainable supply chain on the other. By the late 1980s, 40 per cent of Americans were eating out at least once a week, mostly at fast-food and related chain restaurants (Jakle and Sculle 1999). Between 1968 and 2002, the number of McDonalds outlets in that country grew from around 1,000 to more than 28,000 (Schlosser 2002). Of course, car ascendency as a locational consideration is not confined to the design and planning of fast-food outlets alone. Not only car-focused fast food, but mall-based food courts and car-centred business and office parks have become increasingly the settings for food consumption and meals out-of-home in America and elsewhere. Typically, the need to focus on accessibility by car is a given, whereas walking or cycling are not mentioned either generally or specifically in relation to food access for particular sites. As Baraban and Durocher (2010: 7) demonstrate in their United States–focused book on 'successful restaurant design', assumptions about car use are normatively built into the spatial analysis:

> patterns should be taken into consideration when picking a location for a new restaurant. For example, a restaurant that targets breakfast customers is best located on the inbound side of the road heading into the city or other concentrated employment area, such as an industrial park. Conversely, a restaurant that targets dinner guests or people looking to pick up a home meal replacement after work

can benefit from a location on the outbound side of the roadway. The ease with which drivers can exit and return to limited-access highways is also important. (Baraban and Durocher 2010: 7)

FOOD'S RETREAT FROM THE PUBLIC REALM

That the private realm has flourished at the expense of public space is a truism of the rise of the post-urban region. One of megalopolis's salient characteristics is that foodscapes and practices are often disconnected from the public realm or civic engagement; in part because the spaces for that engagement have been excised. As argued at length in earlier chapters, this situation is associated with a rejection of design principles that govern traditional cities, which have over thousands of years given them a 'pattern book vernacular' (Scruton 1987: 23). Such vernacular offered cities a vocabulary of forms and parts that made up a public realm-focused whole, and 'the resulting street invariably has the character of a public space, in which people are encouraged to linger at will, for it has the character of civility' (Scruton 1987: 23). With the civic domain in decline, 'gradually the will to control and shape the public order eroded, and people put more emphasis on protecting themselves from it' (Sennett 1974, in Glazer and Lilla 1987). Greater stress was instead placed on 'privacy, retreat, personal comfort, private consumption, and security' (Tibbalds 2012: 1). As embedded knowledge about space shaping was lost, in the twentieth century, the necessary design requirements for public space primacy likewise disappeared (Trancik 1986). Place production in megalopolitan regions thus occurs within a kind of historical amnesia (Reeve 1993). The resulting rejection of the public realm calls into question the notion that the two halves of a city's character – its physical form and its cultural significance – come together at the public centre (Nicholson-Lord 1987).

With the rise of privately owned 'public' spaces, what really constitutes public space in relation to food is often blurred or elided (Kohn 2004). From an architectural perspective, Karrholm (2012) notes how the spatial territory of shopping has grown in relation to public space, while architects Gastil and Ryan (2004: 9) argue that we cannot 'ignore the inevitable' but need to accept that these are 'the real conditions of public space' today: spaces that may cost to enter, or only be open for part of the day. For the purposes of this book the focus is the way that food spaces are described as public but are in fact part of the private domain (Kayden 2000). As was described in relation to super-regional malls in Chapter 5, there is an emphasis on security and management of perceived risks (Atkinson 2003; Minton 2006, 2012). In this way the atrium café of a business court building could be described as a public space, as could a food court in a privately owned megamall. However, without a truly public realm, the basis for a sense of shared responsibility for convivial social relations and sustainability stewardship could also be compromised. Freedom was traditionally associated with democratic rights exercised within the civil space of the public realm, and food was at the physical and social centre of this spatiality (Parham 1990, 1992, 1993, 2012). However, this engagement may have become no more than 'the power to choose the image that sanctions a place within social space' (Biddulf 1993: 38). This power is expressed through the creation of symbolic capital, that is, the private consumption of goods that signify or associate the owner with distinction and respectability. There is a connection between the accumulation of such symbolic capital and location, with this interplay expressing strong cultural values which may have negative sustainability consequences. Scholars exploring such space in the 1990s argued that at least some of the foodscapes of post-urbanity offered a 'McDisneyized' (Ritzer and Liska 1997) 'postmodern kitsch-scape' (Wilson 1991: 136) in which

> Postmodern aesthetics makes a virtue out of 'ugliness', often appearing to celebrate the very debris that threatens to suffocate all meaning. So we celebrate Las Vegas and the architecture of the fast food outlet, find an eerie pleasure in ... the shopping mall. ... Their onslaught on the received meaning of city life, its order, regularity and grace, are precisely what makes them radical and exciting; they 'make strange' the perhaps too familiar, outmoded structures of 'classic' city life.

Such perspectives continue to have resonance when considering the privately owned, internalized food spaces of super-regional malls, for example, which can be described as the 'product of stage craft, proscenia for the enactment of consumption-oriented lifestyles' (Knox 1992: 208). Such foodscapes offer architectural expressions to be exploited in the struggle for status which is involved in choosing the right things to eat, and places to dine. 'A central issue ... is how, in the scramble for social distinction, to avoid being caught in a compromising position with *declassé* people, objects and places' (Knox 1992: 223). In other words, packaged architectural settings continue to be important backdrops for stylish materialism (Knox 1992; Bourdieu 1984). More recent work suggests that there may be opportunities for food spaces in rescued malls to be reconfigured in more convivial ways. This may also be possible in other post-urban food settings such as developing cultural quarters (Crewe and Beaverstock 1998). These may offer 'hybrid hospitality' which mixes both commercial and other more unmediated food elements to produce a convivial ecology (Shorthose 2004; Bell and Binnie 2005; Bell 2007: 19). Yet the predominant post-urban foodscapes reviewed here offer few reasons to be hopeful.

GASTRONOMIC MARGINALIZATION: FAST FOOD, FOOD DESERTS AND OBESOGENIC ENVIRONMENTS

The developing landscapes of megalopolitan space have created both winners and losers in food terms. Economic restructuring driving spatial trends to decentralization and deindustrialization in cities tends to privilege some in food terms while marginalizing the food cultures and opportunities of other communities and individuals. Of course, gastronomic marginalization does not only arise in peripheral areas, and much of the literature about food marginalization refers to inner-city poverty (as in Talukdar 2008). It is also clear that post-urban spaces contain a wide mix of income groups and varying levels of housing and living quality so some will enjoy excellent food access. There may also be a gap between residents' attitudes towards the space of megalopolis and its actual, wider effects on conviviality and sustainability. Some gated community dwellers, for instance, who do not see food poverty and marginalization at close hand or personally experience the effects of environmental damage to food systems may find it easier to distance themselves from these externalized effects of post-urbanization.

In the 1990s, this writer explored the paradox of places that were fat in the sense of providing rich gastronomic resources to their inhabitants but thin in relation to the health and longevity of their residents (Parham 1998). The conclusion drawn from fieldwork was that spatial design and social mores in regard to food came together in a positive interplay in places such as Bologna, the Città Grassa in Emilia Romagna. This was in distinct contrast to the shaping of food access in megalopolitan regions which have a complicated interplay with food poverty, food deserts and obesogenic environments. Rising levels of obesity have been correlated with changing foodscapes including the rise of out-of-home food outlets (Burgoine et al. 2009). Poverty, food insecurity and obesity can be identified at a post-urban scale, through the regionalized nature of obesity prevalence in parts of Europe (Shaw 2006: 113; Greenwood and Standford 2008). Thus,

> Obesity, the frequent immediate end-result of a diet low in fruit and vegetables, is also linked to low socio-economic status; this effect shows up at a regional level in several European countries in studies published by Obesity Reviews. Intra-state, obesity is highest in eastern Poland (Milewicz et al. 2005) where the provinces with lowest disposable income are located (*The Economist*, 2005). In Spain obesity rates are highest in southern and central Spain (Martinez 2004), which corresponds to the Spanish provinces with the lowest per capita GDP. (*The Economist*, 2008) (Martinez 2004)

In fact, the last few years have seen a sharp rise in academic interest in spatial factors in health generally and obesity in particular (Corburn 2009). Obesity and post-urban spatiality need to be considered within a wider context provided by a posited global epidemic or international pandemic of 'globesity' identified by the World Health Organization that requires attention to its geographical aspects (Eberwine 2003; Gilman 2008; Popkin 2010; Pearce and Witten 2010). This has identified that around two billion adults are overweight or obese, and this is connected to the development in response of a healthy city paradigm (Corburn 2009). A strong strand in the discussion of post-urban space thus concerns its argued role in causing or supporting obesity through the creation of obesogenic environments (Swinburn et al. 1999; Swinburn and Egger 2002; Lake and Townshend 2006, 2010; Ludwig et al. 2011). In this conception, the level of obesogenity is defined as 'the sum of influences that the surroundings, opportunities, or conditions of life have on promoting obesity in individuals or populations' (Swinburn and Egger 2002). There is a body of evidence that suggests 'a link between the built environment, physical activity, obesity and chronic disease' (Lake and Townshend 2006: 263; Lopez 2004). Place design is critical in determining obesogenity because it affects the propensity to walk and cycle:

> Much in the literature points to a consistent link between urban design, walking and cycling. Research has suggested a number of factors within the built environment that appear to correlate with people's propensity to undertake physical activity and

thereby improved health outcomes: increased residential densities; neighbourhood design features, such as historic structures; land-use mix, in particular local shops, services and schools within primarily residential neighbourhoods; the presence and quality of pavements and footpaths; enjoyable scenery; perceptions of safety; and the presence of others, have all been cited as encouraging walking and cycling. (Lake and Townshend 2006: 263)

The concept has been challenged and arguments made that in certain places the prevalence of, for example, individual fast-food outlets reflect the need for local employment and culturally acceptable places to be, as in inner east London (Food Standards Agency 2004; Bagwell 2011). However, the broader links to modernist city design that undercuts opportunities for active travel on foot or by bicycle, and the increasing prevalence of fast food have also been recognized by some food and health theorists as at least implicated in obesity production (Frumkin et al. 2004). As takeaway and fast-food forms an increasing proportion of food intake, with foods sold being more energy dense and poorer nutritionally than home-cooked alternatives, more frequent consumption in turn is associated with higher levels of being overweight or obese (Branca, et al. 2007; Greenwood and Standford 2008; Wang et al. 2008). Discussed earlier in this chapter and in Chapter 5, the spatiality of this trend matters: the rise of the car as a form of mass transport saw urban space reshaped to favour its use. Compact, walkable, proximity was dispensed with, and this change has had an enormous impact on imposing sedentary lives on more urban dwellers (Christian et al. 2011: 136). In tracing the rise of obesity in Europe since 1800, it has become clear that 'city design in the twentieth century has favoured obesity ... *all* modern cities to a certain extent have built-in discouragements to exercise' (Oddy et al. 2009: 2). Design-related features implicated in relation to the creation of obesogenic environments include a number of elements familiar to those studying post-urban space, such as

> poor perceptions of safety, especially due to street crime; poor walkability of streets (street connectivity, traffic, pavements, dog mess); poor access to upper floors of high-rise residential buildings; high density of fast-food restaurants; lack of public parks and playing fields; few gyms or sports clubs; local retail grocers emphasising cheap, energy-dense foods in what are sometimes called 'food deserts'; housing with inadequate facilities for the preparation, cooking and storage of perishable foods such as fruit and vegetables; low levels of social networking and support, and most common, low disposable incomes. (Oddy et al. 2009: 2)

Such analyses appear to be borne out by specific research in conurbation settings. As Guthman (2011: 77) notes of her fieldwork sites in megalopolitan California, the nature of the place is implicated in the levels of obesity experienced by her participants:

> Although Tracy is a relatively old town in California's Central Valley, it has grown quickly in the last twenty years, primarily because it is within commuting distance of the San Francisco Bay Area, and the high-tech industries of Silicon Valley in particular. It has many of the characteristics that are typically quantified in studies of obesogenic environments: long commute times, sprawling residential housing

without walking linkages to public space, a dearth of nearby sites for outdoor recreation, and new, auto-oriented retail districts of big box stores, malls, and fast food and chain restaurants.

Links between poverty and poor food access had been noted since at least the 1980s (Cole-Hamilton and Lang 1986). However, it was from the 1990s that a considerable amount of attention began to be drawn to the related ideas of food security and food deserts (Whitehead 1998). The food desert metaphor has often been used to describe places in inner cities where residents are unable to access cheap, nutritious food, especially fresh fruits and vegetables, but have to make do with highly processed, highly priced products from convenience shops and stores (Morland et al. 2002) as in Figure 8.2. A number of constraints have been identified for individuals living in food deserts (Whelan et al. 2002). Such communities' disconnection from better-quality and better-priced food is understood as related to their lack of cars and thus ability to reach edge-of-town and out-of-town supermarkets in more affluent areas (Cummins and Macintyre 2002: 2115). This is a trend emphasized by supermarkets' previously noted redlining practices through which they avoid locating in certain neighbourhoods judged unlikely to generate sufficient profits (Eisenhauer 2001). Various studies showed that the presence of supermarkets is correlated with low levels of obesity (such as Morland et al. 2006) and case studies of attempts to bring supermarkets into low-income areas have been reported (as in Pongracz 2004).

A number of theorists have explored issues of race, class and gender playing out in relation to food poverty, food deserts and food shop and supermarket accessibility in the United Kingdom, the United States and Canada, among other places (Caraher et al. 1998; Wrigley et al. 2003; Smoyer-Tomic et al. 2006, 2008; Fulfrost and Howard 2006; Larsen and Gilliland 2008). Work from California has found that

FIGURE 8.2: 'Food desert' located shop, LA, USA.
Photo: Susan Parham.

food insecurity is associated with higher levels of obesity among women (Adams et al. 2003). Researchers have foregrounded links to post-industrial decline sharpened by neoliberal economic regimes, as well as issues of rurality where the village store may remain iconic of community but offer very limited food choices (McClintock 2011; Furey et al. 2002; Smith and Morton 2009; Scarpello et al. 2009). In work exploring the possible presence of food deserts in certain San Francisco neighbourhoods, Short, Guthman and Raskin (2007) find a mixed picture and suggest that 'further work is needed [on] unique economic development histories and [that] cultural politics of neighborhoods affect food availability'. In relation to the post-Katrina environment of New Orleans, for instance, food deserts have been characterized rather as 'food swamps' where there are many more snack food than healthy food choices available (Rose et al. 2009: 16). There are proposals for financial and other interventions to bring supermarkets into food desert areas (such as Fife 2012).

Used to considering food shortages and hunger as problems of developing-world countries, safely far from immediate concern, a troubling and politically contentious development in many Western cities has been the rise in food scarcity and the emergence of food banks and other projects and programmes to respond to this urbanized effect of the inequitable nature of the modern food system. In London, for example, regional government recently adopted a 'zero hunger' strategy to address growing food poverty in its urban population. The Trussell Trust (2014: npr), a charity which runs food banks in the United Kingdom, notes that in '2012-13 foodbanks fed 346,992 people nationwide. Of those helped, 126,889 were children. Rising costs of food and fuel combined with static income, high unemployment and changes to benefits are causing more and more people to come to foodbanks for help'.

Food banks are understood by some to be problematic over the long term, including allowing capital to avoid paying a living wage. In Toronto, for example, community food activists argue that food banks 'despite the apparent goodwill and generosity of those operating, volunteering for and donating to them, might actually contribute to the problems they purport to solve' (McBride 2009: 166). The need for food banks has also been disputed by some politicians on the right who appear uncomfortable with the attention drawn to increasing levels of poverty and punitive welfare arrangements. Instead of arguably 'top-down' interventions, the more radical critique of food banks suggests that grassroots action, including community kitchens, community dining, food education centres, cooking classes, community gardens, greenhouses and healthy grocery stalls in food deserts, constitutes a more robust strategic approach (McBride 2009: 169). An interesting (conceptual) architectural design approach along the same lines comes from Elliott (2010: npr) in her 'Five Points Community Food Center' which attempts to integrate 'multiple food activities into a single architectural entity ... as a tool for place making by expressing a local identity through food culture while improving the social and economic fabric'.

In spatial terms, while, as noted above, food deserts were originally conceptualized as occurring in urban neighbourhoods that had been left behind by transforming urbanized space, they have also been found in suburban areas, rural locations and megalopolitan regions (Clarke et al. 2002). Food deserts and so-called micro-food deserts began to be identified in spatial settings including sprawling urban peripheries, as in the 'unbounded territory' of Nantes in France (Shaw 2006: 235); and rural environments (Blanchard and Lyson 2002; Smoyer-Tomic et al. 2006: 308) such as Nebraska's Sandhills region (Thierolf 2012). Seen as primarily about lack of access to supermarkets, the food desert metaphor could well benefit from being stretched to encompass separation from fresh food markets, high-quality individual food shops and online food box and grocery ordering territories.

Although any exploration of possible megalopolitan locations for food deserts seems underdeveloped, some food desert research is suggestive in relation to the negative impact of design that reflects post-urban spatiality (Thomas 2010). For instance, proximity indicators used to define the existence of food deserts may fail to acknowledge 'elements of the landscape such as a dangerous neighbourhood or a highway or road without sidewalks [which] may create barriers not apparent in typical food desert research methods' (Thomas 2010). The results of Thomas's (2010) study, based on a neighbourhood in Lansing, Michigan, demonstrate that the shaping of space further affects those already suffering food insecurity: short distances to food shops 'as the crow flies' are often translated into much longer journeys because of the indirect, hierarchical street layouts of sprawl-designed development. For car drivers this has problematic time implications; for non-drivers accessibility is very much more compromised. 'While these straight line distances appear rather small, they translate into much larger distances that people must travel following roads and sidewalks to reach the nearest food retailer … distance can create an additional burden on segments of the population that are already struggling to obtain food' (Thomas 2010: 25). Even if access can be gained relatively easily to the supercentres of post-urban space, this does not ensure a healthy relationship with food. Just as car-dependent supermarket development (discussed in Chapter 5) was followed by Wal-Mart style supercentre concentrations in post-urban space, such supercentres have also been found to be associated with localized increases in body mass index and obesity (Thomas 2010: 20). Thus,

> We find evidence that Supercenters increase both BMI and obesity, with effects that are largest for women, low-income married individuals, and those living in the least populous counties. The estimates imply that the proliferation of Walmart Supercenters explains 10.5% of the rise in obesity since the late 1980s, but that the increase in medical expenditures offsets only 5.6% of consumers' savings from shopping at Walmart. (Courtemanche and Carden 2011: 177)

SUSTAINABILITY AND URBAN DESIGN: RETROFITTING SPRAWL FOOD SPACE

The discussion above tends to suggest that megalopolitan space is broadly negative for food and that design is deeply implicated in its gastronomic problems. Post-urban design certainly reflects structural forces impinging on urban spatiality, with often deleterious consequences for conviviality. These, in turn, result in lifestyles that cause sustainability issues and are one of the major areas for criticism in such urban development (Fishman 1987). For those places where land use patterns have already created the kind of splintered urbanism that Marvin (2001) describes, health theorists and designers have proposed techniques to remodel various sprawl conditions and create places that are more civilized and convivial, essentially referencing principles of urbanism set out in the book's Introduction (Duany et al. 2000: 136; Cervero 1998; Friedmann et al. 2002; Frumkin et al. 2004; Corburn 2004, 2009; Farr 2008; Talen 2008; Marshall 2009; Dunham-Jones and Williamson 2009; Tachieva 2010; Parham and McCabe 2014).

There is a need, among other things, to overcome the 'tyranny of excessive roadway and parking requirements' and to build in a regional scale for planning 'to manage urban growth at the scale of people's daily lives' (Duany et al. 2000: 138). Dunham-Jones and Williamson (2009) offer specific proposals for redesigning a range of post-urban spaces to improve individual outcomes including achieving obesity reduction and to institute sustainable and convivial urbanism more broadly. Their design approaches include for regional mall reuse to create public space focused downtowns; edge city infill to repair fragmentation and improve walkability and interconnectivity; and office and industrial park retrofits to mend car-dependent, land-wasting spatiality (Dunham-Jones and Williamson 2009). So, for example, reviving shopping 'ghost boxes' created by dead or dysfunctional malls for adaptive reuse re-cycles big-box buildings and car parks and turns the spaces inside out to re-address the public realm of the street (Smiley 2002). In so doing, it builds in opportunities for convivial social interactions in public space and offers finer-grain ground floor areas for small food shops (Dunham-Jones and Williamson 2009: 67). Retrofitting design proposals in relation to food production, distribution, consumption, waste and clean-up can take in a wide range of sprawl conditions, including urbanizing big-box-based malls (Tachieva 2010). Recent work undertaken in the town of Hatfield in the United Kingdom has shown through a series of visualizations a number of retrofitting design scenarios for introducing edible space into post-urban business park landscapes (Parham and McCabe 2014).

Given the vast and growing expanse of megalopolitan space, design-based retrofitting proposals seem at least part of what is required to repair the damage caused by over-scaled, coarse-grained and unsustainable foodscapes. It is ironic

FIGURE 8.3: Chain store 'urban' food shop.
Photo: Susan Parham.

that just as this form of spatiality has reached an apparent universality, the food retail giants themselves have now moved into the smaller scale and more local. As noted in Chapter 5, transnational retailers have established a series of smaller urban-based food stores, with plans for many more (see Figure 8.3). 'Wal-Mart's interest in smaller grocery stores is also an indication that the company which built itself on a density strategy believes smaller grocery stores remain profitable, despite the daunting scale of supercenters' (Ellickson and Grieco 2012: 13).

THE MEGALOPOLITAN FOOD REALM IN REVIEW

The past fifty years have seen the development of enormous sprawling regions around cities, and this scale of development both challenges our notion of what constitutes urban space and presents some difficult food issues in design terms. Driven by a variety of demographic, economic and cultural changes, megalopolitan settlement patterns are the setting for many peoples' interaction with food, yet much of the dispersed, fragmented and splintered food space of the post-urban region is problematic in terms of both conviviality and sustainability. Arguably representing a crisis of place, loss of connection to location may be offset by new ways of expressing belonging in food terms. However, the so-called McDonaldization of food space evidenced through megamall food courts, gated communities, business parks and distribution centres, among other foodscapes of megalopolis, has created sites for interaction that have turned their back on the public realm, are predicated on most unequal economic relationships and may be judged as uncivil and unsustainable in relation to food as a result. The retreat from engagement with the reality of the need for food resilience by choosing instead the enclave urbanism of privatopia seems particularly egregious, but represents strongly held notions about modernity, and an expression of power to withdraw from externalities caused by its spatiality.

Although not traditionally researched as locations for food poverty, food deserts and obesogenic environments, megalopolitan spatial design is implicated in their development, and thus substantially contributes to the pandemic of 'globesity' identified by the United Nations, which is set to cause massive social and economic disruption and is already blighting many individual lives. All this makes processes of retrofitting food space along convivial and sustainable urbanist lines seem particularly important. Design proposals that remake the typical spaces of megalopolis towards more gastronomic ends are to be welcomed as a positive response to food problems generated at the post-urban scale. In the next chapter we move on to consider the scale of the region as a food space, as here too issues and responses to megalopolitan and larger-scale food problems can be found.

CHAPTER NINE

The Critical Food Region

INTRODUCTION

Up until now this book's focus has been at the scale of urban and peri-urban foodscapes of various kinds, but in this chapter attention shifts to the urbanism of the wider food region. Food's regional attributes are considered in urbanist terms by examining critically notions of the sustainable bioregion and approaches to food production that stress low food miles, deployment of food strategies, support for small-scale food operations and the development of regional and alternative food networks in opposition to more conventionally based food systems. The chapter explores ways in which convivial design and planning for food at the regional scale can actively contribute to a distinctive regionally based urbanism, where inherent natural character is shaped into a sustainable cultural landscape by human activity (Hough 1984). In so doing it contemplates both the diversity of experience of individual regions and the way that economic and cultural relations between regions at the global scale may have impacts on designing for food at this scale. Calthorpe (2010: 4) has argued that in order to develop sustainable urbanism we must consider and work at the regional scale, declaring (to paraphrase) that metropolitan regions have become obese on their high-carbon diet. While food is employed metaphorically to describe this problematic situation, and to propose remedies, in this chapter, food's contribution to designing the critical region is explored in a material rather than metaphorical way.

At the same time, positive design and social practice connections between food and regionality are not presented as simply a normative aspect of urbanism. It is acknowledged that the 'region' as a conceptual and physical space does not offer universal explanatory power in food system terms, as neither food production nor consumption has necessarily been organized or framed in regional terms (Kneafsey 2010). The differing experiences in the global north and south, moreover, demonstrate that reregionalization in high-income countries may be very different in its effects than the more mixed results in relation to issues such as food security and resilience in poorer countries. Thus a fundamental question for an equitable approach to food and urbanism remains whether reregionalized food systems can transcend social inequalities or will more often serve to emphasize them (Donald et al. 2010). As Donald et al. (2010: 174) note, the food system begets unequal societal relations that are very hard to dislodge. In its focus on food's spatiality at this scale, the chapter therefore considers whether regionalized food systems offer only 'cosmetic solutions' or can be more profoundly redistributive towards those in poverty or otherwise marginalized (Donald et al. 2010: 174).

SITUATING FOOD REGIONS HISTORICALLY IN SPATIAL PRACTICE

As explored in Chapter 6, settled agriculture is an ancient human practice, traced back to Neolithic times, with a long period of 'proto-agriculture' thought to exist well before that in which human groups watered plants, weeded, stored food, burned fields and sowed seeds (Tauger 2011: introduction). Domesticated livestock and crops, including wheat, barley, millet and rice, began to emerge in places including the eastern Mediterranean littoral, China, South-east Asia and the Americas (Tauger 2011: introduction). As Tauger (2011) notes, 'agriculture was prior to and a prerequisite to civilisation' as only through the production of a food surplus were other activities and elements given space to emerge. Despite farming's fundamental importance to the creation and maintenance of civilization, it has suffered a 'dual subordination': conditional throughout its history on both the vagaries of the natural environment and the rule of external political agencies (Tauger 2011: 12).

Documentation of farming practices in antiquity suggest that in Greek city-states, for example, crop staples, fruits, vegetables, pulses and livestock were all grown or raised; wine presses and olive oil mills were used; and to deal with shortages, states were trading in and importing food (Tauger 2011: 19). Rome itself began as an agrarian city-state in which long-term military expansion was based on attacking enemy agriculture and producing a vast agricultural stock of land with which to feed large urban populations (Tauger 2011: 21). Although most farming in the Roman Empire remained small scale, the patrician estates or *latifundia* used 'slave labour to produce wine and olive oil for local markets. In more remote conquered regions, such as Sicily and North Africa, large estates employed slaves to grow grain for the cities' (Tauger 2011: 21). In China, meanwhile, peasant farming was ubiquitous, although farmers also lost land to wealthy landowners and agricultural stability was challenged by monsoon weather patterns, leading to devastating crop losses and famines.

By the post-Classical period, farming saw both the decline of the Roman Empire and the rise (and decline) of the European manorial system. Crops originally grown in Asia began to migrate west and agricultural intensification occurred in China and elsewhere. Regional food historians suggest that in the medieval period, while Europe had few large cities, most regions had some kind of commercial food infrastructure of markets and fairs, to trade agricultural surplus that was produced over and above farmers' subsistence requirements (Britnell 2008). Its numerous small towns 'were significant centres of demand for food, fuel, and raw materials' (Britnell 2008: 4). Later innovations such as the introduction of animal-drawn ploughs in eleventh- and twelfth-century Europe, and the mechanical revolution of the past 200 years, in which capital replaced labour, saw agriculture again transformed.

Particular climatic and social conditions still influenced food production practices. The Black Death plagues of 1348, 1349 and beyond, for example, had enormous effects on Europe's fourteenth- and fifteenth-century agriculture (Dodds and Britnell 2008). Early modern agriculture in Europe was fundamentally affected by the long-term effects of the Little Ice Age. Systems of subordination similar to serfdom, meanwhile, were sources of ongoing conflict in Europe, in the Ottoman agrarian system and those of Mughal India, China and Japan (Tauger 2011: 60). By the early modern period, Western Europe was agriculturally underproductive, with servile farming systems that produced regular rebellions and was dependent on food produced by Eastern European serfs. 'The cultural flowering of the Renaissance, the Reformation and the wars of religion, the scientific revolution and the Enlightenment, all were fed in large part by these subordinated farmers' (Tauger 2011: 69).

In the Americas, meanwhile, the early modern period saw two indigenous cultures, the Aztecs and the Incas, with food systems based on American foods: maize, beans, tubers including potatoes, capsicums, peppers and some livestock. The regional nature of food production was in pre-modern times clearly reflected in regional diets. In Mexico, for example, the fish heavy diet of Baja California, was considerably different from that of central Mexico which was based on its rich stock of meat and meat products, citrus fruits and herbs including coriander, and this was different again from the food of the Isthmus of Tehuantepec, with its black turtle beans and various *moles* (Long-Solis and Vargas 2005). With colonial invasion, the development of plantation agriculture, based on Iberian medieval farming systems, again 'dramatically altered world agriculture' (Tauger 2011: 74). European demand for sugar (as well as other luxury crops such as coffee and tobacco) drove the expansion of highly profitable sugar plantations based on grossly exploitative and cruel slave labour (Mintz 1985). Thus 'this pattern of exploiting cheap labour and public taste to make a profit for a few governments, entrepreneurs and share-holders anticipated the twentieth- and early-twenty-first-century agribusiness food system' (Tauger 2011: 79).

THE EXAMPLE OF TUSCAN *MEZZADRIA*

While some of the spatialized food effects of modernity in agriculture are traced below, in certain places over the course of the twentieth century the physically and culturally configured gastronomic landscape of the traditional food region has continued to allow 'local food suppliers, institutions, rural animators and brokers' to operate in a complex, productive interplay (Hardy 1993, 1994; O'Neill and Whatmore 2000; Miele and Murdoch 2002: 322). Tuscany is so often invoked as the food region, par excellence, that its undoubted strengths can be obscured by its pre-eminent positioning as a desirable gastronomic tourism destination. Yet, in reality, Tuscany has regional food attributes very much worth investigating, as were touched on in peri-urban terms in Chapter 7. It has enjoyed a long tradition of intensive viticultural and horticultural production, in which the region's polycentric territorial structure reflects its medieval inheritance of city-states, with Florence historically the dominant city. As paradigmatic of trends elsewhere, and now a region valorized and ascribed considerable meaning in regional food terms, it seems worth pausing to consider these changes in some detail.

In Italy, with a history of deep spatial fragmentation, it is argued that agriculture has changed more in the latter part of the twentieth century than in the preceding 3,000 years (Roden 1989: 2). As Roden (1989: 2) points out, across Italy agriculture using archaic methods was based on sharecropping (*mezzadria*), with large estates divided into fields (*poderi*) in which landlords took half the produce as rent, and the hard grind of peasant life and cooking changed little over the long term. Within Tuscany, agriculture traditionally supported a relatively high rural population whose settlement was centred upon groups of farmhouses, as in the rest of Italy, in the middle of *mezzadria* holdings (Casanova and Memoli 2010: 99). The *mezzadria* system, based on very asymmetrical profit sharing between proprietor and cultivator, was first recorded in the eighth century, while wider societal connections to farming have been historically broad (Roden 1989). With increasing wealth, individuals would invest in land and a country villa for the summer months, while every twenty to thirty holdings were owned by a landowner, usually aristocratic. A network of villa-farmhouses for the nobility also grew up, with associated buildings *(fattoria)* housing the farm manager *(fattore)* and farm facilities. The landscape thus exhibited a strong built form, as well as a linked spatial and economic structure over an extended period. Despite certain incursions, the grouping of the sumptuous villa, farm manager's house, cellars for wine and olive mashing and storage facilities has remained as part of the regional landscape up until the early twenty-first century.

The desire of the Florentine noble class to maintain their traditional structures and way of life into the twentieth century meant that these agricultural arrangements were preserved, despite the advent of the modern administrative state in Italy. Equally, Florentine aristocratic influence within centralized national government also allowed provincial autonomy to be maintained within the national political system (Leonardi 1990). Certain external factors shifted nonetheless: as the 1950s saw considerable changes to Italy's socio-economic system. In particular, *mezzadria* suffered a crisis and a large number of sharecroppers left for the cities, with the reduction in rural population mainly from scattered housing sites, while small and medium sized towns retained population. Large cities, meanwhile, experienced an increase in population at their peripheries and the development of an urbanized countryside that is now a major development pattern within the Tuscan region (Leonardi 1990).

During this phase, the region's urban polycentrism was maintained and consolidated, within a context of tumultuous economic growth, since defined as Italy's economic miracle – which in fact comprised the diffusion of light industry, using skills that had been released from share cropping into a diversity of small and medium sized enterprises (Leonardi 1990). There also began to be a re-aggregation of depopulated areas based on the reuse of abandoned farmhouses as second residences for the urban middle classes, rather than improved farm viability alone. Farmers continued to decline as a proportion of the rural population and the structure of land ownership underwent considerable change that is important not least because it reflects the rise of the urbanized countryside (Leonardi 1990). This was a development trend that resulted from many private, individual, unregulated decisions. 'On the ground, the structure of the urbanized countryside has produced a network which is at times so dense with new roads, houses and factories producing a mix of urban and rural elements that it has not always produced positive results' (Becattini 1975, quoted in Leonardi 1990).

The traditional regional landscape of Tuscany has altered considerably as a result of these changes. The combination of the decline of *mezzadria*, the loss of rural workers, the introduction of mechanization and the changing structure of farm ownership all had an effect on the gastronomic landscape by the 1960s, generally for the worse. 'Trees were cut down, different fields were made into one, bushes were separated, to allow machines through. A few small farmers continue in the old archaic way of varied mixed cultivation, and their bit of landscape has remained like the bits of background in Renaissance paintings, but the rest has changed' (Roden 1989: 95). Thus despite the image of Tuscany as a landscape of gastronomic perfection, discussed in Chapter 7, the reality is more awkward and mixed in both food and design terms, an issue to which we will return later in the chapter in the discussion of the urbanism of regional food tourism.

SITUATING TODAY'S FOOD REGIONS AND REGIONALISM

Regions as food territories have always concerned gastronomers and urbanists, forming a location for food production, a food-centred spatiality and practices of cooking resting on these place-based qualities. Equally, regions are understood as being far from the 'bucolic fantasies' of more simplistic readings as discussed in Parasecoli (2004) in relation to Italy. These cultural food traditions are richly reflected in 'outsider' books like Elizabeth David's *French Provincial Cooking* (1960) or Paula Wolfert's *The Cooking of South West France* (1983), as well as those more embedded in these traditions, such as Ada Boni's (1969) *Italian Regional Cooking* or Lorenza de'Medici's (1990) *The Heritage of Italian Cooking*. Regionality is also critical to the less self-conscious collections of 'recettes paysannes' produced in relation to French regions like Lot-et-Garonne (Lavialle, Lavialle and Béziat 1999) and Perigord (Bonis 1997). Willan (1991), meanwhile, notes that France still describes itself as *profondément rurale* and 'little more than an hour from Notre Dame you'll still find peasants living off the land, growing their own fruit and vegetables, raising rabbits, poultry and eggs, brewing wine or cider and "eau-de-vie", the water of life', a peasant heritage that Luard (1986) shows is alive elsewhere in Europe too. The Roux brothers (1989: 6) similarly argue in their *French Country Cooking* that 'with its temperate climate and fertile soil, France offers a unique and inexhaustible range of produce. Its rich and varied provincial cooking reflects the diverse ethnic origins, tastes and cultures of its inhabitants and the abundant resources of its land'. In introducing their review of various of these regional gastronomies, they point out that France's 'regional, homely' cooking is 'both rustic and sophisticated' and 'rooted in tradition' (Roux brothers 1989: 6), reminding them of their childhood in a family of *charcutiers*, living over the shop in the regional town of Charolles. 'With a twinge of nostalgia, we remember the white wood tables and pull-out trestles sagging under the weight of cooked meats, Lyonnais tripe with onions, *choucroute*, breaded pigs' trotters and a multitude of sausages, *boudins*, pates made from pork, poultry and game and pike quenelles' (Roux brothers 1989: 6).

For Claudia Roden (1989: 1), likewise, part of Italy's regional 'fascination' is in its diversity of landscape, vegetation and climate, with substantial changes in architecture, ambience, behaviour and cooking demonstrated from region to region and town to town. As previously pointed out by Elizabeth David in her *Italian Food* of 1954, and in Waverley Root's (1971) *Food of Italy*, Roden (1989) notes that there 'is no such thing as Italian cooking, only Sicilian, Piedmontese, Neapolitan, Venetian, Florentine, Genoese and so on'. Understood as at once a rich cultural heritage and one that reflects the history and gastronomic character of its twenty-three regions, 'there has long existed an extraordinary variety of regional dishes in Italy, unique among European cuisines for their sharp diversity' (De' Medici 1990: 11).

Of course, while the countries of Southern Europe, including Italy, France, Spain and Greece, offer obvious examples of gastronomic regionality (see Figure 9.1), with

FIGURE 9.1: Regional cheese display, France.
Photo: Susan Parham.

places like Sicily notably, unlike anywhere else, many other places, too, show strong regional food connections (Casas 1982; Gray 1987; Simeti 1991). The Barossa Valley, north of Adelaide in South Australia, for example, settled by German-speaking Prussians in the mid-nineteenth century, has evolved a very particular regionality in food terms (Heuzenroeder 1999: preface). Such immigrant-related, regionally based, and often transforming gastronomies are a feature of New World countries. In the Appalachians, similarly, Long (2010: 4) notes that mountain cuisine is not fixed or clichéd (ie moonshine, cornbread and squirrels), rather, culinary tourism and local foodways interconnect in an iterative way, so that this

> emerging Asheville cuisine emphasizes local natural resources and a specific contemporary ethos blending aesthetic innovation and socially conscious food production and consumption. Tourism highlights this new cuisine and provides a market for it, but this cuisine is also integrated into the lives of local residents, many of whom are recent arrivals.

In Canada, similarly, as Hashimoto and Telfer (2006: 31) point out, regional food is more than simply maple syrup and smoked salmon. 'Canadians are taking part in an international movement towards the development of regional cuisine using indigenous or locally cultivated products. Products such as Atlantic seafood and Alberta beef are culinary attractions; however, there is increasing use of specialized products such as Saskatoon berries, Quebec cheeses and Arctic muskox' (Deneault 2002).

REGIONAL FOOD POLICY – PLACE-BASED APPROACHES

Turning to regionality considered in food-policy terms, it is clear that contemporary regions are subject to the dynamics of both globalization and relocalization (Marsden and Smith 2005). While today's renewed interest in regional planning and food at the regional scale may seem a recent preoccupation, concern for the regional qualities of the food system has been apparent in urban planning since at least the early twentieth century (Wheeler 2002). Ideas about food-centred regionalism were grounded then in an acknowledgement of the locational limitations on production, which more truly reflected the parameters of place. In 1925, for example, the Regional Planning Association of America proposed a national plan that would involve

> regions delimited on the basis of natural geographic entities; a maximum of foodstuffs, textiles and housing materials grown and manufactured in the home region; a minimum of interregional exchanges based only on such products as the home region cannot economically produce; plus regional power plants, short hauls by truck and a decentralised distribution of population. (Chase, in Hall 1992: 151)

That proposal failed to make the transition to public policy but similar tropes are evident in the ecologically and gastronomically situated programmes of bio-regionalism today which would shrink food catchments and acknowledge food webs (Marsden 2010). In so doing, this revisits some classic planning theory of the 1920s, promulgated by Mumford and others in the American Planning Association, and articulating Geddes' even earlier, but less widely known views that rejected the idea of large cities and embraced an ecologically based regional settlement pattern ((Meller 1994; Talen 2005). A growing interest and concern for problems of regional food development today can be clearly recognized in the work on alternative food geographies and networks (Parrot et al. 2002; Goodman 2004; Watts et al. 2005; Maye et al. 2007; Jarosz 2008; Sims 2009). This gives due recognition to the place-specific nature of economic and environmental outcomes in relation to food, as well as food networks' social embededness (Sage 2003). These themes are considered in some detail later in the chapter.

In the period since the mid-to-late nineteenth century, theorists of agricultural systems speak of changing food regimes whereby agriculture has been linked to forms of capitalist accumulation. The post–Second World War era saw agriculture shift from a productivist to a post-productivist mode in which constant modernization and industrialization was often replaced by reduced farm outputs and greater integration with non-farm activities, in line with wider economic and environmental objectives (Ilbery and Bowler 1998). Marked by trends to concentration, intensification and specialization the productivist phase gave rise to externalized food and spatial consequences that helped set the stage for a model characterized by 'extensification', dispersion and diversification (Ilbery and Bowler 1998: 70). This in turn ushered in alternative food networks, of which more later in this chapter.

By the early 2000s, interest in regional food policy was again evident including in the United States where movements such as those for regional equity began to articulate the region as the appropriate scale for dealing with social justice issues, including food security and combating obesity (Pavel 2009). In 2007, the American Planning Association developed a *Community and Regional Food Planning Policy Guide*. This set out two goals for planners in relation to food at a regional level, to 'help build stronger, sustainable, and more self-reliant community and regional food systems [and] suggest ways the industrial food system may interact with communities and regions to enhance benefits such as economic vitality, public health, ecological sustainability, social equity, and cultural diversity' (Pothukuchi 2009: 352). The Guide focused on policies to undertake comprehensive regional food planning and strengthen regional food economies that would be both regionally sustainable and protect regionally based foods and cultures. It showcased regionally supportive examples that included more direct local relationships between farms and markets, such as community-supported agriculture, support for small-scale, organic production and protection of regional agricultural land. In Australia, meanwhile, a *Food Sensitive Planning and Design Guide* (2011: 25) has been produced for the metropolitan region of Melbourne and argues for a much greater strategic focus for food planning at this scale. As Pothukuchi (2009: 350) notes, the 'growing planning interest in community and regional food systems is of course bolstered by, and in turn supports, a broader social movement for sustainable and just food systems, an emphasis on localism in food, and public health concerns related to the global obesity crisis'. Although support for conviviality is not articulated, it appears to be implied by the thrust of these strategic policy documents.

ALTERNATIVE FOOD NETWORKS – GLOBAL CONNECTIONS AND LOCAL FOOD MOVEMENTS

Turning to regionalism considered in relation to the spatiality of food consumption, Peace (2006: 51) suggests that the very idea of a regional cuisine is challenged by the breakdown of traditional regional structures given the ease with which regional boundaries are now crossed. 'Can any other commodities rival foodstuffs and drinks in the ease with which they are shipped from one place to another? The more one takes a hard look at the very idea of a regional cuisine, the more improbable it sounds. At the least it warrants closer inspection' (Peace 2006: 51). Yet the push to reregionalize the food system has emerged from a number of different but linked directions set out by Donald et al. (2010), including

> broader concerns that the conventional agro-industrial food system has not effectively provided a nutritious, sustainable and equitable supply of food to the world's population. Technological innovations have provided cheap food to millions, but there are costs of such a system in terms of soil and water depletion, food safety scares, animal welfare, declining rural communities, rising obesity and diet-related health problems, as well as growing food insecurity.

The late twentieth century saw a renewed focus in food terms on local communities and regional units of analysis and action, as the basis for the development of a new geography of food, built on linked approaches to elements including sustainable agriculture, food justice, resilience and obesity reduction. Parrot et al. (2002: 242) argue, for example, that as the previously mentioned productivist model has come into question, the 'protection of cultural patrimony in the south and the fragmentation of the food supply system in the north' both represent expressions of resistance to the legitimacy and hegemony of the global agrifood system. Globalization and its counter-movement thus give rise to what Whatmore and Thorne (1997) describe as an 'alternative geography of food'. Yet we are cautioned against understanding the local in relation to food in an unreflexive way. Case study research in France and Norway suggests that the local in food terms can be quite 'concrete and physical', that is, merely about closeness, but can also reflect strategies of relocalization related to food product quality (Fort et al. 2007: npr).

The term alternative food networks (AFNs) emerged as part of the 'quality turn' away from global 'agrifood' experienced by both those who produce and consume food, although what constitutes 'quality' remains contested (Goodman 2004; Harvey et al. 2004) as does 'authenticity' (Pratt 2007). Similarly, food's aesthetic aspects may have been under-explored (Murdoch and Miele 2004). However defined, increases in quality did not start here: they were always a part of food production and 'the turn to quality has no single defining set of characteristics based around local ecologies' (Winter 2003: 25) as in Figure 9.2. Rather, high food quality has come to be seen as associated with alternatives to dominant production modes and connected to embeddedness in particular locations through alternative food networks and producer groupings (Ilbery and Kneafsey 2000). Such networks cover newly emerging combinations of producers, consumers and other actors who embody alternatives to the more standardized industrial mode of food supply (Murdoch et al. 2000, in Renting et al. 2003: 394). 'Variously and loosely defined in terms of "quality", "transparency", and "locality", such newly emerging networks are (somewhat contentiously) signalling a shift away from the industrialized and conventional food sector, towards a re-localized food and farming regime' (Sonnino and Marsden 2006: 181). These changes are closely related to the way that trust in conventional agriculture has diminished, with food scares and scandals suggesting mainstream methods are not fit for purpose. 'Since the late 1970s the public image of agriculture has become dominated by an ongoing stream of "food scandals" ranging from salmonella and bovine spongiform encephalopathy (BSE) to dioxine residues in milk' (Renting et al. 2003: 395).

It has been argued that alternative food networks emerged as an idea to help niche products in regions marginal to (or marginalized by) global commodity chains and regions highly integrated into global food markets using modern production methods and market structures (Renting et al. 2003: 395). As it has become apparent to some that the dominant discourse of continued expansion of such mainstream arrangements is mistaken and cannot deliver either sustainable agriculture or rural development, 'increasingly we are witnessing an impressive growth of a variety of new food-production and trade circuits falling outside the conventional model of agriculture' (Renting et al. 2003: 395). Among direct producer-consumer techniques

FIGURE 9.2: Local food products, Porto.
Photo: Susan Parham.

are Pick Your Own arrangements, roadside food stalls, vegetable boxes, mail order and Community Supported Agriculture schemes. It has been noted that carbon emissions from vegetable box schemes are likely to be lower than driving to farm shops (Coley et al. 2009).

Planning regimes have sometimes had difficulty in coping with these new entrants to farming and retailing (Nichol 2003). Driven by changing consumer attitudes which have hardened into a distrust of industrialized farming methods, as well as by pressure on farm incomes, the implications of such practices are a challenge to productivist approaches to rural food development, with their stress on continued increases in the volume of production, buttressed by technical and regulatory treadmills that also increase costs (Nichol 2003). At the same time, the very notion of food quality has become contested:

> a comparative analysis makes clear that there is a diversity of competing definitions of quality along these food supply chains, both between and within countries. This is exemplified by the very different ways in which consumer demands and new producer supplies are articulated to specific (organic, integrated, regional, artisanal, etc) production 'codes'. These differences result from a diversity in farming systems and territorial settings, different cultural and gastronomic traditions, a diversity in the organisational structures of food supply chains, variations in consumer perceptions, and also from substantial differences in institutional and policy support. (Renting et al. 2003: 394)

Alternative views about quality have spatial implications, and although work on food chains in the Scottish Borders by Ilbery and Maye (2005) cautions against conflating terms such as 'local', 'alternative', 'speciality' and 'sustainable', shorter supply chains can allow food to be respatialized and re-localized, with particular possibilities related to public procurement (Morgan 2008). These shorter chains act to create new linkages between a range of actors in agriculture and society, bringing closer food producers and consumers (Renting et al. 2003: 394). One of the things that these supply chains do is allow consumers to 'make new value judgments about the relative desirability of foods on the basis of their own knowledge, experience, or perceived imagery. Commonly, such foods are defined by either the locality or even the specific farm where they are produced; and they serve to draw upon an image of the farm and/or region as a source of quality' (Renting et al. 2003: 398). These conceptualizations also reflect concern for authenticity and quality in food that is very much tied to place (Feagan 2007). Rather than signalling the 'invisible hand' of the elusive free market, short food supply chains (sometimes given the acronym, SFSCs) are constructed by the networking actions of farmers, distributors, processors and consumers. 'In countries such as Italy, Spain, and France SFSC-development to a large extent appears to centre around activities of regional quality production and direct selling, which built on long-lasting cultural and gastronomic traditions' (Renting et al. 2003: 406).

Yet there is a heterogeneity in relation to alternative food networks that means their nuances need to be explored, in part because the relationship between conventional and alternative food chains is complicated by the nature of relocalization processes (Renting et al. 2003: 406). If conventional agriculture has been all about delocalizing, the current relocalization that alternative food networks represents is likely to be highly contested. Despite the arguments made by Donald (2008: 1259) that a more 'firm centred' approach can help with 'the challenges facing local food communities' and in developing a 'post-Fordist hybrid food economy', it appears that there will be increasing pressure brought to bear from those promoting genetically modified technologies, involved in the internationalization of corporate retailing and developing market share for retailer own brands, with a range of spatial effects (Sonnino and Marsden 2006: 195). At the same time, a more heterodox agrifood paradigm may be emerging. One argument is that Michael Storper's (1997a) '"Holy Trinity" of new regional economics: technological changes, organizations, and territorial transformations' may be reconstituted and evolved in a more regionally embedded way (Sonnino and Marsden 2006: 194).

Equally, it is worth remembering that alternative food networks have only been partially successful. Food consumers remain very wedded to mainstream food systems which reflect cultural and economic norms in relation to consumption. Work from Ireland on so-called consumer lifestyle segments suggests that there are some 'extremely uninvolved consumers' who would not in any case be likely to develop more adventurous food practices (Ryan et al. 2004). Sage's (2003: 49) account of the difficulties of establishing a farmer's market in south-west Ireland 'in the face of official and private developer hostility' is a case in point. While Sage (2010) and others have offered suggestions for ways to reconnect food production and consumption in more positive ways, with alternative food networks in an embryonic state, any argument that a paradigm shift is well advanced from mainstream to alternative would be premature. It is possible thus that 'quality food production is destined to remain as a narrow "class diet" of privileged income groups in the absence of consumer price subsidies and related instructional changes'. It seems probable that only if greater attention is paid to the consumption side of the food chain will poorer consumers also benefit from the safe, nutritious food offered by alternative food networks. In the next section we explore how enlightened forms of urbanism could assist in this process.

CASE STUDIES OF REGIONAL FOOD AND CONVIVIAL URBANISM

How regional relationships play out in real places can be traced through case studies of the interplay of food and urbanism. This section briefly reviews findings from primary research in two regional food contexts, one in the Italian region of Tuscany, centred on Florence and the other in the southern French region of Languedoc Roussillon, around Montpellier (Parham 1996). The first of these two sites was previously discussed in Chapter 7 in relation to the gastronomic landscapes of the urban edge and touched on above in the discussion of *mezzadria* and its impact on shaping the region's gastronomic landscape. In this part of the chapter regional food implications in contemporary practice are foregrounded in both study areas. In each region, wealth has traditionally come from agricultural and viticultural production. These are activities which after some indications of decline have regenerated by responding successfully to challenging market conditions, both by improving product quality and through clever spatial planning and design. The ways in which valued landscapes, ways of life, local ecologies and regional economies are being protected and enhanced in the face of rapid urban growth and economic change may make them useful examples for other regions facing similar food-related contradictions and imperatives. Both regions are dealing with powerful global economic forces: the internationalization of capital, the dominance of multinational food and wine companies and the decline of manufacturing and traditional markets in the face of low-cost, high-volume products. At the same time they have sought to protect their gastronomic advantages in part through attention to urbanism which stresses convivial and sustainable elements.

Of course, one obvious question is whether such place-specific experiences of regional success can be transplanted elsewhere. For example, one claim made in this area is that Europeans have been somewhat reluctant to embrace the very technically driven wine-making methods of the New World that it is claimed discount *terroir*, while New World wine makers undervalue and even disparage European experience as somehow elitist and inappropriate to viticultural conditions in the Americas, South Africa, Australia and New Zealand. This may be both an oversimplification and a false dichotomy. Differences in the intensity of cultivation can be expected given varying landscape conditions but are not in themselves sufficient to overshadow the range of points of comparison. Obvious similarities between the regions include the relationship between high-quality agricultural and viticultural production and upmarket tourism and the landscape and ecology of rural terrains with cool microclimates.

At the time primary research was undertaken (1996), both Tuscany and Languedoc Roussillon had been affected by difficult economic changes occurring globally. However, in each region rural spaces were being transformed without losing their strengths from a gastronomic perspective. Both places revitalized the economic capacity of rural food and wine production as sources of innovation, while maintaining or renewing the natural, cultural and built environment. New farm-related activity and new non-farm land uses, evolving in two quite different styles in the two regions studied exploited the associated benefits of food and wine production and agritourism, and each melded rural and urban business activity in a way that supported the regional economy, culture and environment. Both showed that urban and rural land uses need not undermine one another but can produce a symbiotic urbanism relationship that plays a key part in revitalization, innovation and conviviality.

The fieldwork suggested that both the Tuscan and Languedoc regions have worked with their physical terrain and tried to reestablish a fine grain of land uses, recognizing especially where the agricultural region intersects with city fringes that this grain is very important for diversity, which is, in turn, positive both ecologically and gastronomically. By concentrating on producing and marketing the best of their high value-adding produce in food and wine at home and beyond the region, they avoided joining the race to the bottom represented by mass volume production.

Both regions concentrated on linking agriculture, viticulture and landscape quality, making the regions more attractive to invest and live in. Both linked their built and cultural heritage, urban and rural, to food production and gastronomic tourism, from artisan producer to traditional central city market place.

In each case, the region was flexible about mixing farm and non-farm uses on rural holdings, supporting 'boutique' producers, increasing food and wine product quality and using regional and locational guarantees to ensure recognition of these high-quality products. Those interviewed were very opposed to generic brand-based labelling, and considered appellation contrôlée or DOCG systems critical to economic success, cultural stability and landscape ecology. Languedoc, in particular, linked high technology industry sectors spatially and economically to the increasing quality of its agricultural base; thus, for example, people who came to Montpellier to work in the computer and other technology industries were also investing in boutique viticulture. Tuscany, meanwhile, was notable in protecting its olive trees, which were seen as not only a gastronomic but also a cultural and ecological landscape device, essential to regional rural character.

Both case studies demonstrated that the wider benefits of maintaining rural and rural-urban fringe agriculture in a region could not be adequately defined in terms of strict sectoral accounting, although even within the agricultural sector value-added commodities were already increasingly critical to regional economies. Against the conventional wisdom of agribusiness, support for supposedly marginal, small-scale production was seen to be important because of its flow-on effects in terms of food-centred tourism, landscape protection, social cohesiveness and regional ecology. Perhaps the key finding from the research was that a very high level of awareness in government and among other regional players meant that food and urbanism were consciously brought together to valuable regional effect.

SLOW FOOD AS A CONTESTED EXPRESSION OF FOOD-CENTRED REGIONALISM

The Slow Food movement has increasingly come to symbolize and embody the approaches to regionality in food explored in the cases above (Petrini et al. 2007) but has been equally subject to critical readings. Founded in 1986 in the Italian town of Bra in Piedmont, the Slow Food movement (and more latterly that of the related Slow Cities or *Cittàslow*) are embedded in notions of sustainability and conviviality for both food systems and places, focusing in on the 'right to pleasure' but problematized as a middle-class discourse (Sassatelli and Davolio 2010). In focusing on liveability and quality of life, these linked movements have been situated as offering an alternative, unorthodox approach to urban economic development which provide a viable model 'that is especially sensitive and responsive to the complicated interdependencies between the goals for economic development, environmental protection, and social equity' (Mayer and Knox 2006: 321). As Pietrykowski (2004: 307) argues, Slow Food manages to both focus on the material pleasure of food consumption and to 'promote a socially and environmentally conscious stance' towards such consumption. Yet van Otterloo (2005: 259) cautions against thinking of this as a movement solely emanating from the last two decades of the twentieth century: rather Slow Food is intimately connected to the rise and development of the modern-industrial food chain since the 1890s which has been marked by 'recurrent manifestations of discontent and protest' about refined, factory-produced foods (van Otterloo 2005: 268).

Slow Food has famously offered a holistic alternative (and challenge) to mainstream thinking and practice in the regional development of food systems as it spread globally to some 80,000 members in over forty convivia by 2002 (Miele and Murdoch 2002; Petrini et al. 2007). Similarly to alternative food networks reviewed above, these have been described as Alternative Agro-Food Networks (AAFNs) in which producers and consumers attempt to find ways to circumvent the extreme lengthening of food chains and re-emphasize quality (van Otterloo 2005: 259). Both Slow Food and Slow Cities offer ideas for city planners on how to pursue a convivial food-centred urban and regional agenda that emphasizes distinctiveness and valorizes place identity (Mayer and Knox 2006: 322). Thus,

> Slow Food's emphasis on the way food is produced and consumed and its normative goal of promoting organic, seasonal, traditional, and distinctive food highlights characteristics such as high quality, asset specificity, sensitivity to local history and culture, as well as crafts orientation and sustainability. Local sensitivity and authenticity seem to be an important component of the alternative agenda. (Mayer and Knox 2006: 322)

Drawing on urban regime theory, Mayer and Knox (2006) distinguish the ways in which Slow Food departs economically from the mainstream of regional and city-based development strategies (including for food), which are predicated on corporate understandings of what constitutes success in an increasingly globalized place-based competition. Instead, Slow Food (and Slow Cities, of which more below) offer opportunities to build alternative economic spaces that stress territoriality based traditional foods and food practices, grassroots efforts in localizing food systems, and food- and place-related distinctiveness. Slow Food's regionality both arises out of its core principles and is reflected in specific regional affiliations, including the biannual *Salone del Gusto* which is supported by the Regional Authority of Piedmont and offers opportunities to showcase regional food products and increase consumer awareness. This, in turn, is intended to garner support for aspects of everyday life under threat, like Italy's traditional *osterias* and *trattorias* (Mayer and Knox 2006: 326) that retain close links to local food production systems (Miele and Murdoch 2002: 317). Slow Food produces a guide to *Osterias* which also covers a range of other traditional regional food businesses including cafés, artisan *gelaterias*, bakeries, delicatessens and makers of traditional food products. Central to Slow Food's arguments are specific cases in which regional food and wine products, cultures and landscapes at risk can be retrieved from oblivion by action to protect and enhance them, through creating or reviving positive feedback loops between production and consumption. The case of the Cinque Terre's characteristic wine is one example:

> The region is known for its steep terraced hills along the Mediterranean coast of Northwestern Italy. Wine production on these steep hills became almost extinct and the cultural landscape was in danger. Slow food promoted the protection of

the vineyards by emphasizing the quality of the locally produced wine, the so-called Sciacchetra wine. Higher quality means higher prices for the wine, which in turn makes it worthwhile for young people in the villages to become vintners. This made it more appealing to younger generations to continue caring for the vineyards and thereby cultivating the landscape. (Mayer and Knox 2006: 327)

Another example of the interplay of Slow Food and regional space is found in Leitch's (2003) analysis of Slow Food and the politics of pork fat in Carrara where the locally celebrated *lardo* of Colonnata has become increasingly widely known by gastronomers. In a wider regional context by which artisanal regional specialities (and their associated cultural traditions) were being threatened by European Union food safety requirements (as noted, for example, in Otiman 2008, and discussions of identity and distinctiveness in Castellanos and Bergstresser 2006), the phenomenology of Colonnata's *lardo* (Castellanos and Bergstresser 2006) came to be associated with questions of regional identity:

> by the late 1990s, Colonnata had become a major destination for international culinary tourism. Venanzio, a restaurant named eponymously after its owner, a local gourmet and lardo purveyor, was one attraction, but Colonnata's pork fat was also being promoted with great acclaim by Pecks, Milan's epicurean mecca. Moreover, it had even been nominated as a delicious, albeit exotic, delicacy by writers as far afield as the food columns of *The New York Times*. (Leitch 2003: 438)

In approaching Slow Food's rurality, Miele and Murdoch (2002) meanwhile foreground its gastronomically based aesthetics: reclaiming the centrality of food in its season and terroir from a mainstream of mass-produced standardization in which food consumption environments rather than the food itself have taken precedence. The notion of terroir is explored in the next part of the chapter but looking in detail at Bagnoli, a restaurant located in rural Tuscany that is thought to be representative of the *osteria* that Slow Food seeks to protect, Miele and Murdoch (2002: 318) find that it embodies 'slow' principles through typical cuisine, materials, products and conviviality, as well as how 'a practical aesthetic becomes bound into the everyday practices of food production and consumption'.

Not all are equally impressed, however, and critiques of Slow Food often focus on perceived weaknesses in what has been conceptualized as the 'intermediate field' between consumption and production that Slow Food occupies in its attempts to condense the food chain (van Otterloo 2005: 273). It has certainly been difficult for Slow Food to escape the charges of representing various individuals' elite habitus in Bourdieu's terms. It has also been argued to be complicit in processes of gentrification through the gastronomic tourism that its work perforce underpins. As Chrzan (2006) notes, the so-called Tuscan Experience in which upper-middle-class visitors perform certain dietary and health identities perhaps 'represents new concepts of the idealized-self created through dietary renewal, lifestyle management, and rituals of imagined tradition and community'. Yet it should also be noted that work about rural slow tourism in Japan and in the Brazilian coastal town of Paraty does not fit this analysis but suggests more authentic relationships and experiences rooted in real traditions (Murayama et al. 2012; Parasecoli and de Abreu e Lima 2012).

Clearly irritated by Slow Food's claims for authenticity and conviviality, Peace (2006: 51) meanwhile, presents the experience of a Slow Food event in the Barossa Valley region of rural South Australia, to support the thesis that in a regional location argued to be fully integrated into global agribusiness, such an event is more about 'myth manufacturing' than an expression of genuinely regional cuisine. This appears to connect to ideas about food nostalgia discussed in Chapter 3 and features in a rather less sour analysis of Slow Food's translation to American circumstances, where Paxson (2005: 15) concludes that the movement could help Americans work their way through food's social and ethical concerns. However, Paxson (2005: 15) argues, this capacity has been stripped of its radical political edge, and, by its celebration of capitalist consumption, been denatured. If understood in these terms, 'Slow Food may be viewed as an exercise in what anthropologist Renato Rosaldo has called "imperialist nostalgia", romantic regret for selectively recollected conditions whose loss one's own society has condoned if not orchestrated. Harboring Native American wild rice does little to address the poverty-based malnutrition afflicting Native American communities' (Paxson 2005: 15).

Notwithstanding these divergent perspectives on Slow Food, its influence on thinking about regionality in food terms has been significant and its consideration of the spatiality of 'slow' was part of the impetus for the related *Cittàslow* (Slow Cities) movement, established in 1999 by mayors of three Italian towns. By 2006, *Cittàslow* covered a network of over forty towns, mostly in Italy, but also Waldkirch, Hersbruck, Schwarzenbruck and Uberlingen in Germany, Levanger and Sokndal in Norway and Aylsham, Ludlow and Diss in the United Kingdom, while Seferihisar and Gokceada have emerged as a *Cittàslow* destinations in Turkey (Mayer and Knox 2006; Yurtseven and Kaya 2011; Yurtseven and Karakas 2013). To join the network, such Slow Cities must house fewer than 50,000 inhabitants and 'comply with a list of criteria covering environmental policies, urban design, support for local products, conviviality, and hospitality' (Mayer and Knox 2006; Yurtseven and Kaya 2011; Yurtseven and Karakas 2013). A Slow City charter encapsulates these slow principles (Mayer and Knox 2009). For example, the Slow City of Waldkirch in the Black Forest, home to some 20,000 residents and located in Freiburg's agricultural and recreation hinterland, has emphasized the connections between social sustainability and 'slow' qualities through a number of food-centred projects and initiatives (Mayer and Knox 2006: 328). The interplay of design and social life of this food-focused activity connects to points made in Chapter 3 about food's role in the spatial and social structure of everyday life:

> a strong sense of place in Waldkirch's city center is maintained by the tradition of conducting the main farmers' market on the prominent central square. Twice a week, the market attracts local residents as well as visitors from outside. Because the local square is automobile free, vendors and visitors use the square without being disturbed by cars. Market visitors typically take time to sample produce and to interact with friends and acquaintances. (Mayer and Knox 2006: 330)

Andrews (2008: 152) notes that Slow Food argues for a kind of virtuous globalization which critiques neoliberalism, but instead of rejecting globalizing food trends outright, it seeks to offer alternative forms of globalization which support and involve small producers and poorer countries. In a Slow Food–influenced food system, networks of local economies would support small farmers, through economic assistance, the revival of artisan food skills and practices and biodiversity protection, including that represented in food terms through the 'Ark of Taste'. Terre Madre, the 'World Meeting of Food Communities', held by Slow Food in 2004 is argued to have provided the basis for this turn in the movement's direction (Andrews 2008: 48). Of interest in relation to conviviality is the argument that instead of this being a concept rooted in abundance, it may be closer to the working-class convivial practices noted by Bourdieu and those of 'traditional cultures [which] had strong histories of gastronomic expertise, including in places like India where malnutrition was rife ... Pleasure, Petrini argued, was prevalent in these traditions and their convivial ways of producing food which had been denied by "monocultural production"' (Andrews 2008: 158).

REGIONAL FOOD AND TERROIR

The notion of 'slow' is closely tied to the idea of terroir to which we now turn. In the face of expanding waves of globalization in agriculture, the homogenization and hybridization of cuisine and increasing uniformity in urban life, support for regional distinctiveness in food (and especially) wine has been linked to the concept of terroir (Hall and Mitchell 2002; Van Leeuwen and Seguin 2006). Parrott et al. (2002) situate this as reflecting a north-south divide in European terms, with terroir largely associated with food cultures of the south of Europe rather than its north, something which is reflected in significantly more place-based protective appellations (which are discussed below). Derived from the Latin territoire, terroir is argued to be a 'near mystical' French term that, despite certain semantic shifts, 'signifies an elusive combination of the effects of sun, soil, weather and history' (Deloire et al. 2008: 1). As Tomasik (2001: 523) notes in his study of de Certeau's use of the term,

> Etymologically, the term is a conflation of tioroer (1198) and tieroir (1212), both deriving from the popular-Latin terratorium, a Gallo-Roman variation on territorium, from which we get territoire, territory. Terroir, originally denoting land or an expanse of land, had begun by the end of the thirteenth century to be associated with agricultural land, specifically in terms of soil qualities that affected the flavor of wine.

In modern times, terroir is situated as 'an area or terrain, usually rather small, whose soil and micro-climate impart distinctive qualities to food products' (Barham 2003: 131). It has been defined by UNESCO in the following way:

> Terroirs constitute a responsible alliance of man and his territory encompassed by know-how: production, culture, landscape and heritage. By this token, they are the fount of great human biological and cultural diversity. Terroirs are expressed by

products, typicality, originality and the recognition associated with them. They create value and richness. A terroir is a living and innovative space, where groups of people draw on their heritage to construct viable and sustainable development. Terroirs contribute to the response to consumer expectations in terms of diversity, authenticity, nutritional culture and balance and health. (INRA, INAO, UNESCO 2005)

Terroir can be understood as a central element in defining a region's gastronomic identity, but one that is also subject to critical analysis. The concept can oscillate from the particular to the general, as slippage from *produit du terroir* to *produits de terroir*, which can refer to products from more than one region (Tomasik 2001: 524). For some, terroir has more broadly become a kind of code word speaking to the valorization of local food production in opposition to globalizing agricultural practices, while for others more pragmatically it is a way of signifying quality to consumers (Aurier et al. 2005). Barham (2001) argues that it may thus be liable to be misunderstood and misappropriated and used to reflect 'invented traditions' rather than actual ones (Trubek et al. 2010, quoting Hobsbawm and Ranger 1983) although as Hobsbawm and Ranger have made clear traditions are always undergoing a process of invention and reinvention. In France, it is suggested, terroir can articulate the notion of a stable and ahistorical culinary paradise of authentic and natural places and products that stand in opposition to urbanization and homogeneity (Tomasick, quoting Poulain 2002). Terroir has also been conceptualized as embodied in Bourdieu's sense, reflecting high degrees of symbolic capital and tied to forms of rural elite formation (Brunori 2006).

However, at the level of the individual *produit du terroir*, these rather critical readings appear to sit uneasily with the often-passionate advocacy by growers of the quality and regional specificity of their products. The *Ratte du Touquet*, for example, is a particular kind of small potato grown in the Touquet region of Nord-Pas-de-Calais, which is publicized at the point of purchase by leaflets that accompany the straw baskets of these *pommes-de-terre*. These leaflets offer information about the producers (Dominque, Olivier, Benoît, Audouin, Éric, and Marc-Antoine), as well as recipes which emphasize the *Ratte's* gastronomic quality by combining it with other prized ingredients including *foie gras* and truffles. More information can be found through a producers' website and a smart phone quick response code. It is even possible to now *suivez la Ratte du Touquet sur Facebook*. The growers explain on the website set up to publicise the potato that over a forty-five-year period they had seen production of the potato variety reborn and then gain momentum: 'gradually integrated into a land that it fits very well. More importantly, it has taught us seven producers to adapt to its quirks and its low yield, and continually rethink our approach to progress'. As the growers' blog notes, the producers have attempted to regenerate this potato varietal as a part of France's culinary heritage, and to reflect its terroir:

> Depuis 1965, et l'année où mon beau-père a régénéré la Ratte du Touquet, cette petite pomme de terre a parcouru un long chemin, jusqu'à être aujourd'hui reconnue, il me semble, comme un véritable produit de terroir: une variété ancienne à la saveur atypique, appartenant au patrimoine culinaire français. (http://www.larattedutouquet.com/)

[Since 1965, the year my stepfather regenerated the *Ratte du Touquet*, this small potato has come a long way, to be recognised today, it seems to me a genuine local product of the terroir: an old variety with an atypical flavour belonging to French culinary heritage.] (author's translation)

Development of *goût de terroir* (the taste of place) has also become a conscious programme in certain locations outside France (as in Figure 9.3) where food producers work to 'create desired places through promoting an artisanal taskscape of farm-based agricultural food processing' (Paxson 2010: 444). Trubek and Bowen (2008: 28) argue that, on the basis of their case study of Comté cheese, it would be possible to create the cultural and political capital necessary in the United States to localize production. At the level of practice, the concept of terroir is being invoked as a way of reverse engineering agrarian, social, gastronomic and environmental stewardship of place through the development of particular food products such as artisan cheeses (Paxson 2010: 444). Paxson's ethnographically based exploration of this phenomenon indicates that

> Terroir is also being translated to suggest that the gustatory values that make artisan cheeses taste good to consumers are rooted in moral values that make the cheeses ethically good for producers to make. Drawing on the holism of terroir – what one cheesemaker described as 'everything that goes into the cheese' – artisans argue that the commercial value of their cheese is derived from underlying assets that cheese sales also protect: independent family farms, unconfined dairy animals, and working landscapes.

FIGURE 9.3: Regional meat products stall, South Australia.
Photo: Susan Parham.

APPELLATION AND OTHER PLACE-SPECIFIC BRANDING SYSTEMS FOR FOOD

Attempts to retrofit terroir need to be approached in an informed way. In France, the concept of terroir is closely, and somewhat messily, connected to the Appellation d'Origine Contrôlée (AOC) system of which more below (Barham 2001: 1). Some major food-producing countries have strongly opposed AOC-style labelling and have made attempts to ensure this is not accepted within the World Trade Organization with whom ultimate power resides. This difference of view is made manifest in conflict between the United States and the European Union in trade negotiations, over what, if any, level of protection or recognition should be given to geographical indications (GIs), which would in turn give terroir legal expression (Josling 2006: 338). It was reflected in protests in 1999 led by French sheep farmers and activists Jose Bové and Francois Dufour, against American trade barriers to Roquefort cheese, imposed in retaliation against European barriers to the import of beef growth hormones. This advocacy for ethically conscious regionality and against technocratic, placeless functionalism in food production also struck a chord with urban consumers (van Otterloo 2005: 259). Gastronomic heritage can thus be understood as a form of intellectual property requiring protection, and this may also underpin the capacity to strengthen food systems on the basis of terroir in future (Ravenscroft and van Westering 2002: 163). However,

> Many food activists, particularly those from 'northern' and settler countries such as Britain, the United States and Australia, know little or nothing of the details of the French AOC system, or similar systems of other 'southern' countries of Europe (Spain, Italy, Portugal and Greece). By failing to recognize the connection between terroir and AOC labeling, sustainable agriculture activists may miss the opportunity of pursuing similar systems at home that have some administrative 'backbone'. (Barham 2001: 3)

In France, as noted above, the AOC guarantees the quality and authenticity of products such as *Puy* lentils and *Bresse* chickens, but there are also *Le Label Rouge* covering some 254 food products: *le label régional* which classifies foods of specific regions as 'products highlighting local character', *l'appellation Montagne* covering mountain food products and *les certificats de conformité* which show that foods included have been processed within certain quality norms (Bessière 1998: 25). In Brittany, for example, foods including seafood, artichokes, cauliflowers, onions, potatoes, carrots, endive and lettuce are seen to typify the region (Roux and Roux 1989: 37). Products such as certain Breton ciders are labelled as *Produit en Bretagne* or fall within an Appellation d'Origine Contrôlée (Barkat and Vermignon 2006). Equally, protected status for mountain region foods has emerged in a number of countries given ancient traditions such as transhumance where livestock are moved seasonally between valley and mountain pastures. Protected status differentiates and reflects the fragility of such agricultural systems and products in the face of food commercialization (Meiberger and Weichbold 2010: 1626). In Austria, for example, this supports small-scale economic health, gastronomic landscape preservation, ecology and regional food tourism:

> Mountain farms guarantee a small scale interesting cultivated landscape by means of the farmers care with little costs for tourism management. At the same time tourists are prospective customers for mountain food and farm lodging. Thus, besides economic arguments, reasons to maintain cultivated land in the mountains are social: to prevent rural exodus, and ecological: to preserve biodiversity. (Meiberger and Weichbold 2010: 1626)

In the United Kingdom, early adopters of EU quality marks have done so to differentiate their food products from those of inferior quality while in Italy, some typical Italian foods have been given official recognition as Protected Denomination of Origin (PDO) or Protected Geographical Indication (PGI) products under European Regulation 2081/92 (de Roest and Menghi 2000: 439; Ilbery and Kneafsey 2000). Other places in Eastern Europe have also started employing regional branding systems to certify the quality of food products in place-based terms (Spilková and Fialová 2013). In Italy's case, it is estimated that around '6,000 firms are involved in processing protected products and that 300,000 people are directly or indirectly involved in typical product supply chains' (Spilková and Fialová 2013). The story of *parmigiano reggiano* (Parmesan cheese) offers an instructive example of the economic, environmental and cultural strengths of a place-based product and production system, as well as an insight into threats from inferior products and an industrialized race to the bottom for food products (Marsden 1998). Like *prosciutto crudo*, another long-lived, regionally based food, with which Parmesan production is interconnected, at base, *parmigiano* reflects the ecological requirements to preserve food through Italy's long hot summers and cold winters and was first produced by Benedictine monks in an area of Emilia in the thirteenth century (de Roest and Menghi 2000: 439; Pacciani and Italiani 2005). Although Benedictine lands were increasingly expropriated by the Este family, Parmesan production continued, with cheese dairies emerging in the eighteenth century and co-operative dairies in the late nineteenth, to become a characteristic feature of Emilio Reggiano's countryside (de Roest and Menghi 2000). By the mid-1930s, a regional organization was established – the Voluntary Consortium of Typical Grana – which was revived, reconstituted and renamed post-war as the Consorzio del Formaggio Parmigiano Reggiano (de Roest and Menghi 2000: 440). The geographical area for Parmesan production was set and remains critical to its successful gastronomic functioning:

> In 1955, an official decree was issued defining the *Parmigiano Reggiano* production area as the Provinces of Parma, Reggio Emilia and Modena and parts of the Provinces of Bologna and Mantua. The entire production area covers 1.02 million ha and contains 550,000 ha of Utilized Agricultural Area. Three main sub-areas can be distinguished in the *Parmigiano Reggiano* region, the mountains, hills and plains. Each sub-area has a different altitude, rainfall, climate and farming conditions. Farms in the Apennine mountains are at altitudes of between 800 and

1,200 metres above sea level and upland areas, some 400 metres above sea level, dominate the hilly landscape. (de Roest and Menghi 2000: 441)

With around 73,000 farmers supplying numerous small-scale co-operative dairies and cheese ripening firms, the supply chain is marked by strong, stable relationships, which are thought to be at the core of a mature food system. The traditional methods used are labour intensive, so that Parmesan production supports greater numbers of workers than does mainstream agriculture. The methods used are also more ecologically benign, causing lower levels of nitrogen loss from soil than conventional dairying, and more broadly, protecting a valued landscape:

> The positive effect on employment is generated both in the plains and in the less favoured mountainous Apennines region. The continued use of steep prairie areas for agriculture contributes to the preservation of its characteristic landscape. In this respect the hills and mountains within the Parmigiano Reggiano production area are better off than those of neighbouring areas not involved in its production. In the latter, land has been abandoned which has lead to problems of soil instability that can only be reduced by costly forestation programmes. (de Roest and Menghi 2000: 441)

Parmesan cheese, as a delicious accompaniment or eaten alone (dug from the block with a specifically designed gouging implement) continues to be popular in Italy, and markets for it are expanding globally as Italian gastronomy increases its cultural reach. However, threats remain, including inferior products sold as 'parmesan' outside Europe, increasing labour cost, new technologies for feed systems and, as de Roest and Menghi (2000: 441) note,

> A fundamental issue that will have to be dealt with in the near future is the extent to which the drive towards cost reduction compromises the final quality of the cheese. Some large dairy farmers may feel less integrated into the still strong Parmigiano Reggiano actor's system with its commonly shared values and norms and may be attracted to the more industrial dairy farming system where they have more freedom to introduce cost reducing techniques.

REGIONAL FOOD TOURISM

Since travelling for pleasure emerged as a mass pursuit, especially in the twentieth century, to the seaside, the countryside and to cities, an important area for the intersection of food, design and regional space has arisen through various kinds of food-centred tourism (Segreto et al. 2009; Urry and Larsen 2011). Rural tourism and gastronomy are increasingly seen as a good fit (Sidali et al. 2011). Food, culinary or gastronomic tourism forms have been identified (Hjalager and Corigliano 2000; Hjalager and Richards 2004; Cánoves and de Morais 2011; Long 2012; Parasecoli and de Abreu e Lima 2012). These are now sometimes situated as slow tourism (Fullagar et al. 2012). Some work has also been done on developing sustainable gastronomy as a tourism product (as in Scarpato's 2002 collective case study). Hjalager (2002: 21), for example, has developed a typology of what she terms 'gastronomy tourism' from an economic point of view, but clearly this does not capture (nor seek to) some of its multifaceted nature in design terms. Such tourism may start out with green tourism activities such as renting out rooms, as a side operation to mainstream agricultural activities, to growth in the form of diversified rural agritourism, organic agritourism and ecotourism activities and places (Epler Wood 2002; Ecker et al. 2010). As Clarke et al. (2010) have noted, small, diversified farms are well represented and clearly work for agritourism and food tourism activities.

This kind of tourism is argued to be not only a set of practices but also a social construct that reflects, and offers respite from, a largely urbanized and industrialized society and rests on an idealized image of the countryside (Bunce 1994). Obscure regions may thus be understood as offering greater authenticity in tourism experiences (MacCannell 1976). In such places culinary tourism is increasingly important (Hashimoto and Telfer 2006). This offers among other things the chance to demonstrate cosmopolitanism and taste 'scary' foods as in the example of *Smalahove* (salted, smoked and cooked sheep's head) from Voss in Norway (Molz 2007; Gyimóthy and Mykletun 2009: 2). Food tourism may become correlated with retention and development of regional identity and territorial protection, as research in Cornwall has demonstrated (Everett and Aitchison 2008; Montanari and Staniscia 2009; Timothy and Ron 2013). As shown in Canadian research, those attracted to visit are not an undifferentiated cohort but a market that can be segmented and profiled in quite subtle ways. Activities might include visiting farmers' fairs, markets, pick-your-own farms or participating in harvesting; shopping for regional foods in shops or farms; dining in internationally celebrated restaurants or those with well-regarded regional or local cooking; touring and staying at a region's wineries; staying at a cooking or wine-tasting school or at a restaurant with boarding and lodging (Ignatov and Smith 2006: 241).

This kind of tourism has also been seen as a response to the post-productive rural landscape (Nielsen et al. 2010) and even a form of mass tourism (Weaver 2001). The most salient point is that with the rise of so-called eco or responsible tourists, instead of deprecating rural people and pursuits, the countryside becomes a place of meaning and authenticity (see Figure 9.4); with food and gastronomy a rich source for regional place-identity formation and representation for destination marketing (Bessière 1998: 23; Goodwin and Francis 2003; Kivela and Crotts 2006; Moginon et al. 2012). Coverage is uneven: with places like Sweden having very small agritourism sectors (Nielsen et al. 2010: 5). In countries including Italy, France and Spain, and in particular regions where proximity to urban areas offers the likelihood of more visitors, this is a substantial area of rural activity (Cánoves et al. 2004; Dettori et al. 2004). One example is the previously mentioned regional town of Paraty, Brazil, in which an explicit programme defined as sustainable gastronomy is founded on traditional *cascara* cuisine (Parasecoli and de Abreu e Lima 2012). Its experience suggests that not only are local food traditions being revived, but that the public-realm-focused spatiality of these traditions, including the Festival of Cahaca (sugar cane spirit), perhaps unsurprisingly, is highly attractive to tourists (Parasecoli and de Abreu e Lima 2012).

FIGURE 9.4: Traditional agricultural landscapes, Transylvania.
Photo: Susan Parham.

REGIONAL FOOD TOURISM'S CONNECTIONS TO SPATIALITY AND DESIGN

Starting at the level of food, the *products du terroir* discussed in the previous section have played an important part in many regions' economic, cultural and sustainability trajectory and connected to intriguing spatial design elements. In France, for instance, the Inventaire du patrimoine culinaire de la France, started in 1990 by the Conseil National des Arts Culinaire, is at a national governmental level an attempt to reassert a regional basis for cuisine (Tomasik 2001: 525). In direct support of regional food as an identity marker, that inventory includes a list of a hundred most outstanding food sites across France's regions; the Sites Remarquables du Goût (Bessière 1998: 21). On the ground, such moves have been reflected regionally in ecotourism and agritourism operations where by trying local specialities tourists take part in, rather than merely observe, local food culture. 'By eating a so-called natural or traditional product, the eater seems to incorporate, in addition to nutritional and psycho-sensorial characteristics of the food, certain symbolic characteristics: one appropriates and embodies the nature, culture, and identity of an area' (Bessière 1998: 25).

The spatiality and design settings for this gastronomy matter. Even within a rather technocratic 'tourism studies' mode it is argued that a spatial tourist district 'characterized by a common local identity based on historical, social and economic elements' is required to contain 'a range of enterprises producing and distributing the local tourist product' (Corigliano 2002: 172). Examples of regions boasting such districts in Italy include the Marche, Sicily and Piedmont (Corigliano 2002:

181). Here and elsewhere, interaction with regional foods may take place in traditional buildings and spaces such as renovated *fattoria* as well as in the bespoke architecture of restaurants tied to place-specific food and wine production, such as (to offer a few examples from many) Urban at O. Fournier in Mendoza, Argentina; Folio Enoteca and Microwinery in California's Napa Valley; Magill Estate on the edge of Adelaide, in South Australia; or Marqués de Riscal at Elciego, in Rioja. Food product museums and producer institutes also play a part, often supported by producer tours, gastronomic itineraries and cooking classes (Bessière 1998: 30). The Academia Barilla in Parma, for example, is located in a former pasta factory owned by the firm and is the site for a wide variety of gastronomic activity, much of it educational. In Denmark, meanwhile, manor houses and larger farms act as sites for tourism development with special events and food products (Nielsen et al. 2010: 15).

Regional restaurateurs may demonstrate the typicality of their cuisine through the specifics of their restaurant design as well as the food served. Thus the Ristarante Bagnoli in Tuscany, which began as an extremely basic farmhouse restaurant serving very simple local dishes, eventually became a specifically Slow Food restaurant in which local ingredients and traditional recipes and cooking methods attract diners from local towns, as well as gastronomically inclined visitors from further afield, in part because of the ambience of the location, buildings and dining space (Miele and Murdoch 2002). Explored by Boyne et al. (2002) in relation to the 'Arran taste trail', another typical example of this kind of tourism approach comes from the Burren area of north-west County Clare in Ireland, where food production (salmon smoking) is explicitly linked to the spatiality and economy of the area, including farms and accommodation:

> This week will see the official launch of the Burren Eco-Tourism Network (BEN). The network links very different businesses in their efforts to offer the gentle approach to tourism. The Burren Smokehouse is part of this network, and also innovative ideas like the Clare Farm Heritage Tours Co-Op which offers tours through working farms in the area, and Heart of Burren Walks for guided tours through our spectacular Burren area. Other suppliers are high-quality accommodation, the Cliffs of Moher and many more.

FOOD BUILDINGS AND SPACES

Farm and other rural buildings and spaces are important settings in other ways too. Food-focused farm stays and visits of various kinds have increasingly become the sites for regional food tourism (Sidali 2011). As Bessière (1998: 30) argues, these spaces offer direct ways that tourists can tap into a region's gastronomic and culinary heritage, however problematized that may be. In France, for example, there is a well-established system of *fermes-auberges* (farmstead inns), *tables d'hotes* (family inns with accreditation granted by the Gîtes de France organization), *fermes de séjour* (stays on the farm) and *goûters à la ferme* (snacks on the farm) (Bessière 1998: 30). The *bienvenue à la ferme* promotional efforts refer to farm holidays in which 'real food with flavor' is the objective (Barkat and Vermignon 2006: npr). Certain countries like the Czech Republic have chosen to adopt regionally based branding schemes as quality markers for such spaces, and these, it is argued, could also support culinary tourism (Spilková and Fialová 2012). In Spain, although with a somewhat less formalized system at national level, there is national 'green Spain' branding, and in particular regions like Catalonia, 'spatial' products such as Gites de Catalunya have been created which integrate 'elegant country houses in harmony with their surroundings, with a variety of accommodation options, marketed in various categories by different numbers of ears of corn (as in France)' (Cánoves et al. 2004: 763). In Denmark, meanwhile, the *bondegårdsferie* (farm holiday) or *landboferie* (rural holiday) constitute a well-known tourism form, although the number of farms is steadily declining (Nielsen et al. 2010: 8). *Agritourismo* in Italy has by contrast grown as a developed form of regional tourism provision, again reflecting the cultural attitudes discussed above (Privitera 2009). An account from Abruzzo demonstrates such provision keys into the same kinds of visitor preferences as in Spain, for small-scale, traditional, vernacular architecture, with close ties to food production in beautiful rural built and landscape settings (Roddy 2013).

These 'on farm' spaces are not the only sites for regional food tourism; tourists and second homeowners can take up regional and rural housing stock, squeezing out local populations (Butler and Clark 1992). Thus, for instance, many of Canada's rich stock of rural cottages have been converted into year-round homes, generating change and demands on rural host communities and services (Halseth 1998: 4). In regions suffering from depopulation and decline, however, more positive impacts come from the use of public spaces and leftover building stock in villages and towns, where these are employed in supporting regional gastronomic tourism for accommodation and food services. While farmers sell their products directly on their farms and through mail order, and regional food markets in local villages and regional towns are similarly a feature of the rural food landscape that is highly attractive to food tourists, it would be wrong to suggest that such markets only exist to cater for these visitors. Part of their appeal is that they are forms that cater to local populations and thus again contribute to a sense of authenticity and directness in the food chain. This not only appeals immensely to visitors but, it has more recently been argued, also offers keystones for remaking regional food systems (Gillespie et al. 2007).

Alongside regular weekly markets, may be found specialist markets connected to particular products (such as Bresse poultry) and annual food festivals, fairs and shows (Bessière 1998: 30). Regional food festivals may be situated in the tourism studies literature as about place branding (Lee and Arcodia 2011). They may be approached more sociologically as about reifying ethnicity in inauthentic ways, as was discussed in Chapter 4 in relation to the gastronomic townscape. However, it is clear that regional festivals focused on particular foods often have extremely strong cultural associations to place and history (Field 1990). Thus, in the French region of Brittany, this would include the Fête de la Confiture in la Chapelle Fougeretz or the Chestnut festival in Redon (Barkat and Vermignon 2006: npr). Place and product branding also occur in diverse regional circumstances. Work on food festivals and fairs in Portugal demonstrates a consciously gastronomic strategy of regional product promotion and branding of regional food culture (Beer et al. 2002: 217). From Hong Kong's New Territories region comes the intriguing morphing of *poonchoi* (separately cooked food items served together in a large bowl) from festive rural village food to trendy urban dish (Chan 2007).

RISE OF REGIONAL FOOD POLICY AND DESIGN GUIDANCE

Increasingly, regional approaches to food space are reflected in public policy guidance: a development manifesting growing acknowledgement of issues including climate change and obesity and reflecting an implied interest in more nebulous conceptual areas including conviviality. Given the particular urban edge space tensions explored in Chapter 7, regional food plans often focus on areas linking rural and urban domains, as in the plan for the Greater Melbourne region of Victoria, in Australia (Donovan et al. 2011). There are also examples of broader spatial coverage to be found in a range of regional settings. Examples include the previously mentioned *Community and Regional Food Planning Policy Guide* (2007) in the United States, which offers principles applicable across diverse regions, while in Canada, the Waterloo region's *Healthy Community Food System Plan*, published in 2007, proposes ambitious objectives for the regional food system in that part of Ontario. Paraphrasing Pothukuchi (2009: 354), in the latter case, these objectives encompass increasing availability of healthy food so that healthy choices are easier to make, increasing the viability of farms that sell food to local markets to preserve rural communities and culture, strengthening the food economy, preserving the region's agricultural lands, ensuring all residents can afford to buy the type of food they need to sustain their health, strengthening food-related knowledge and skills among consumers and forging dynamic partnerships to implement the plan.

Similar regionally focused food plans in other places include that for Marin County in California where the plan components read as a tool box for alternative food systems planning, with a stress on short food supply chains and a respatialization to smaller scale, mixed farming arrangements (Pothukuchi 2009: 354). Thus the plan encompasses aspects that (again paraphrasing Pothukuchi 2009: 355) promote organic certification, support local, organic and grass-fed agriculture, small-scale diversification, local processing, marketing of local products (including through the development of a permanent public market), increasing knowledge of agriculture and underpinning intergenerational transfer of agricultural land.

There have also been proposals to 'borrow' labelling approaches used successfully in certain regions to emphasize the regionality and quality of particular foodstuffs, as described above. For instance, a system similar to the red label for poultry in France could well be employed in the United States to deal with a highly dualistic poultry production system that raises chickens in ways which create serious quality, ethical and sustainability issues (Stevenson and Born 2007: 145). 'The "red label" sector of the French poultry industry demonstrates how public policy and private agribusiness strategy are joined to result in high-quality poultry products for French consumers and sustainable economic returns to midsize French farmers' (Stevenson and Born 2007: 145). While such food policy advances and marks of quality have only limited application thus far, they seem hopeful signs in bringing together spatiality and food practice in positive ways for a regionally grounded urbanism.

THE CRITICAL FOOD REGION IN REVIEW

This chapter has been about the region as food space, arguing that this scale of the interplay of food and urbanism is of critical importance to creating a convivial and sustainable future. The region's food practices have been interrogated from a variety of angles, starting with a brief review of the way agricultural practices have shaped rural space over millennia, with systems like Italy's *mezzadria* considered in spatial terms. This set the context for exploring today's food regions and regionalism through place-based ideas and policies and foregrounds the rise of alternative food geographies and food networks. As noted, the late twentieth century saw a renewed focus in food terms on local communities and regional units of analysis and action, based on elements including sustainable agriculture, food justice, resilience and obesity reduction. Case studies of regional food and urbanism focused on Languedoc Roussillon and Tuscany demonstrate how gastronomic landscapes are supported by both rural and urban activities in food and farming, with particular, positive spatial effects.

The chapter has considered Slow Food's claims to offer a holistic alternative (and challenge) to mainstream thinking and practice in the regional development of food systems and has looked at specific products such as the *lardo* of Colonnata to explore both its food-centred regionalism and critiques of that approach. Some attention has also been paid to the related movement of *Cittàslow* (Slow Cities), to consider how it developed slow ideas about food in a spatialized way. The chapter linked this discussion to the related concept of terroir, in support of regional distinctiveness in food and wine terms and as a basis for claims for authenticity, but argued that at the level of the individual *produit du terroir*, some rather critical readings sit uneasily with the high quality and regional specificity of food products and places. It suggested that various forms of geographical indication of food quality are critical to supporting gastronomic landscapes, as the example of *parmigiano reggiano* demonstrates. The chapter finally turned its attention to food tourism at regional scale, showing how this is both based upon and can help maintain particular food-related buildings and spaces that reinforce regionally sustainable and convivial urbanism.

Having looked at food and urbanism over nine linked chapters, the next and final part of the book reviews themes explored from the scale of the table to that of the region, and offers some final thoughts about the interplay of food and urbanism at all scales.

Conclusion: Food and Urbanism in Review

A REMINDER ABOUT SOME KEY THEMES

Food and Urbanism has sought to explore some of the critical ways that food interconnects with place, and how that affects, for good or ill, possibilities for conviviality and sustainability in a largely urbanized future. As set out in the book's Introduction, urbanism's focus is on human settlement, and food is understood as central to supporting vital, well-shaped places that put the public realm first (Talen 2005). Over nine chapters, the book has considered the interplay of food and urbanism at and across a series of spatial scales: from the intimacy of the table to the breadth of the rural region. It has examined a huge diversity of food spaces including kitchens, gardens, markets, food shops, street food, cafes, restaurants, supermarkets, shopping malls, fast-food outlets, allotments, community gardens and farms, among others. By working from the small to the very large scale, it has been possible to delve into some of the history of food's design interplay with urban form, from the traditional city to the post-urban megalopolis that now confronts us. This has also been a chance to survey contemporary food and urbanism practices and to foreshadow possible futures in a spatially structured way. However, given the sheer range of concerns about food and place, completeness has not been the intention. Instead, there has been an effort to explore particular food and urbanism themes using a diversity of examples from primary and secondary research in some of the very wide range of disciplines that are integral to this topic.

While undertaking research for this book, a recurrent theme has been the way that food is still often considered within narrow discipline boundaries that limit the possibility for making useful urbanism connections, especially at the intersection of design and social science. While there is an abundance of excellent research in food history, gastronomy, sociology, anthropology, economics and geography, among other disciplines, there are also gaps, and consideration of the spatial design and urbanism implied by these explorations has often appeared severely limited. Although this is starting to change, place is still often situated as no more than a static backdrop for playing out food processes, if it is regarded at all. Food theories and practices, understandably defined within disciplinary terms and bodies of knowledge, are nonetheless approached in largely aspatial ways, without reference to principles that govern place shaping. Similarly, while there are exceptions, food has too often been relegated to the margins of the design disciplines, as a taken-for-granted aspect of place, narrowly conceived as offering a surface gloss of vitality or applied as a kind of pleasant afterthought in spatial design terms. Thus, a central task for this book has been making (or retrieving) more fundamental connections between food and design as an overlooked or marginalized aspect of the interplay between physical form and social, economic and cultural practice.

In this book, the nature of food's relationship to spatial design has been argued to have a significant capacity to either support or undermine making sustainable and convivial places. A fundamental theme following on from this central contention is the argued importance of food to public space and its decline. The loss of public space primacy has been discussed as both a fundamental urbanism issue and a by-product of ideology, with modernism argued to have had largely deleterious effects on settlement form in relation to food. At the same time, in numerous examples, the book has explored physical co-ordinates for successful place shaping for food. In design terms this has been shown to respond to the need to make human-scaled space, in the public domain configured as outdoor rooms, with the right balance between positive and negative space, not only to achieve spatial enclosure but also to create resilient food-centred social and economic space. The book has not only focused on the spatial, but also emphasized the complicated interplay between physical space and the practices of everyday life. It is acknowledged that the urbanism of food responds to structural forces that impinge on how food's spatiality plays out. The connections between food and urbanism reflect a globalized food system associated with particular forms of economics, urbanization and agriculture, among other influencing elements. One of the most critical roles for urbanism is to make these spatial and non-spatial elements work in a positive, intertwined way for a healthy food system. This has been explored at each of the book's scales.

As suggested in the Introduction, taking food seriously is sometimes read as valorizing gastronomy as an elitist cultural obsession in ways that reinforce or ignore inequality. Food may be viewed as simply an aspect of pretentious individual identity formation, or as a frivolous topic compared with the real issues of the day. In such perspectives, the radicalism of alternative food networks may be reduced to no more than a personal expression of a particular habitus, self-indulgently expressed by those who do not have to grapple with more fundamental problems of existence. Yet such views may fail to come to grips with a great deal of evidence about how important food is in place shaping globally, and the critical nature of ordinary and everyday food activities for a sustainable and convivial future.

The notion of the convivial city, in fact, has offered a framework for considering food and space: focusing on fleeting, sociable, everyday pleasures by sharing food and understood as capable of nourishing wider civil society (Peattie 1998). In a considerable number of examples, an ethics of conviviality has been seen to govern food's relationship with place, and convivial ecologies have been observed to either exist already or have been developed in a variety of ways (Bell and Binnie 2005; Bell 2007). Similarly, food's spatiality has been shown to have an extremely important role to play in achieving sustainability, in a context set out in the Introduction of increasingly unsustainable urban forms, which are being exacerbated by climate change. The book has shown how negative sustainability effects exist all along the food chain, with over-consumption of resources and over-production of waste contributing to a negative feedback loop in which urban development has alienated agricultural land and made intensive demands on environmental resources in urban foodsheds (Rudlin and Falk 2001). Just as some enjoy a year-round dietary summer, others do not have enough to eat, and aspects of this untenable food resilience and security situation have been explored in spatial design terms through a number of chapters.

LOOKING CHAPTER BY CHAPTER

In the initial section, 'Food, domesticity and design', the book considered food's spatiality in the private, domestic domain, considering the table, kitchen, dining room, house and garden as food space in terms of both conviviality and sustainability. It has pointed out that urbanism does not cease at the front door but is as important to understanding food's spatiality in the private domain as it is in public space – categories which are in any case significantly blurred and intertwined in food terms. The section's two chapters covered issues associated with the spatiality of the table or its equivalents, design aspects of the changing roles of the kitchen with the rise of modernism, the putative decline of home dining and the impact on the design of domestic life of the rise of ready prepared meals and fast food at home. The section considered the once productive nature of private gardens and their design transformations in an era of mass urbanization, as places primarily for food consumption and symbolic display, to re-emerge in some cases (and be reshaped again) as productive urban food spaces within sometimes more conscious gastronomic strategies for cities.

Chapter 1, 'Starting with the table', explored the meaning and centrality of sharing food at the table (or equivalent) as a near universal expression of conviviality. Its spatialized expressions over time have been considered, as well as aspects of alienation from that state. At the micro scale, the design elements of the tablescape and platescape have been assessed in terms of design's impact on both food consumption and social relations at the table. Evolving kitchen design (including kitchenless flats and houses) has been given particular attention through a review of twentieth- and twenty-first-century house plans. The book concludes that in food terms kitchens must be understood as design models of technological change, as well as expressing the way food space intersects with gender roles, informing both sense of self and self-denial. The rise of Taylorist notions of efficiency, with kitchen design to match, has been a focus for the design analysis of kitchen transformations, reflecting wider societal norms about food and, in particular, women's roles in relation to it. As both deskilling and conspicuous competence have again transformed the design of kitchen space, changing the 'food axis' (Cromley 1996: 8), the chapter moved to issues associated more broadly with the argued decline of food at home, expressed in design changes between kitchen and other rooms for dining and reflecting the impact of the culinary transition (Lang and Caraher 2001: 2).

In Chapter 2, 'The garden and gastronomy', the central role that private gardens have played over the very long term in urban food productivity, and as sites for dining, was explored through a range of examples of productive domestic urbanism including the Classical *xystus*, the Roman *hortus*, the medieval *potager*, the early modern town garden, the allotment and the *jardin ouvrier*. The chapter considered domestic gardens' predominant loss of productive status in twentieth-century cities and investigated design aspects of garden revival for informal food production, as well as more collective urban agriculture, with particular attention to largely positive implications for urban conviviality and sustainability. In recent and contemporary spatial practice, the way that the productive, utilitarian garden was broadly superseded by private garden space for leisure has been explored through a range of urban settings. Themes include the contrast between the decline of garden productivity associated with the advent of modernist tropes in city design and highly productive immigrant gardening experience as an exception to this broader spatial and social shift.

The garden as designed space for food consumption has also been reviewed in both historical terms and in relation to current practices through which it became predominantly the spatial extension of the domestic consumption space of the house. Outdoor dining, as a very long-term practice and design focus, has been seen today to connect to a growing informalization in family relations and domestic spaces, while reflecting stereotypical gender assumptions. Changing garden mores have been seen to be reflected in particular design features apparent in house plan analysis and accounts. The chapter also gave attention to the reanimation of the productive garden (while acknowledging that in some places this never disappeared), and investigated this in spatial design terms as part of a sometimes conscious programme of urban food productivity.

Part 2, 'Gastronomy and the urbanism of public space', both broadened the discussion in scale terms and shifted from the private to the public domain. The focus moved away from the design and urbanism of the individual, private spaces of domestic household production and consumption to aspects of food and public space in cities, elements of which have been explored at increasing scales over four chapters. At one level this is a narrative of the development of the city in food terms from its traditional structure to its suburban expansion in the nineteenth and twentieth centuries, prior to the more recent transformations at urban edges that are considered in Part 3. Across the four chapters, topics included the outdoor room of the food market; the gastronomic townscape of related food spaces; the particular food landscape of suburban growth, including the supermarket and mall; and the role and function of green areas in cities for food production, exchange, consumption and waste.

In 'Food's outdoor room' (Chapter 3), the focus was on food markets as fundamental elements in urban life over very long settlement history. Markets have been seen to be central to everyday social and economic practices, convivially shaping city space and contributing to the economic functioning of towns and cities from the Greek agora and the classical 'hypermarket', to the contemporary market hall, street or square. Their location and design as outdoor rooms at the heart of public space has been shown to be a consistent convivial thread running through urban history, despite a diversity of cultural expressions in spatial terms. The development of market hall typologies has been counterpointed by the decline of wholesale and retail food markets in many twentieth-century urban circumstances. The wanton destruction of Baltard's superb *Les Halles* in the 1970s, and its egregious replacement with a failed underground mall complex, for example, was contrasted with the continued resilience and renewal of market halls and market squares in other places where the food market strongly supported a richly convivial public realm. The chapter also considered the emergence of hybrid markets which reflect new food consumption forms associated with both a hip individual habitus and convivial spaces conceptualized as gastronomic quarters and food quarters (Parham 2005, 2012). Tied to the renewal, reconfiguration (and sometimes gentrification) of traditional market spaces and quarters and the development of some new market halls, it has been argued that these spaces reflect urbanist principles of human scale, compactness, mix and fine grain, and these design qualities were further explored through case studies of a series of extant food markets in Italy, France and Australia.

Chapter 4, 'The gastronomic townscape', widened out the scale from the food market to the broader townscape of food streets, street food, cafés, restaurants and third spaces, considering among other things the decline and re-imagining of traditional high-street food retailing and the related role of street food and food quarters. It was argued that the very long tradition of small food shops has been challenged and largely undermined by predominant trends in twentieth- and twenty-first-century urban spatiality, despite contestation about this evidence. Street food, similarly a marker of place identity and connected to certain street, place and vending infrastructure, has been seen to have undergone a gastronomic renaissance that is also contended by some to signify a particular habitus of personal distinction in Bourdieu's (1984) terms. Design for food buying on food streets and at hybrid markets has been a focus, as has been the connection to ordinary, everyday and 'slow' places, often located within traditional town forms: a spatial connection that has been seen to be a good fit in design terms with convivial approaches to place shaping. The spatial nature of food court dining within 'traditional' townscapes has similarly been interrogated, before the chapter moved on to consider the nature of the café and snack bar as food space: with one element the emergence of food-related third spaces, located somewhere between public and private realms across a diversity of urban circumstances and reflecting melding of actual and virtual worlds.

The development, decline and rise of café culture as convivial (if contested) hospitable space has been explored, including in the detailed discussion of the Singaporean *kopitiam*. The chapter linked this discussion to the restaurant as an important design element in the gastronomic townscape, yet one that has held a curiously ambivalent place of exclusion as well as conviviality. Other semi-public dining forms, including the office canteen and its recent fashionable offshoots, were similarly considered in spatial terms, to explore their complicated contribution to urbanism. The chapter concluded with consideration of dining forms that have taken over public space, as in traditional festivals and through more recent expressions of urban celebration, to consider the part spatiality plays in what may be forms of commodification of space and identity for place branding, but also sometimes acting as examples of food led conviviality.

In Chapter 5, 'Ambivalent suburbia', the scale again expanded outwards to consider in food terms the rise and decline of suburbia, including of 'new urban frontier' cities of the Americas and Australia (Frost 1993; Sudjic 1991; Rowe 1991; Garreau 1991) and explored its spatially expressed ambivalence about food space. Tracing suburban development in food terms through early examples, the chapter focused in particular on Ebenezer Howard's detailed proposals for designing in space for food production, exchange and consumption in his ideas for the Garden City. It considered the implications for food's spatiality of suburban development post-war, including the decline of high-street-based individual food shops, the development of shopping strips and supermarkets and the emergence of shopping malls. The chapter explored the rise of car-oriented and private-space-focused food forms, including fast-food, mall-based consumption and interiorized office 'streets'. Attention was given to the food aspects of mono-functional suburban development forms dependent on car-based urban shaping. Contemporary debates about suburbia were touched on where these illuminated food themes including health and obesity. Spatial design aspects of the uneven nature of suburban development in food terms, including food wealth and poverty, were briefly reviewed, prefiguring the more in-depth discussion in relation to food poverty, food deserts and obesogenic environments in Chapter 8.

Chapter 6, 'Convivial green space', enlarged the focus to the city-wide scale, and concentrated on the transforming role and function of designed and informal green spaces connected with urban food production and consumption. Situating food-focused green space as an ancient part of city design and functioning, and one with particular roles to play in supporting conviviality and sustainability, the chapter explored some of its spatial expressions historically, before turning its attention to contemporary practice, including immigrant market-gardening experience. Urban agriculture was explored as a long-term resilience and food security strategy to deal with poverty and express cultural food preferences, and while seen as a practice in decline in some cities, was also considered in spatial terms as a more recent revival, in part responding to climate change. Examining urban food growing today, the chapter surveyed green spaces including informal vegetable gardens, market gardens and community gardens, and explored the rebirth of the allotment movement, the burgeoning of front yard farms, urban orchards and urban bee-keeping, as well as foraging from street trees and public gardens. In so doing, attention was paid to socio-spatial practices and theoretical perspectives on design emerging in relation to convivial green space, edible landscapes and agrarian urbanism. Allotments were considered as an example and a kind of test case for design-led approaches, and the chapter concluded with a discussion of recent advances in convivial green space as public and design policy.

In Part 3, 'Food space and urbanism on the edge' three chapters again widened out the scale of the spatial analysis to consider some of the ways food has connected to the relatively recent urbanism of the megalopolis as a dominant urban development mode. Foreshadowing a likely spatial future for many, it argued that the advent of massive urbanization and many of its complicated spatial forms and design approaches not only reflected the triumph of modernism but have also had considerable and often negative implications for conviviality and sustainability in food terms. With examples drawn from a wide range of megalopolitan sources, the final third of the book explored places that have kept or re-invigorated the food landscapes of their peripheries and considered the complex urban to rural economics and politics behind this trend, including arguments for food regionalism, a stress on low food miles, food strategies, support for small-scale production and regional food networks such as Slow Food and *Cittàslow* despite their unpopularity among academic geographers.

Chapter 7, 'The productive periphery', focused on places that have maintained and strengthened the gastronomic landscape of their urban edges, and contrasted such success with other less positive experience of peri-urban food space. The complex story of the broad decline of food production around cities was considered, with the presumption of primacy for urban development to the fore. Urban edge food resilience has been understood to be the result of a complex interplay between critical shifts in the nature of urban expansion and also of changes that are internal to the evolution of productive landscapes themselves. The chapter reviewed a rich history from biblical times onwards of productive edge spatiality to contextualize the nature of more recent practices in relation to peri-urban food spaces connected to production, exchange and consumption. A diversity of examples, from Asia, South America, Europe and Australia have shown how edge space has been traditionally shaped in food terms for both pleasure and production, and how in contemporary practice its value in economic, social and environmental terms has often been underestimated. The chapter explored the complex attempts to capture theoretically a diversity of edge spatiality in relation to food and interrogated concepts which connect sustainability and design, including the ecological footprint, the foodshed and the transect. It discussed the kinds of planning and design tools which may help protect and sustain peri-urban food production into the future, before moving on to consider the urban edge as a food tourism location (foreshadowing the longer discussion in Chapter 9). Detailed case studies of the urbanism of edge space gastronomy were explored, and the chapter concluded by considering design responses that support peri-urban food space.

Chapter 8, 'The megalopolitan food realm', expanded to a yet larger scale, to explore the relatively recent rise of vast post-urban regions and delve into their food-related design and urbanism implications. A brief survey of the huge range of research into trends in urbanization, including global, regional and location-specific shifts in population, offered a necessary backdrop to the more focused discussion of the sustainability and conviviality implications for food consumption, distribution and production of massive megalopolis-style urbanism. Applying a food-centred design analysis, the chapter considered some of the competing prescriptions for dealing with this new urban form, from broadly welcoming dystopia to more vernacular, place-specific solutions. The chapter showed that within a broad context of 'ad hoc laissez-faire urbanism' (Marshall 2009: 243) food landscapes and services have been radically reconfigured from their more traditional expressions as the outdoor rooms of the gastronomic townscape that were explored in Chapters 3 and 4. While place-based interest in food might be thought to suffer, the chapter showed that actual outcomes of edge and edgeless cities in food terms have offered a more complex picture. Food spaces explored included gated communities, distribution and customer fulfilment centres, business and office parks, 'big box' food stores and hypermarkets, fast-food outlets and chain restaurants, petrol station 'road pantries' and the food courts of megamalls. The food aspects of the conflict between standardized, large-scale food-retailing formats and the finer-grained spatiality of more convivial cultural norms in relation to shopping have been discussed in relation to food retailers' global expansion.

Megamalls and super-regional malls as archetypical spaces in this megalopolitan spatiality have been explored in some detail, with a particular focus on the rise, decline and transformation of shopping malls, including their food court designs. Food has been considered as the starting-off point for the McDonaldization thesis in relation to post-urban space, with intriguing examples of ways in which these standardized spatial and economic formats have been subverted into more convivial social spaces by certain cultures and groups. Food has also been considered in relation to employment spaces in landscapes including business parks, distribution centres and dark stores. The chapter considered the close connections between food and the predominance of car-centred spaces and the related retreat of food from a public realm largely shaped for car use. Space was also given to issues of food marginalization this spatiality has fostered, the food deserts and obesogenic environments associated with megalopolitan landscapes and designs. The chapter concluded with a discussion of more positive interconnections between sustainability and urban design possible through the retrofitting of sprawl-determined food space.

In Chapter 9, 'The critical food region', the focus once again shifted outwards, this time to food at the wider regional scale beyond the peri-urban and the megalopolitan. The chapter refracted arguments for food regionalism through the lens of design and urbanism for food. Food regions were first situated historically as forms of spatial practice, from the Roman *latifundia* to the farming innovations of the European medieval period and some of the spatial design implications for food of colonial expansion. The spatiality of long-term agricultural systems such as the Tuscan *mezzadria* were explored in urbanism terms, demonstrating their profound impact on gastronomic landscape stability (and rural inequality). Such traditional systems were later compared with the post-productivist model that has dominated agricultural spatiality in the post-war period and is now being challenged in various ways.

The chapter went on to consider today's food regions and explore the close cultural connections between urban and rural space, including France's self-described status as a place which is *profondément rurale*. The food-centred nature of regionality was interrogated in spatial terms, and in relation to various food policy and regulatory attempts to assert that place matters for food. On the food consumption side the chapter explored the rise of both an alternative geography of food and the emergence of alternative food networks as part of the quality turn away from global agrifood. It offered case studies of regional food and urbanism drawn from primary research in Italy and France. Slow Food and *Cittàslow* were critically considered in regional spatial terms, focusing in part on the example of Carrara's *lardo* (pork fat) as a Slow Food product, as a precursor to the wider discussion of concepts and practice of *terroir* and Appellation d'Origine Contrôlée and related systems. These were understood as defining regional gastronomic identity and strongly influencing its spatiality, including through 'reverse engineering' agrarianism into places lacking these qualities. The case of Parmesan cheese was used to explore some of the spatial design implications of food regionality in these terms. The chapter then turned again to consider regional food tourism and the convivial urbanism it has sometimes influenced, and concluded with a deliberation of the role that regional food policy and design guidance might have in shaping food and urbanism at the regional scale in future.

SOME FINAL THOUGHTS

Clearly, this book could not hope to do justice to the huge range of relevant research now beginning to link food and place. For instance, a planned chapter considering the urbanism implications of the relationship between the global north and south in food terms was omitted for lack of space. The book has, however, set out some fundamental theoretical considerations for connecting food and urbanism, filled in some gaps in thinking about the still under-researched theme of the interplay of food and design, and offered useful evidence in support from a diversity of relevant sources. In many cases this has required a kind of detective work, to retrieve knowledge from the 'edges' of research in a range of disciplines in which food's spatiality and design is present but not always very clearly acknowledged or framed as a research concern.

Notwithstanding this marginality, there are a number of both tentative and more certain conclusions that can be drawn from what has been considered over these nine chapters. Perhaps the most obvious overarching inference from this exploration is that making convivial and sustainable space requires conscious attention to design for food at every scale. As argued in the Introduction, urbanism is at the heart of this activity, and urbanists should take a central role in this task. This is especially the case given the urgency for action following climate change effects on settlement space. As Calthorpe (2010) notes, the scales for action need to be considered in a nested way so that positive activity on food's spatiality is integrated across each scale, and due regard is again given to the regional level for action. And as Talen (2005) points out, this is a global as well as a local, metropolitan and national issue, reinforcing the need for food and urbanism to be thought about and acted upon at each of these scales.

Despite the kind of contestation discussed in the book's Introduction, which situates food as a frivolous or self-indulgent subject for the entitled, the book's findings suggest that increasingly for many theorists and practitioners food does matter, just as it self-evidently does for those trying to achieve food resilience for themselves and others. As a fundamental aspect of the design and practice of everyday life generally undertaken in ordinary spaces, food is the material basis for more convivial lives in both public and private realms and in their interstices. This is very clear from the range of examples offered throughout the book of the way that people and food have interacted at home and in public space over millennia. Given the increasingly obvious relationship between food and personal and global health, with obesity and food insecurity evident for many, making and remaking settlements in ways that put food at the centre seems an extremely necessary way forward. It is equally certain that much more work is needed to understand these dynamics at every scale, so that a positive interplay between food and urbanism can form the basis for a sustainable and convivial future.

BIBLIOGRAPHY

Aalders, Inge and Morrice, Jane (2012), 'Land use change and ecosystem delivery by green networks'. *Land Management: Potential, Problems and Stumbling Blocks.* Land Management 21.
Abbott, Marylyn (2001), *Gardens of Plenty.* London: Kyle Cathie.
Aben, R. and De Wit, S. (1999), *The Enclosed Garden: History and Development of the Hortus Conclusus and its Reintroduction into the Present-day Urban Landscape.* Rotterdam: 010 Publishers.
Ackard, Diann M. and Neumark-Sztainer, Dianne (2001), 'Family mealtime while growing up: Associations with symptoms of bulimia nervosa'. *Eating Disorders* 9.3, pp. 239–49.
Adams, Elizabeth J., Grummer-Strawn, Laurence and Chavez, Gilberto (2003), 'Food insecurity is associated with increased risk of obesity in California women'. *The Journal of Nutrition* 133.4, pp. 1070–4.
Adell, Germán (1999), 'Theories and models of the peri-urban interface: A changing conceptual landscape'. London: Development Planning Unit, UCL.
Adema, Pauline (2006), 'Festive foodscapes: Iconizing food and the shaping of identity and place'. Dissertation, The University of Texas at Austin.
Agnelli, M., Pietromarchi, L., Bright, R. E. and Forquet, F. (1987), *Gardens of the Italian Villas.* New York: Rizzoli.
Aguilar, Adrián G., Ward, Peter M. and Smith Sr, C. B. (2003), 'Globalization, regional development, and mega-city expansion in Latin America: Analyzing Mexico City's peri-urban hinterland'. *Cities* 20.1, pp. 3–21.
Alanen, Arnold R. (1990), 'Immigrant gardens on a mining frontier', in Mark Francis and Randolph Hester (eds), *The Meaning of Gardens: Idea, Place and Action.* Cambridge, MA: MIT Press, pp. 160–5.
Alcock, Joan (2006), *Food in the Ancient World.* Westport, CT; London: Greenwood Press.
Albala, Ken (2003), *Food in Early Modern Europe.* Westport, CT; London: Greenwood Press.
Albala, Ken (2006), *Cooking in Europe, 1250-1650.* Westport, CT: Greenwood Press; Oxford: Harcourt Education.
Albala, Ken (2007), *The Banquet: Dining in the Great Courts of Late Renaissance Europe.* Urbana and Chicago: University of Illinois Press.
Alexander, A., Nell, D., Bailey, A. R. and Shaw, G. (2009), 'The co-creation of a retail innovation: Shoppers and the early supermarket in Britain'. *Enterprise and Society* 10.3, pp. 529–58.
Alexander, C., Ishikawa, S. and Silverstein, M. (1977), *A Pattern Language: Towns, Buildings, Construction.* New York: Oxford University Press.

Alexander, Nicholas and Akehurst, Gary (1998), 'Introduction: The emergence of modern retailing, 1750–1950'. *Business History* 40.4, pp. 1–15.

Alkon, Alison (2008), 'Paradise or pavement: The social constructions of the environment in two urban farmers' markets and their implications for environmental justice and sustainability'. *Local Environment* 13.3, pp. 271–89.

Alkon, Alison Hope and Agyeman, Julian (eds) (2011), *Cultivating Food Justice: Race, Class, and Sustainability*. Cambridge, MA; London: MIT Press.

Alkon, Alison Hope and Mares, Teresa Marie (2012), 'Food sovereignty in US food movements: Radical visions and neoliberal constraints'. *Agriculture and Human Values* 29.3, pp. 347–59.

Alkon, Alison Hope and Norgaard, Kari Marie (2009), 'Breaking the food chains: An investigation of food justice activism'. *Sociological Inquiry* 79.3, pp. 289–305.

Allen, Adriana (2003), 'Environmental planning and management of the peri-urban interface: Perspectives on an emerging field'. *Environment and Urbanization* 15.1, pp. 135–48.

Allen, Patricia (1993), *Food for the Future: Conditions and Contradictions of Sustainability*. New York; Chichester: John Wiley and Sons.

Allison, Anne (1991). 'Japanese mothers and obentōs: The lunch-box as ideological state apparatus'. *Anthropological Quarterly*, pp. 195–208.

Alsayyad, Nezar (1994), 'The street between the two palaces', in Zeynep Çelik, Diane Favro and Richard Ingersoll (eds), *Streets: Critical Perspectives on Public Space*. Berkeley, CA: University of California Press.

Altieri, Miguel A. et al. (1999), 'The greening of the "barrios": Urban agriculture for food security in Cuba'. *Agriculture and Human Values* 16.2, pp. 131–40.

Amery, Colin and Curran Jr., Brian (2002), *The Lost World of Pompeii*. London: Frances Lincoln.

Amilien, V., Torjusen, H. and Vittersø, G. (2005), 'From local food to terroir product? Some views about Tjukkmjølk, the traditional thick sour milk from Røros, Norway'. *Anthropology of Food* 4.

Amin, A. and Thrift, N. (1992), 'Neo-Marshallian nodes in global networks'. *International Journal of Urban and Regional Research* 16, pp. 571–87.

Amin, A. and Thrift, N. (2002), *Cities: Reimagining the Urban*. Cambridge, MA: Polity Press.

Amin, Ash and Graham, Stephen (1997), 'The ordinary city'. *Transactions of the Institute of British Geographers* 22.4, December, pp. 411–29.

Amine, Abdelmajid and Lazzaoui, Najoua (2011), 'Shoppers' reactions to modern food retailing systems in an emerging country: The case of Morocco'. *International Journal of Retail & Distribution Management* 39.8, pp. 562–81.

Anderson, Eugene Newton (2005), *Everyone Eats: Understanding Food and Culture*. New York: NYU Press.

Anderson, Kay (2005), 'Introduction: After sprawl: Post-suburban Sydney', in Kay Anderson, Reena Dobson, Fiona Allon and Brett Neilson (eds), 'After Sprawl: Post-Suburban Sydney'. E-Proceedings of 'Post-Suburban Sydney: The City in Transformation' Conference, 22–23 November 2005.

Andrews, Geoff (2008), *The Slow Food Story: Politics and Pleasure*. Montreal: McGill-Queen's University Press; Sidmouth: Chase.

Anjaria, J. S. (2006), 'Street hawkers and public space in Mumbai'. *Economic and Political Weekly*, pp. 2140–6.
Architecture Foundation (2011), *The Union Street Urban Orchard: A Case Study of Creative Interim Use*. http://www.architecturefoundation.org.uk/programme/2010/london-festival-of-architecture-2010/the-union-street-urban-orchard.
Ashkenazi, Michael and Jacob, Jeanne (2000), *The Essence of Japanese Cuisine: An Essay on Food and Culture*. Richmond and Surrey: Curzon Press, p. 252.
Ashkenazi, Michael and Jacob, Jeanne (2003), *Food Culture in Japan*. Westport, CT: Greenwood Press; Oxford: Harcourt Education.
Ashley, Bob, Hollows, Joanne, Jones, Steve and Taylor, Ben (2004), *Food and Cultural Studies: Studies in Consumption and Markets*. London and New York: Routledge.
Atkins, P. J. and Oddy, D. J. (2007), 'Food and the city', in P. J. Atkins, P. Lummel and D. J. Oddy (eds), *Food and the City in Europe since 1800*. Aldershot, UK: Ashgate Publishing.
Atkins, Peter and Bowler, Ian (2001), *Food in Society: Economy, Culture, Geography*. London: Arnold.
Atkinson, Rowland (2003), 'Domestication by cappuccino or a revenge on urban space? Control and empowerment in the management of public spaces'. *Urban Studies* 40.9, pp. 1829–43.
Atkinson, Rowland and Flint, John (2004), 'Fortress UK? Gated communities, the spatial revolt of the elites and time–space trajectories of segregation'. *Housing Studies* 19.6, pp. 875–92.
Attfield, Judy (1989), 'Inside pram town', in Judy Attfield and Pat Kirkham (eds), *A View from the Interior: Feminism, Women and Design*. London: The Women's Press, pp. 215–38.
Attfield, Judy (1999), 'Bringing modernity home. Open plan in the British domestic interior', in Irene Cieraad (ed.), *At Home: An Anthropology of Domestic Space*. New York: Syracuse University Press, pp. 73–82.
Audirac, I. (1999), 'Unsettled views about the fringe: Rural-urban or urban-rural frontiers?', in O. J. Furuseth and M. B. Lapping (eds), *Contested Countryside: The Rural Urban Fringe in North America*. Aldershot, UK: Ashgate, pp. 7–32.
Augé, Marc (1995), *Non-Places: Introduction to an Anthropology of Supermodernity*. London: Verso.
Aurier, P., Fort, F. and Sirieix, L. (2005), 'Exploring terroir product meanings for the consumer'. *Anthropology of Food* 4.
Austin, S. B., Melly, S. J., Sanchez, B. N., Patel, A., Buka, S. and Gortmaker, S. L. (2005), 'Clustering of fast-food restaurants around schools: A novel application of spatial statistics to the study of food environments'. *Journal Information* 95.9, p. 1575.
Bacon, Edmund (1982), *Design of Cities*. London: Thames and Hudson.
Bagwell, S. (2011), 'The role of independent fast-food outlets in obesogenic environments: A case study of east london in the UK'. *Environment and Planning-Part A* 43.9, pp. 2217–36.
Bah, M., Cissé, S., Diyamett, B., Diallo, G., Lerise, F., Okali, D. ... and Tacoli, C. (2003), 'Changing rural–urban linkages in Mali, Nigeria and Tanzania'. *Environment and Urbanization* 15.1, pp. 13–24.

Bailis, Rob, Cowan, Amanda, Berrueta, Victor and Masera, Omar (2009), 'Arresting the killer in the kitchen: The promises and pitfalls of commercializing improved cookstoves'. *World Development* 37.10, pp. 1694–705.

Baker, Lauren E. (2004), 'Tending cultural landscapes and food citizenship in Toronto's community gardens'. *Geographical Review* 94.3, pp. 305–25.

Ballantyne, Glenda and Benny, Helen (2009), 'Kitchen matters: Ethno-cuisines and contemporary foodways'. In *The Future of Sociology: The Annual Conference of The Australian Sociological Association*, pp. 1–12.

Banerjee, Tridib (2001), 'The future of public space: Beyond invented streets and reinvented places'. *Journal of the American Planning Association* 67.1, pp. 9–24.

Banta, M. (1993), *Taylored Lives: Narrative Productions in the Age of Taylor, Veblen, and Ford*. Chicago: University of Chicago Press.

Baraban, Regina S. and Durocher, Joseph F. (2010), *Successful Restaurant Design*. Hoboken, NJ: Wiley.

Barber, Benjamin R. (2002), 'Civic space', in David J. Smiley (ed.), *Sprawl and Public Space Redressing the Mall*. Washington, DC: National Endowment for the Arts, pp. 31–6.

Bardo, Matt and Warwicker, Michelle (2012), *Does Farmers' Market Food Taste Better?* BBC News, 23 June.

Barham, Elizabeth (2001), 'Translating "terroir": Social movement appropriation of a French concept', in *Workshop on International Perspectives on Alternative Agro-Food Networks, Quality, Embeddedness and Bio-Politics*. Santa Cruz: University of California, October, pp. 12–13.

Barham, Elizabeth (2003), 'Translating terroir: The global challenge of French AOC labeling'. *Journal of Rural Studies* 19.1, pp. 127–38.

Barkat, S. M. and Vermignon, V. (2006), *Gastronomy Tourism: A Comparative Study of Two French Regions: Brittany and La Martinique*. Presentation at the Sustainable Tourism with Special Reference to Islands and Small States conference Malta, 25–27 May.

Barnett, Jonathan (1995), *The Fractured Metropolis: Improving the New City, Restoring the Old City, Reshaping the Region*. New York: Harper Collins.

Bartley, J. R. (2006), *Designing the New Kitchen Garden: An American Potager Handbook*. Portland: Timber Press.

Barton, H. (ed.) (2000), *Sustainable Communities. The Potential for Eco-Neighbourhoods*. London: Earthscan.

Barton, H., Grant, M. and Guise, R. (2003), *Shaping Neighbourhoods: A Guide for Health, Sustainability and Vitality*. London: Taylor & Francis.

Basker, Emek, Klimek, Shawn and Pham, Hoang Van (2012), 'Supersize it: The growth of retail chains and the rise of the "Big Box" store'. *Journal of Economics & Management Strategy* 21.3, pp. 541–82.

BBC (2013), 'Elderly malnutrition highlighted in government project', 22 December 2013. Last updated at 18:14. http://www.bbc.co.uk/news/health-25463483.

Beardsworth, Alan and Keil, Teresa (1990), 'Putting the menu on the agenda'. *Sociology* 24.1, pp. 139–51.

Beardsworth, Alan and Keil, Teresa (1997), *Sociology on the Menu: An Invitation to the Study of Food and Society*. London; New York: Routledge.

Beatley, Timothy (1997), *The Ecology of Place: Planning for Environment, Economy, and Community*. Washington, DC: Island Press.
Beatley, Timothy (2000), *Green Urbanism: Learning from European Cities*. Washington, DC: Island Press.
Beatley, Timothy (2012), *Green Cities of Europe: Global Lessons on Green Urbanism*. Washington, DC: Island Press.
Beatley, Timothy (2004), *Native to Nowhere: Sustaining Home and Community in a Global Age*. Washington, DC: Island Press.
Beatley, Timothy (2010), *Biophilic Cities: Integrating Nature into Urban Design and Planning*. Washington, DC: Island Press.
Beatley, Timothy and Manning, Kirsty (1997), *The Ecology of Place: Planning for Environment, Economy and Community*. Washington, DC: Island Press.
Beatley, Timothy and Newman, Peter (2009), *Green Urbanism Down Under*. Washington, DC: Island Press.
Beck, U., Bonss, W. and Lau, C. (2003), 'The theory of reflexive modernization: Problematic, hypotheses and research programme'. *Theory, Culture & Society* 20.2, pp. 1–33.
Becker, Franklin D. and Mayo, Clara (1971), 'Delineating personal distance and territoriality'. *Environment and Behavior* 3.4, December, pp. 375–81.
Beecher, Mary Anne (2001), 'Promoting the "unit idea": Manufactured kitchen cabinets (1900-1950)'. *APT Bulletin* 32.2/3, pp. 27–37.
Beer, Sean, Edwards, Jonathan, Fernandes, Carlos and Sampaio, Francisco (2002), 'Regional Food Cultures: Integral to the rural tourism product?', in Hjalager Anne-Mette and Greg Richards (eds), *Tourism and Gastronomy*. Hillsdale, NJ: Psychology Press, pp. 1207–23.
Beeton, Isabella (1861), *Beeton's Book of Household Management*. First Edition Facsimile. London: Thomas Nelson.
Beirne, Valerie (2011), 'Union street urban orchard', in Moira Lascelles (ed.), *The Union Street Urban Orchard: A Case Study of Creative Interim Use*. Architecture Foundation.
Bell, David (2007), 'The hospitable city: Social relations in commercial spaces'. *Progress in Human Geography* 31.1, February, pp. 7–22.
Bell, David and Binnie, J. (2005), 'What's eating Manchester? Gastro-culture and urban regeneration', in K. Franck (ed.), *Food and the City: Architectural Design* 75.3, pp. 78–85.
Bell, David and Hollows, Joanne (eds) (2005), *Ordinary Lifestyles: Popular Media, Consumption and Taste*. Maidenhead: Open University Press.
Bell, David and Valentine, Gill (1997), *Consuming Geographies: We are Where We Eat*. London: Routledge.
Bendiner, K. (2004), *Food in Painting: From the Renaissance to the Present*. London: Reaktion Books.
Benedictus, Leo (2014), 'Inside the supermarkets' dark stores'. *The Guardian*, 7 January 2014. Accessed 10 March 2014.
Benfield, Kaid (2013), 'Malls and big boxes continue to lose their grip – will this mean more abandoned properties?'. http://switchboard.nrdc.org/blogs/kbenfield/malls_and_big_boxes_continue_t.html.

Bennett, S. (1992), 'Combining good business and good works'. *Progress: Grocer* 71.12, pp. 65–9.
Bentley, Ian et al. (ed.) (1985), *Responsive Environments: A Manual for Designers*. Oxford: Architectural Press.
Bessière, Jacinthe (1998), 'Local development and heritage: Traditional food and cuisine as tourist attractions in rural areas'. *Sociologia Ruralis* 38.1, pp. 21–34.
Betts, P. (2004), *The Authority of Everyday Objects: A Cultural History of West German Industrial Design* (vol. 34). Berkeley, CA and Los Angeles: University of California Press.
Bhatti, Mark and Church, Andrew (2000), 'I never promised you a rose garden': Gender, leisure and home-making'. *Leisure Studies* 19.3, pp. 183–97.
Bhatti, Mark and Church, Andrew (2001), 'Cultivating natures: Homes and gardens in late modernity'. *Sociology* 35.02, pp. 365–83.
Bhatti, Mark and Church, Andrew (2004), 'Home, the culture of nature and meanings of gardens in late modernity'. *Housing Studies* 19.1, pp. 37–51.
Biddulf, Michael (1993), 'Consuming the sign value of Urban Form', in Richard Hayward and Sue McGlynn (eds), *Making Better Places: Urban Design Now*. Oxford: Butterworth-Heinemann, pp. 35–41.
Biggs, Steven (2009), 'High off the hog: Higtown as food processing hub', in Christina Palassio and Alana Wilcox (eds), *The Edible City Toronto's Food from Farm to Fork*. Toronto: Coach House Books, pp. 32–7.
Binns, Tony and Lynch, Kenneth (1998), 'Feeding Africa's growing cities into the 21st century: The potential of urban agriculture'. *Journal of International Development* 10.6, pp. 777–93.
Birch, Eugenie L. and Wacher, Susan M. (eds) (2008), *Growing Greener Cities Urban Sustainability in the Twenty-First Century*. Philadelphia: University of Pennsylvania Press.
Bisogni, Carole A., Falk, Laura Winter, Madore, Elizabeth, Blake, Christine E., Jastran, Margaret, Sobal, Jeffery and Devine, Carol M. (2006), 'Dimensions of everyday eating and drinking episodes'. Research Report, *Appetite*.
Björklund, Annika (2010), 'Historical urban agriculture: Food production and access to land in Swedish towns before 1900', Stockholm University, Faculty of Social Sciences, Department of Human Geography.
Blais, Pamela (2010), *Perverse Cities: Hidden Subsidies, Wonky Policy, and Urban Sprawl*. Vancouver: UBC Press.
Blake, Megan Kathleen (2013), 'Ordinary food spaces in a global city: Hong Kong'. *Streetnotes* 21.21, pp. 1–12.
Blake, Megan Kathleen, Mellor, Jody and Crane, Lucy (2010), 'Buying local food: Shopping practices, place, and consumption networks in defining food as "local"'. *Annals of the Association of American Geographers* 100.2, pp. 409–26.
Blakeley, E. J. and Snyder, M. G. (eds) (1997), *Fortress America: Gated Communities in the United States*. Washington, DC: Brookings Institution Press.
Blanchard, T. and Lyson, T. (2002), 'Access to low cost groceries in nonmetropolitan counties: Large retailers and the creation of food deserts', in *Measuring Rural Diversity Conference Proceedings*. Washington, DC: Economic Research Service, US Department of Agriculture.

Blinnikov, M., Shanin, A., Sobolev, N. and Volkova, L. (2006), 'Gated communities of the Moscow green belt: Newly segregated landscapes and the suburban Russian environment'. *GeoJournal* 66.1–2, pp. 65–81.

Bloom, Jonathan (2013), 'American wasteland: How America throws away nearly half of its food (and what we can do about it)'. *Journal of Progressive Human Services* 24.2, pp. 165–71.

Blunt, Alison and Dowling, Robyn (2006), *Home*. London: Routledge.

Blythman, Joanna (2006), *Bad Food Britain*. London: Fourth Estate.

Bohl, C. C. and Plater-Zyberk, E. (2006), 'Building community across the rural-to-urban transect'. *Places* 18.1, pp. 1–14.

Bonfiglio, Olga (2009), 'Delicious in Detroit-Vacant land can be an asset, as Olga Bonfiglio reports. Sidebars on montreal and on LA's vertical gardens'. *Planning* 75.8, p. 32.

Boni, Ada (1969), *Italian Regional Cooking*. London: Godfrey Cave Associates.

Boniface, Priscilla (2003), *Tasting Tourism: Travelling for Food and Drink*. Aldershot; Burlington, VT: Ashgate Publishing.

Bonis, Françoise (1997), *Les recettes Périgourdines de Tante Célestine*. Rennes: Editions Ouest-France.

Bontje, Marco (2004), 'From suburbia to post-suburbia in the Netherlands: Potentials and threats for sustainable regional development'. *Journal of Housing and the Built Environment* 19.1, pp. 25–47.

Bontje, Marco and Burdack, Joachim (2005), 'Edge cities, European-style: Examples from Paris and the Randstad'. *Cities* 22.4, pp. 317–30.

Born, B. (2006), 'Avoiding the local trap, scale and food systems in planning research'. *Journal of Planning Education and Research* 26.2, pp. 195–207.

Bose, C. E., Bereano, P. L. and Malloy, M. (1984), 'Household technology and the social construction of housework'. *Technology and Culture* 25.1, pp. 53–82.

Bosio, Andrea (2013), 'Food places through the visual media: Building gastronomic cartographies between Italy and Australia'. http://www.inter-disciplinary.net/probing-the-boundaries/wp-content/uploads/2013/02/Food1_wpaper_Bosio.pdf.

Bost, Peter Grael (2010), 'The creation and maintenance of Community Gardens', in Moira Lascelles (ed.), *The Union Street Urban Orchard: A Case Study of Creative Interim Use*. Architecture Foundation.

Bourdieu, Pierre (1984), *Distinction: A Social Critique of the Judgement of Taste*, trans. Richard Nice. London: Routledge and Kegan Paul.

Bourdieu, Pierre (1993), *Sociology in Question*. London; Thousand Oaks: Sage.

Bourne, L., Bunce, M., Taylor, L., Luka, N. and Maurer, J. (2003), 'Contested ground: The dynamics of peri-urban growth in the Toronto region'. *Canadian Journal of Regional Science* 26.2/3, pp. 251–70.

Bowen Jr., John T. (2008), 'Moving places: The geography of warehousing in the US'. *Journal of Transport Geography* 16.6, pp. 379–87.

Bowes, Kimberly Diane (2010), *Houses and Society in the Later Roman Empire*. London: Duckworth.

Bowlby, Rachel (1985), *Just Looking: Consumer Culture in Dreiser, Gissing and Zola*. New York and London: Methuen.

Boyne, Steven, Williams, Fiona and Hall, Derek (2002), 'On the trail of regional success: Tourism, food production and the *Isle of Arran Taste Trail*', in Anne-Mette Hjalager and Greg Richards (eds), *Tourism and Gastronomy*. London: Routledge, pp. 91–114.

Boyne, Steven, Williams, Fiona and Hall, Derek (2003), 'Policy, support and promotion for food-related tourism initiatives: A marketing approach to regional development'. *Journal of Travel & Tourism Marketing* 14.3–4, pp. 131–54.

Bramall, R. (2011), 'Dig for victory! Anti-consumerism, austerity and new historical subjectivities'. *Subjectivity* 4.1, pp. 68–86.

Braun, B. (2005), 'Environmental issues: Writing a more-than-human urban geography'. *Progress in Human Geography* 29.5, pp. 635–50.

Braudel, F. (1988), *The Identity of France*, trans. S. Reynolds. London: Collins (original work published 1986).

Branca, F., Nikogosian, H. and Lobstein, T. (eds) (2007), *The Challenge of Obesity in the WHO European Region and the Strategies for Response: Summary*. World Health Organization.

Braverman, Harry (1974), *Labor and Monopoly Capital: The Degradation of Work in the Twentieth Century*. New York: Monthly Review Press.

Bray, Tamara L. (ed.) (2003), *The Archaeology and Politics of Food and Feasting in Early States and Empires*. London: Springer.

Brears, Peter et al. (1993), *A Taste of History: 10,000 Years of Food in Britain*, intro. Maggie Black. London: English Heritage in association with the British Museum.

Brennan, Thomas (1984), 'Beyond the barriers: Popular culture and Parisian guinguettes'. *Eighteenth-Century Studies* 18.2, pp. 153–69.

Brennan, Thomas (2005), 'Taverns in the public sphere in 18th-century Paris'. *Contemporary Drug Problems* 32, p. 29.

Bridge, Gary and Dowling, Robyn (2001), 'Microgeographies of retailing and gentrification'. *Australian Geographer* 32.1, pp. 93–107.

Briggs, John and Mwamfupe, Davis (2000), 'Peri-urban development in an era of structural adjustment in Africa: The city of Dar es Salaam, Tanzania'. *Urban Studies* 37.4, pp. 797–809.

Briquel, Vincent and Collicard, Jean-Jacques (2005), 'Diversity in the rural hinterlands of European cities', in Keith Hoggart (ed.), *The City's Hinterland: Dynamism and Divergence in Europe's Peri-urban Territories*. Aldershot: Ashgate, pp. 19–40.

Broadbent, Geoffrey (1990), *Emerging Concepts in Urban Space Design*. London; New York: Van Nostrand Reinhold.

Bromley, R. D. and Thomas, C. J. (2002), 'Food shopping and town centre vitality: Exploring the link'. *The International Review of Retail, Distribution and Consumer Research* 12.2, pp. 109–30.

Brook, Robert M. and Dávila, Julio D. (eds) (2000), *The Peri-urban Interface: A Tale of Two Cities*. London: School of Agricultural and Forest Sciences, University of Wales.

Brooks, Karen, with Bosker, Gideon and Gelber, Terry (2012), *The Mighty Gastropolis Portland A Journey through the Centre of America's New Food Revolution*. San Francisco: Chronicle Books.

Brown, Kate H. and Jameton, Andrew L. (2000), 'Public health implications of urban agriculture'. *Journal of Public Health Policy* 21.1, pp. 20–39.

Brownlie, D., Hewer, P. and Horne, S. (2005), 'Culinary tourism: An exploratory reading of contemporary representations of cooking'. *Consumption Markets & Culture* 8.1, pp. 7–26.

Bruce, Nigel, Neufeld, Lynnette, Boy, Erick and West, Chris (1998), 'Indoor biofuel air pollution and respiratory health: The role of confounding factors among women in highland Guatemala'. *International Journal of Epidemiology* 27, pp. 454–8.

Brueckner, Jan K. (2000), 'Urban sprawl: Diagnosis and remedies'. *International Regional Science Review* 23.2, pp. 160–71.

Bruegmann, Robert (2005), *Sprawl: A Compact History*. Chicago, IL; London: University of Chicago Press.

Brüel, M. (2012), 'Copenhagen, Denmark: Green City amid the Finger Metropolis', in *Green Cities of Europe*. Washington, DC: Island Press/Center for Resource Economics, pp. 83–108.

Brunori, G. (2006), 'Post-rural processes in wealthy rural areas: Hybrid networks and symbolic capital'. *Research in Rural Sociology and Development* 12, pp. 121–45.

Bryant, C. R. and Johnston, T. R. R. (1992), *Agriculture in the City's Countryside*. London: Belhaven Press.

Buckingham, Susan (2005), 'Women (re) construct the plot: The regen(d)eration of urban food growing'. *Area* 37.2, pp. 171–9.

Bullock, Nicholas (1988), 'First the kitchen: Then the Façade'. *Journal of Design History* 1.3/4, pp. 177–92.

Bunce, Michael (1994), *The Countryside Ideal: Anglo-American Images of Landscape*. London: Routledge.

Bunker, Raymond (2002), 'In the shadow of the city: The fringe around the Australian metropolis in the 1950s'. *Planning Perspectives* 17.1, pp. 61–82.

Bunker, Raymond and Holloway, Darren (2001), 'Fringe city and contested countryside: Population trends and policy developments around Sydney'. Issues Paper No. 6, *Urban Frontiers Program*. University of Western Sydney.

Bunker, Raymond and Houston, Peter (1992), 'At and beyond the fringe: Planning around the Australian city with particular reference to Adelaide'. *Urban Policy and Research* 10.3, pp. 23–32.

Bunker, Raymond and Houston, Peter (2003), 'Prospects for the rural-urban fringe in Australia: Observations from a brief history of the landscapes around Sydney and Adelaide'. *Australian Geographical Studies* 41.3, pp. 303–23.

Burchell, Robert W. and Sahan, Mukherji (2003), 'Conventional development versus managed growth: The costs of sprawl'. *American Journal of Public Health* 93.9, pp. 1534–40.

Burgoine, T., Lake, A. A., Stamp, E., Alvanides, S., Mathers, J. C. and Adamson, A. J. (2009), 'Changing foodscapes 1980–2000, using the ASH30 Study'. *Appetite* 53.2, pp. 157–65.

Burke, Matthew (2001), 'The pedestrian behaviour of residents in gated communities'. *Australia: Walking The 21st Century, International Conference, 2001, Perth, Western Australia*, Vol. 2.

Burnett, John (2003), 'Eating in the open air in England, 1830–1914', in Marc Jacobs and Peter Scholliers (eds), *Eating Out in Europe: Picnics, Gourmet Dining and Snacks since the Late Eighteenth Century*. Oxford: Berg, pp. 21–38.

Burnett, John (2004), *England Eats Out: A Social History of Eating Out in England from 1830 to the Present*. Harlow: Longman.

Burnett, Sierra Clark and Ray, Krishnedu (2012), 'Sociology of food', in Jeffrey M. Pilcher (ed.), *The Oxford Handbook of Food History*. Oxford; New York: Oxford University Press.

Burnley, I. H. (2003), 'Population and environment in Australia: Issues in the next half century'. *Australian Geographer* 34.3, pp. 267–80.

Burrichter, Felix (2011), 'Courtyard Canteen'. *New York Times Magazine*, Blogs Retrieved 22 January 2013.

Burridge, Joseph and Barker, M. (2009), 'Food as a medium for emotional management of the family: Avoiding complaint and producing love', in P. Jackson (ed.), *Changing Families, Changing Food*. Basingstoke: Palgrave Macmillan, pp. 146–64.

Busck, A. G., Kristensen, S. P., Præstholm, S., Reenberg, A. and Primdahl, J. (2006), 'Land system changes in the context of urbanisation: Examples from the peri-urban area of Greater Copenhagen'. *Geografisk tidsskrift-Danish Journal of Geography* 106.2, pp. 21–34.

Butler, Richard and Clark, Gordon (1992), 'Tourism in rural areas: Canada and the United Kingdom', in I. R. Bowler, C. R. Bryant and M. D. Nellis (eds), *Contemporary Rural Systems in Transition 2*. Wallingford: CAB International, pp. 166–83.

Butler, Sarah (2014), 'Grocers rush to open "dark stores" as online food shopping expands', *The Guardian*, Monday 6 January 2014, 20.04 GMT. Accessed 10 March 2014.

Buxton, M. and Choy, D. L. (2007), *Change in Peri-urban Australia: Implications for Land Use Policies*. Adelaide, Australia.

Bynum, Caroline Walker (1988), *Holy Feast and Holy Fast: The Religious Significance of Food to Medieval Women*. Berkeley, CA; London: University of California Press.

Cabannes, Y. and Raposo, I. (2013), 'Peri-urban agriculture, social inclusion of migrant population and Right to the City: Practices in Lisbon and London'. *City* 17.2, pp. 235–50.

Cabedoce, Béatrice (1991), 'Jardins ouvriers et banlieue: le bonheur au jardin?', in Alain Faure (ed.), *Les Premiers banlieusards. Aux origines des banlieues de Paris 1860-1940*. Paris: Créaphis, pp. 249–79.

Cabedoce, Béatrice and Pierson, Philippe (eds) (1996), *Cent ans d'histoire des jardins ouvriers, 1896-1996: la Ligue française du coin de terre et du foyer*. Grâne: Créaphis.

Caldeira Teresa, P. R. (2000), *City of Walls: Crime, Segregation, and Citizenship in Sao Paulo*. Berkeley, CA: University of California Press.

Calthorpe, Peter (1993), *The Next American Metropolis: Ecology, Community, and the American Dream*. New York: Princeton Architectural Press.

Calthorpe, Peter (2010), *Urbanism in the Age of Climate Change*. Washington, DC: Island Press.

Calthorpe, Peter (2011), 'Urbanism and Climate Change', in *Urbanism in the Age of Climate Change*. Washington, DC: Island Press/Center for Resource Economics, pp. 7–24.

Camagni, R., Gibelli, M. C. and Rigamonti, P. (2002), 'Urban mobility and urban form: The social and environmental costs of different patterns of urban expansion'. *Ecological Economics* 40.2, pp. 199–216.

Cameron, R. W., Blanuša, T., Taylor, J. E., Salisbury, A., Halstead, A. J., Henricot, B. and Thompson, K. (2012), 'The domestic garden–Its contribution to urban green infrastructure'. *Urban Forestry & Urban Greening* 11.2, pp. 129–37.
Campbell, Denis (2011), 'How to save school dinners – part two'. http://www.guardian.co.uk/education/2011/oct/24/jamie-oliver-fight-healthy-eating-schools, 24 October 2011. Retrieved 22 January 2013.
Campbell, Marcia Caton (2004), 'Building a common table: The role for planning in community food systems'. *Journal of Planning Education and Research* 23.4, pp. 341–55.
Campbell, Susan (1996), *Charleston Kedding History of Kitchen Gardening*. London: Ebury Press.
Campbell, Susan (2005), *A History of Kitchen Gardening*. London: Frances Lincoln.
Campigotto, Rachelle Marie (2010), 'Farmers' markets and their practices concerning income, privilege, and race: A case study of Wychwood Artscape Barns in Toronto'. Dissertation, University of Toronto.
Canavan, Orla, Henchion, Maeve and O'Reilly, Seamus (2007), 'The use of the internet as a marketing channel for Irish speciality food'. *International Journal of Retail & Distribution Management* 35.2, pp. 178–95.
Cánoves, G. and de Morais, R. S. (2011), 'New forms of tourism in Spain: Wine, gastronomic and rural tourism', *Tourism and Agriculture: New Geographies of Consumption, Production and Rural Restructuring*. Abingdon: Oxon; New York: Routledge, p. 205.
Cánoves, G., Villarino, M., Priestley, G. K. and Blanco, A. (2004), 'Rural tourism in Spain: An analysis of recent evolution'. *Geoforum* 35.6, pp. 755–69.
Cantacuzino, S. (1978), *Wells Coates: A Monograph*. London: Gordon and Fraser.
Capon, Anthony G. and Thompson, Susan M. (2011), 'Built environments of the future'. *Making Healthy Places*. Washington, DC: Island Press/Center for Resource Economics, pp. 366–78.
Caraher, M., Dixon, P. and Lang, T. R. (1998), 'Barriers to accessing healthy foods: Differentials by gender, social class, income and mode of transport'. *Health Education Journal* 57.3, pp. 191–201.
Caraher, M., Dixon, P., Lang, T. and Carr-Hill, R. (1999), 'The state of cooking in England: The relationship of cooking skills to food choice'. *British Food Journal* 101.8, pp. 590–609.
Carlin, Martha (2008), '"What say you to a piece of beef and mustard?": The Evolution of Public Dining in Medieval and Tudor London'. *Huntington Library Quarterly* 71.1, pp. 199–217.
Carroll-Spillecke, M. (1992), 'The gardens of Greece from homeric to roman times'. *The Journal of Garden History* 12.2, pp. 84–101.
Carruthers, John I. and Vias, Alexander C. (2005), 'Urban, suburban, and exurban sprawl in the rocky mountain west: Evidence from regional adjustment models'. *Journal of Regional Science* 45.1, pp. 21–48.
Casanova, Anna and Memoli, Paolo (2010), *La Romagna Toscana: Mille Anni di Caccia, Mezzadria e Carbone*. Firenze: Samus.
Casas, Penelope (1982), *The Foods and Wines of Spain*. London: Penguin.

Casey, Emma and Martens, Lydia (eds) (2007), *Gender and Consumption: Domestic Cultures and the Commercialisation of Everyday Life*. Aldershot; Burlington, VT: Ashgate.

Castellanos, Erick and Bergstresser, Sara M. (2006), 'Food fights at the EU table: The gastronomic assertion of Italian distinctiveness'. *European Studies: A Journal of European Culture, History and Politics* 22.1, pp. 179–202.

Catalán, Bibiana, Saurí, David and Serra, Pere (2008), 'Urban sprawl in the Mediterranean?: Patterns of growth and change in the Barcelona Metropolitan Region 1993–2000'. *Landscape and Urban Planning* 85.3, pp. 174–84.

Central Bedfordshire Council and Guise, Richard (2010), *Design in Central Bedfordshire*. Concept and Design by Richard Guise RIBA MRTPI, Context4D with James Webb, Forum Heritage Services in association with officers of Central Bedfordshire Council.

Cervero, Robert (1998), *The Transit Metropolis: A Global Inquiry*. Washington, DC: Island Press.

Chan, Sucheng (1996), *This Bittersweet Soil: The Chinese in California Agriculture, 1860–1910*. Berkeley, CA; London: University of California Press.

Chan, Kwok Shing (2007), 'Poonchoi: The production and popularity of a rural festive cuisine in urban and modern Hong Kong', in Sidney C. H. Cheung and Che-Beng Tan (eds), *Food and Foodways in Asia: Resource, Tradition and Cooking*. London: Routledge, pp. 53–66.

Chapman, M. and Jarnal, A. (1997), 'Acculturation: Cross cultural consumer perceptions and the symbolism of domestic space', in Brucks Merrie and Debbie MacInnis (eds), *Advances in Consumer Research* (vol. 24). Provo, UT: Association for Consumer Research, pp. 138–44.

Charmes, Eric (2010), 'Cul-de-sacs, superblocks and environmental areas as supports of residential territorialization'. *Journal of Urban Design* 15.3, pp. 357–74.

Chase, J., Crawford, M. and Kaliski, J. (1999), *Everyday Urbanism*. Monacelli Press.

Chaves, E., Knox, P. and Bieri, D. (2011), 'The rest landscape of metroburbia', *International Perspectives on Suburbanization*. New York: Palgrave MacMillan, pp. 35–53.

Cheang, Michael (2002), 'Older adults' frequent visits to a fast-food restaurant: Nonobligatory social interaction and the significance of play in a "third place"'. *Journal of Aging Studies* 16.3, pp. 303–21.

Chen, Jie (2007), 'Rapid urbanization in China: A real challenge to soil protection and food security'. *Catena* 69.1, pp. 1–15.

Cherry, Gordon E. (1972), *Urban Change and Planning: A History of Urban Development in Britain since 1750*. Henley-On-Thames: Oxfordshire GT Foulis and Co.

Cheung, Sidney C. H. (2007), 'Fish in the marsh: A case study of freshwater fish farming in Hong Kong', in Sydney Cheung and Che-Beng Tan (eds), *Food and Foodways in Asia: Resource, Tradition and Cooking*. London: Routledge, pp. 37–50.

Chevalier, Sophie (1998), 'From Woollen Carpet to grass carpet: Bridging house and garden in an English suburb', in Daniel Miller (ed.), *Material Cultures: Why Some Things Matter*. London: UCL Press, pp. 47–71.

Chiang, Marylyn (2008), 'Hungry for a serious food production plan in the city of Vancouver'. Graduation Project. School of Community and Regional Planning (SCARP), Vancouver.

Choi, Hwansuk Chris et al. (2010), 'Food hygiene standard satisfaction of Singaporean diners'. *Journal of Foodservice Business Research* 13.3, pp. 156–77.
Choo, Kristin (2013), 'Plowing over: Can urban farming save detroit and other declining cities? Will the law allow it?', http://www.abajournal.com/magazine/article/plowing_over_can_urban_farming_save_detroit_and_other_declining_cities_will. Accessed 13 December 2013.
Choon-Piew, Pow (2009), *Gated Communities in China Class, Privilege and the Moral Politics of the Good Life*. London: Routledge.
Christian, H., Giles-Corti, B., Knuiman, M., Timperio, A. and Foster, S. (2011), 'The influence of the built environment, social environment and health behaviors on body mass index. Results from RESIDE'. *Preventive Medicine* 53.1, pp. 57–60.
Christie, Maria Elisa (2004), 'Kitchenspace, fiestas, and cultural reproduction in Mexican house-lot gardens'. *Geographical Review* 94.3, pp. 368–90.
Chrzan, J. (2006), 'Why study culinary tourism?'. *Expedition* 48.1, pp. 40–1.
Clark, Clifford E. (1987), 'The vision of the dining room: Plan book dreams and middle-class realities', in Kathryn Grover (ed.), *Dining in America: 1850–1900*. Amherst: University of Massachusetts Press, pp. 142–72.
Clark, Samuel and Clark, Samantha (2007), *Moro East*. London: Ebury Press.
Clarke, Ethne (1988), *The Art of the Kitchen Garden*. London: Joseph.
Clarke, Graham, Eyre, Heather and Guy, Cliff (2002), 'Deriving indicators of access to food retail provision in British cities: Studies of Cardiff, Leeds and Bradford'. *Urban Studies* 39.11, pp. 2041–60.
Clarke, R., Cartwright, S., Kancans, R., Please, P. and Binks, B. (2010), *Drivers of Regional Agritourism and Food Tourism in Australia*. Canberra: ABARE-BRS.
Clendenning, Jessica (2011), 'Between empty lots and open pots: Understanding the rise of urban food movements'. Research Paper, The Hague, Erasmus University.
Clery, E. J. (1991), 'Women, publicity and the coffee-house myth'. *Women: A Cultural Review* 2.2, pp. 168–77.
Clinch, J. Peter and O'Neill, Eoin (2010), 'Designing development planning charges: Settlement patterns, cost recovery and public facilities'. *Urban Studies* 47.10, pp. 2149–71.
Clos, Juan (2005), 'Towards a European city model', in *London: Europe's Global City?* Urban Age Conference Newspaper. London: LSE and Alfred Herrhausen Society.
Cobbett, William (1967), *Rural Rides*. London: Penguin (First published 1830).
Cockrall-King, J. (2012), *Food and the City: Urban Agriculture and the New Food Revolution*. Amherst: Prometheus Books.
Coe, N. M. and Lee, Y. S. (2006), 'The strategic localization of transnational retailers: The case of Samsung-Tesco in South Korea'. *Economic Geography* 82.1, pp. 61–88.
Cofie, Olufunke O., Veenhuizen, Rene van and Drechsel, Pay (2003), 'Contribution of urban and peri-urban agriculture to food security in sub-Saharan Africa'. Africa session of 3rd WWF, Kyoto 17.
Cohen, Lizabeth (1986), 'Embellishing a life of labor: An interpretation of the material culture of American working-class homes', in Upton Dell and John Michael Vlach (eds), *Common Places: Readings in Vernacular Architecture*. Athens; London: University of Georgia Press, pp. 261–80.

Cohen, Lizabeth (1996), 'From town center to shopping center: The reconfiguration of community marketplaces in postwar America'. *The American Historical Review*, pp. 1050–81.

Cole-Hamilton, I. and Lang, Tim (1986), *Tightening Belts: A Report of the Impact of Poverty on Food*. London: London Food Commission.

Coley, David, Howard, Mark and Winter, Michael (2009), 'Local food, food miles and carbon emissions: A comparison of farm shop and mass distribution approaches'. *Food Policy* 34.2, pp. 150–5.

Coley, David, Howard, Mark and Winter, Michael (2011), 'Food miles: Time for a re-think?' *British Food Journal* 113.7, pp. 919–34.

Collins, Jock, Gibson, Katherine, Castles, Stephen and Tait, David (1995), *A Shop Full of Dreams: Ethnic Small Business in Australia*. Sydney: Pluto Press.

Commission for Architecture and the Built Environment (2000), *By Design, Urban Design in the Planning System: Towards Better Practice*. London: HMSO.

Conisbee, M. and Murphy, M. (2004), *Clone Town Britain: The Loss of Local Identity on the Nation's High Streets*. London: New Economics Foundation.

Conlin, J. (2008), 'Vauxhall on the boulevard: Pleasure gardens in London and Paris, 1764–1784'. *Urban History* 35.1, pp. 24–47.

Connell, David (2013), *Economic and Social Benefits Study*. British Columbia Association of Farmers Markets.

Conzen, M. (1960), *Alnwick Northumberland: A Study in Town Plan Analysis*. London: Institute of British Geographers.

Cook, N. and Harder, S. (2013), 'By accident or design? Peri-urban planning and the protection of productive land on the urban fringe', in Quentin Farmar-Bowers, Vaughan Higgins and Joanne Millar (eds), *Food Security in Australia*. London: Springer, pp. 413–24.

Cooper, Gail (2008), 'Escaping the house: Comfort and the California garden'. *Building Research & Information* 36.4, pp. 373–80.

Corburn, Jason (2004), 'Confronting the challenges in reconnecting urban planning and public health'. *American Journal of Public Health* 94.4, pp. 541–6.

Corburn, Jason (2009), *Toward the Healthy City: People, Places, and the Politics of Urban Planning*. London: The MIT Press.

Corigliano, Magda Antonioli (2002), 'The route to quality: Italian gastronomy networks in operation', in Anne-Mette Hjalager and Greg Richards (eds), *Tourism and Gastronomy*. Hillsdale, NJ: Psychology Press, pp. 166–85.

Corlett, J. L., Dean, E. A. and Grivetti, L. E. (2003), 'Hmong gardens: Botanical diversity in an urban setting'. *Economic Botany* 57.3, pp. 365–79.

Costley, Debra (2006), 'Master planned communities: Do they offer a solution to urban sprawl or a vehicle for seclusion of the more affluent consumers in Australia?' *Housing, Theory and Society* 23.3, pp. 157–75.

Côté, Raymond P. and Cohen-Rosenthal, Edward (1998), 'Designing eco-industrial parks: A synthesis of some experiences'. *Journal of Cleaner Production* 6.3, pp. 181–8.

Couch, Chris, Leontidou, Lila and Petschel-Held, Gerhard (eds) (2007), *Urban Sprawl in Europe Landscapes, Land-Use Change and Policy*. Blackwell: RICS Research.

Counihan, Carole M. (1999), *The Anthropology of Food and Body: Gender, Meaning and Power*. London: Routledge.
Counihan, Carole M. (2004), *Around the Tuscan Table: Food, Family, and Gender in Twentieth Century Florence*. London: Routledge.
Counihan, Carole M. and Kaplan, Steven L. (eds) (2004), *Food and Gender: Identity and Power*. London: Routledge.
Courtemanche, Charles and Carden, Art (2011), 'Supersizing supercenters? The impact of Walmart Supercenters on body mass index and obesity'. *Journal of Urban Economics* 69.2, pp. 165–81.
Cowan, Brian (2001), 'What was Masculine about the Public Sphere? Gender and the coffeehouse milieu in post-Restoration England'. *History Workshop Journal* 51, pp. 127–57.
Cowan, Brian (2004a), 'Mr Spectator and the coffeehouse public sphere'. *Eighteenth-Century Studies* 37.3, pp. 345–66.
Cowan, Brian (2004b), 'The rise of the coffeehouse reconsidered'. *The Historical Journal* 47.01, pp. 21–46.
Cowan, Brian (2005), *The Social Life of Coffee: Curiosity, Commerce and Civil Society in Early Modern Britain*. London: New Haven.
Cowan, Ruth Schwartz (1983), *More Work for Mother: The Ironies of Household Technology from the Open Hearth to the Microwave*. New York: Basic Books.
Cowan, Ruth Schwartz (1987), 'The consumption junction: A proposal for research strategies in the sociology of technology', *The Social Construction of Technological Systems*, pp. 261–80.
Coy, Martin (2006), 'Gated communities and urban fragmentation in Latin America: The Brazilian experience'. *GeoJournal* 66.1–2, pp. 121–32.
Craig, Amanda (2013), 'Jolly kitchen suppers aren't as casual as you'd think. Owners of the smartest new homes might regret no longer having a dining room', 8:53PM BST 30 September 2013. http://www.telegraph.co.uk/foodanddrink/10344653/Jolly-kitchen-suppers-arent-as-casual-as-youd-think.html.
Craig, Suzanne (2012), 'What restaurants know (about you)', http://www.nytimes.com/2012/09/05/dining/what-restaurants-know-about-you.html?pagewanted=all&_r=0. Published 4 September 2012, Retrieved 21 January 2013.
Crawford, Margaret (1992), 'The world in a shopping mall', in Michael Sorkin (ed.), *Variations on a Theme Park: The New American City and the End of Public Space*. New York: Hill and Wang.
Crawford, Margaret (2002), 'Suburban life and public space', in David J. Smiley (ed.), *Sprawl and Public Space: Redressing The Mall*. Washington, DC: National Endowment for the Arts, pp. 21–30.
Creasy, Rosalind (1982), *The Complete Book of Edible Landscaping*. San Francisco: Sierra Club Books.
Crewe, L. and Beaverstock, J. (1998), 'Fashioning the city: Cultures of consumption in contemporary urban spaces'. *Geoforum* 29.3, pp. 287–308.
Crewe, L. and Lowe, M. (1995), 'Gap on the map? Towards a geography of consumption and identity'. *Environment and Planning A* 27.12, pp. 1877–98.

Cromley, Elizabeth C. (1996), 'Transforming the food axis: Houses, tools, modes of analysis'. *Material Culture Review/Revue de la culture matérielle* 44.1. http://journals.hil.unb.ca/index.php/MCR/article/view/17695/22264.

Cromley, Elizabeth C. (2010), *The Food Axis: Cooking, Eating, and the Architecture of American Houses*. Charlottesville: University of Virginia Press.

Crouch, David and Ward, Colin (1997), *The Allotment: Its Landscape and Culture*. Nottingham: Five Leaves.

Crouch, David and Wiltshire, Richard (2005), 'Designs on the plot: The future for allotments in urban landscapes', in A. Viljoen, K. Bohn and J. Howe (eds), *Continuous Productive Urban Landscapes (CPULs): Designing Urban Agriculture for Sustainable Cities*. Oxford: Architectural Press, pp. 124–30.

Cruickshank, D. and Burton, N. (1990), *Life in the Georgian City*. London: Viking.

Csergo, Julia (2003), 'The picnic in nineteenth-century France. A social event involving food: Both a necessity and a firm of entertainment', in Marc Jacobs and Peter Scholliers (eds), *Eating Out in Europe. Picnics, Gourmet Dining and Snacks since the Late Eighteenth Century*. Oxford: Berg, pp. 39–52.

Cullen, Gordon (1961), *The Concise Townscape*. London: Architectural Press.

Cummins, Steven and Macintyre, Sally (2002), 'A systematic study of an urban foodscape: The price and availability of food in greater Glasgow'. *Urban Studies* 39.11, pp. 2115–30.

Cummins, Steven and Macintyre, Sally (2006), 'Food environments and obesity – neighbourhood or nation?' *International Journal of Epidemiology* 35.1, pp. 100–4.

Daniels, G. D. and Kirkpatrick, J. B. (2006), 'Comparing the characteristics of front and back domestic gardens in Hobart, Tasmania, Australia'. *Landscape and Urban Planning* 78.4, pp. 344–52.

Davey, Peter (2008), 'Hearth and home: Food preparation locations in changing times', in Petra Hagen Hodgson and Rolf Toyka (eds), *The Architect, the Cook, and Good Taste*. Basel: Birkhauser, pp. 100–9.

David, Elizabeth (1954), *Italian Food*. London: Macdonald.

David, Elizabeth (1986), *French Provincial Cooking*. London: Penguin.

Davidson, Alan (ed.) (1981), *National and Regional Styles of Cookery*. Oxford Symposium Proceedings. London: Prospect Books.

Davidson, Alan (ed.) (1991), *The Cook's Room: A Celebration of the Heart of the Home*. Sydney: Doubleday.

Davies, LLewelyn (2000), *Urban Design Compendium*. London: English Partnerships.

Davis, B. and Carpenter, C. (2009), 'Proximity of fast-food restaurants to schools and adolescent obesity'. *American Journal of Public Health* 99.3, p. 505.

Davis, Judy S., Nelson, Arthur C. and Dueker, Kenneth J. (1994), 'The New'Burbs The Exurbs and their implications for planning policy'. *Journal of the American Planning Association* 60.1, pp. 45–59.

Davis, Mike (2005), *Planet of Slums*. London: Verso.

Davis, S. (1996), 'Theme park: Global industry and cultural form'. *Media, Culture and Society* 18, pp. 399–422.

Day, K., Carreon, D. and Stump, C. (2000), 'The therapeutic design of environments for people with dementia a review of the empirical research'. *The Gerontologist* 40.4, pp. 397–416.

De Certeau, Michel (1984), *The Practice of Everyday Life*, trans. S. Rendall. Berkeley, CA: University of California Press.

De Certeau, Michel (1986), *Heterologies: Discourse on the Other*, trans. Brian Massumi; foreword by Wlad Godzich. Manchester: Manchester University Press.

De Certeau, Michel (2000), *The Certeau Reader*, edited by Graham Ward. Oxford: Blackwell.

De Certeau, M., Giard, L. and Mayol, P. (1998), *The Practice of Everyday Life, 2: Living and Cooking*. Minneapolis, MN: University of Minnesota Press.

Deelstra, Tjeerd and Girardet, Herbert (2000), 'Urban agriculture and sustainable cities', in Nico Bakker et al. (eds), *Growing Cities, Growing Food: Urban Agriculture on the Policy Agenda. A Reader on Urban Agriculture*. Feldafing, Germany: DSE, pp. 43–65.

Deener, Andrew (2007), 'Commerce as the structure and symbol of neighborhood life: Reshaping the meaning of community in Venice, California'. *City & Community* 6.4, pp. 291–314.

de la Bruheze, Adri Albert and van Otterloo, Anneke H. (2003), 'Snacks and snack culture in the rise of eating out in the Netherlands in the twentieth century', in Marc Jacobs and Peter Scholliers (eds), *Eating Out in Europe: Picnics, Gourmet Dining and Snacks since the Late Eighteenth Century*. Oxford: Berg.

de Lera, Enrique Ruiz (2012), 'Gastronomy as a key factor in branding Spain'. OECD Studies on Tourism Food and the Tourism Experience. The OECD-Korea Workshop: The OECD-Korea Workshop. OECD Publishing.

DeLind, Laura B. (2002), 'Place, work, and civic agriculture: Common fields for cultivation'. *Agriculture and Human Values* 19.3, pp. 217–24.

DeLind, Laura B. (2011), 'Are local food and the local food movement taking us where we want to go? Or are we hitching our wagons to the wrong stars?' *Agriculture and Human Values* 28, pp. 273–83.

Deloire, A., Prévost, P. and Kelly, M. (2008), 'Unravelling the terroir mystique: An agro-socio-economic perspective'. *Perspectives in Agriculture, Veterinary Science, Nutrition and Natural Resources* 3.32, pp. 1–9.

De'Medici, Lorenza (1990), *The Heritage of Italian Cooking*. Willoughby, NSW: Weldon Russell Publishing.

Dennis, Richard (2008), *Cities in Modernity Representations and Productions of Metropolitan Space, 1840–1930*. Cambridge: Cambridge University Press.

Dentith, Simon (2000), 'From William Morris to the Morris Minor: An alternative suburban history', in R. Webster (ed.), *Expanding Suburbia: Reviewing Suburban Narratives* (vol. 6). London: Berghahn Books, pp. 15–30.

DeSalvo, L. A. and Giunta, E. (eds) (2003), *The Milk of Almonds: Italian American Women Writers on Food and Culture*. New York: Feminist Press at CUNY.

de Roest, K. and Menghi, A. (2000), 'Reconsidering "traditional" food: The case of Parmigiano Reggiano cheese'. *Sociologia Ruralis*, 40, Part 4, pp. 439–51.

Derry, N. (2008), Ridley Road Market, Yelp. http://www.yelp.co.uk/biz/ridley-road-market-london.

Deslauriers, Pierre, Bryant, Christopher and Marois, Claude (1992), 'Farm business restructuring in the urban fringe: The Toronto and Montreal regions', in I. R. Bowler, C. R. Bryant and M. D. Nellis (eds), *Contemporary Rural Systems in Transition 1*. Wallingford: CAB International, pp. 74–86.

Dettori, D. G., Paba, A. and Pulina, M. (2004), *European Rural Tourism: Agrotouristic Firms in Sardinia and their Life Cycle*.

D'Haese, Marijke and Van den Berg, Marrit (2007), 'Who shops at supermarkets? A study of retail patronage in Nicaragua'. *Journal of Food Distribution Research* 38.1, p. 50.

Di, A. (2007), *Consumers' Perceptions toward Retail Stores: A Comparision between Foreign Superstores and Family-Run Stores in Bangkok*. Shinawatra University.

Dixon, Jane (2007), 'Supermarkets as new food authorities', in David Burch and Geoffrey Lawrence (eds), *Supermarkets and Agri-food Supply Chains: Transformations in the Production and Consumption of Foods*. Cheltenham: Edward Elgar Publishing, pp. 29–50.

Dixon, J. M., Donati, K. J., Pike, L. L. and Hattersley, L. (2009), 'Functional foods and urban agriculture: Two responses to climate change-related food insecurity'. *New South Wales Public Health Bulletin* 20.2, pp. 14–18.

DMR Architecture (2013), 'Traditional Malls adopting Lifestyle Center Elements'. http://www.dmrarch.com/traditional-malls-adopting-lifestyle-center-elements/. Accessed 27 August 2013.

Docking, A. (2009), 'Invigorating town centres through alternate local produce marketing strategies'. International Cities Town Centres and Communities Society.

Dodds, B. and Britnell, R. H. (eds) (2008), *Agriculture and Rural Society After the Black Death: Common Themes and Regional Variations* (vol. 6). Hatfield: University of Hertfordshire Press.

Dokmeci, Vedia and Beygo, Cem (1998), 'Analysis of market areas of mega malls in Istanbul'. *ERSA Conference Papers*. European Regional Science Association.

Dokmeci, Vedia, Altunbas, Ufuk and Yazgi, Burcin (2007), 'Revitalisation of the main street of a distinguished old neighbourhood in Istanbul'. *European Planning Studies* 15.1, pp. 153–66.

'Domo + Urbis Now Open' (2013), Homes. Adelaide Matters, Wednesday 13 November: 3.

Domosh, M. (1998), 'Geography and gender: Home, again?'. *Progress in Human Geography* 22.2, pp. 276–82.

Don, Monty (2009), 'As the credit crunch brings an allotment boom, Monty Don explains "my plot to save Britain"', 20 February 2009. http://www.dailymail.co.uk/debate/article-1150805/As-credit-crunch-brings-allotment-boom-Monty-Don-explains-plot-save-Britain.html#ixzz3Hdg8BjM0.

Donald, B. (2008), 'Food systems planning and sustainable cities and regions: The role of the firm in sustainable food capitalism', in Graham Haughton and Kevin Morgan (eds), Sustainable Regions. *Regional Studies* 42.9, pp. 1251–62.

Donald, B. and Blay-Palmer, A. (2006), 'The urban creative-food economy: Producing food for the urban elite or social inclusion opportunity?' *Environment and Planning A* 38.10, pp. 1901–20.

Donald, B., Gertler, M., Gray, M. and Lobao, L. (2010), 'Re-regionalizing the food system?' *Cambridge Journal of Regions, Economy and Society*, rsq020.

Donovan, Jenny, Larsen, Kirsten and McWhinnie, Julie-Anne (2011), *Food-sensitive Planning and Urban Design: A Conceptual Framework for Achieving a Sustainable and Healthy Food System*. David Lock Associates, University of Melbourne and National Heart Foundation of Australia.

Dorian, D. (2007), 'The Good Life-An ornamental and highly productive potager in Sonoma County, California, devoted to organic principles, offers lessons in great garden style'. *Garden Design* 72.

Douglas, Ian (2006), 'Peri-urban ecosystems and societies: Transitional zones and contrasting values', in Duncan McGregor, David Simon and Donald Thompson (eds), *Peri-Urban Interface: Approaches to Sustainable Natural and Human Resource Use*. Abingdon: Oxon; Earthscan, pp. 18–29.

Dovey, Kim (1994), 'Dreams on display: Suburban ideology in the model home', in Sarah Ferber, Chris Healy and Chris McAuliffe (eds), Beasts of Suburbia: Reinterpreting Cultures in Australian Suburbs. *The Australian and New Zealand Journal of Sociology* 31.3, pp. 112–13.

Dresner, S. (2008), *The Principles of Sustainability*. London: Earthscan.

Drew, Jane (c.1944), *Kitchen Planning*. London: Gas Industry.

Dries, Liesbeth, Reardon, Thomas and Swinnen, Johan F. M. (2004), 'The rapid rise of supermarkets in Central and Eastern Europe: Implications for the agrifood sector and rural development', *Development Policy Review* 22.5, pp. 525–56.

Drummer, Randyl (2012), 'The de-malling of America: What's next for hundreds of outmoded malls? Strained by online commerce, changing shopper preferences and trendier competition, many outmoded malls face bleak future', 3 October. http://www.costar.com/News/Article/The-De-Malling-of-America-Whats-Next-for-Hundreds-of-Outmoded-Malls-/141980. Accessed 27 August 2013.

D'Silva, L. and Webster J. (eds) (2010), *The Meat Crisis: Developing more Sustainable Production and Consumption*. London, UK: Earthscan.

Duany, Andrés (2002), 'Introduction to the special issue: The transect'. *Journal of Urban Design* 7.3, pp. 251–60.

Duany, Andrés and DPZ (2011), *Garden Cities: Theory and Practice of Agricultural Urbanism*. London: Duany Plater-Zyberk and Co. and The Prince's Foundation.

Duany, Andrés, Plater-Zyberk, Elizabeth and Speck, Jeff (2000), *Suburban Nation: The Rise of Sprawl and the Decline of the American Dream*. New York: North Point Press.

Dubbeling, M., de Zeeuw, H. and Otto-Zimmermann, K. (2011), 'Urban agriculture and climate change adaptation: Ensuring food security through adaptation', in Konrad Otto-Zimmermann (ed.), *Cities and Adaptation to Climate Change – Proceedings of the Global Forum*. Local Sustainability, vol. 1, Resilient Cities. Accessed 8 May 2013.

Duchin, F. (2005), 'Sustainable consumption of food: A framework for analyzing scenarios about changes in diets'. *Journal of Industrial Ecology* 9.1–2, pp. 99–114.

Duffin Wolfe, Jessica (2009), 'City of snacks', in Christina Palassio and Alana Wilcox (eds), *The Edible City Toronto's Food from Farm to Fork*. Toronto: Coach House Books.

Dummitt, Chris (1998), 'Finding a place for father: Selling the Barbecue in postwar Canada'. *Journal of the Canadian Historical Association/Revue de la Societe historique du Canada* 9.1, pp. 209–23.

Dunbabin, K. M. (2003), *The Roman Banquet: Images of Conviviality*. Cambridge: Cambridge University Press.
Dunham-Jones, E. and Williamson, J. (2009), *Retrofitting Suburbia: Urban Design Solutions for Redesigning Suburbs*. Hoboken: Wiley.
Dunnett, Nigel and Qasim, Muhammad (2000), 'Perceived benefits to human well-being of urban gardens'. *HortTechnology* 10.1, pp. 40–5.
Duruz, J. (1994), 'Suburban gardens: Cultural notes', in Sarah Ferber, Chris Healy and Chris McAuliffe (eds), *Beasts of Suburbia: Reinterpreting Cultures in Australian Suburbs*. Melbourne: MUP.
Duruz, J. (2001), 'Home cooking, nostalgia, and the purchase of tradition'. *Traditional Dwellings and Settlements Review* 12.2, pp. 21–32.
Duruz, J. (2002), 'Rewriting the village: Geographies of food and belonging in Clovelly, Australia'. *Cultural Geographies* 9.4, pp. 373–88.
Duruz, J. (2005), 'Eating at the borders: Culinary journeys'. *Environment and Planning D: Society and Space* 23, pp. 51–69.
Duruz, J., Luckman, S. and Bishop, P. (2011), 'Bazaar encounters: Food, markets, belonging and citizenship in the cosmopolitan city'. *Continuum* 25.5, pp. 599–604.
Dyson, Sue and McShane, Roger (undated), Michel Bras Review. http://www.foodtourist.com/ftguide/content/i3212.htm.
Eberwine, Diane (2003), 'Globesity: A crisis of growing proportions'. *Diabetes Voice* 48.1, pp. 30–3.
Ecker, S., Clarke, R., Cartwright, S., Kancans, R., Please, P., and Binks, B. (2010), *Drivers of Regional Agritourism and Food Tourism in Australia*. Australian Government, Australian Bureau of Agricultural and Resource Economics- Bureau of Rural Sciences.
Edge, J. T. (2012), *The Truck Food Cookbook: 150 Recipes and Ramblings from America's Best Restaurants on Wheels*. Workman Publishing.
Edinburgh Garden Partners Kitchen Garden at Princes Street Gardens Cottage (undated), http://www.edinburghgardenpartners.org.uk/. Accessed 13 December 2013.
Edwards, F., Dixon, J., Friel, S., Hall, G., Larsen, K., Lockie, S. and Hattersley, L. (2011), 'Climate change adaptation at the intersection of food and health'. *Asia-Pacific Journal of Public Health* 23.2 suppl, pp. 91S–104S.
Edwards, John S. A. and Gustafsson, Inga-Britt (2008), 'The room and atmosphere as aspects of the meal: A review'. *Journal of Foodservice* 19.1, pp. 22–34.
Eid, Jean, Overman, Henry G., Puga, Diego and Turner, Matthew A. (2008), 'Fat city: Questioning the relationship between urban sprawl and obesity'. *Journal of Urban Economics* 63.2, pp. 385–404.
Eisenhauer, E. (2001), 'In poor health: Supermarket redlining and urban nutrition'. *Geojournal* 53, pp. 125–33.
Ekanem, Etok O. (1998), 'The street food trade in Africa: Safety and socio-environmental issues'. *Food Control* 9.4, pp. 211–15.
Elias, Norbert (1982), *The Civilising Process*, trans. E. Jephcott. Oxford: Blackwell.
Ellaway, Anne and Sally Macintyre (2000), 'Shopping for food in socially contrasting localities'. *British Food Journal* 102.1, pp. 52–9.

Ellickson, Paul B. (2011), 'The evolution of the supermarket industry: From A&P to Wal-Mart'. University of Rochester – Simon School of Business, 1 April 2011, Simon School Working Paper Series No. FR 11-17.
Ellickson, Paul B. and Grieco, Paul L. E. (2012), 'Wal-Mart and the geography of grocery retailing'. *Journal of Urban Economics* 75, pp. 1–14.
Ellin, Nan (1996), *Postmodern Urbanism*. Cambridge, MA; Oxford: Blackwell.
Elliott, Shannon Brooke (2010), 'Landscape, kitchen, table: Compressing the food axis to serve a food desert'. Master's Thesis. University of Tennessee. http://trace.tennessee.edu/utk_gradthes/793.
Ellis, Markman (2004), 'Pasqua Rosee's coffee-house, 1652–1666'. *The London Journal* 29.1, pp. 1–24.
Ellis, Markman (2011), *The Coffee-House: A Cultural History*. UK: Hachette.
Elms, Jonathan, Canning, C., de Kervenoae, R., Whysall, P. and Hallsworth, A. (2010), '30 years of retail change: Where (and how) do you shop?' *International Journal of Retail & Distribution Management* 38.11/12, pp. 817–27.
Eng, L. A. (2010), 'The Kopitiam in Singapore: An evolving story about migration and cultural diversity'. *Asia Research Institute: Working paper Series No 132; Singapore*.
Engelhard, Benjamin (2010), 'Rooftop to tabletop: Repurposing urban roofs for food production'. Dissertation, University of Washington.
Engler-Stringer, Rachel (2010), 'The domestic foodscapes of young low-income women in Montreal: Cooking practices in the context of an increasingly processed food supply'. *Health Education & Behavior* 37.2, pp. 211–26.
Erlanger, Steven (2008), http://seattletimes.com/html/nationworld/2008425346_cafes23.html. Retrieved 21 January 2013.
Esman, Marjorie R. (1982), 'Festivals, change, and unity: The celebration of ethnic identity among Louisiana Cajuns'. *Anthropological Quarterly* 55.4, pp. 199–210.
Esperdy, Gabrielle (2002), 'Edible urbanism'. *Architectural Design* 72, pp. 44–50.
Essex, Stephen J. and Brown, Graham P. (1997), 'The emergence of post-suburban landscapes on the North Coast of New South Wales: A case study of contested space'. *International Journal of Urban and Regional Research* 21.2, pp. 259–87.
Esteva, G. and Prakash, M. S. (1998), *Grassroots Post-Modernism: Remaking the Soil of Cultures*. New York: Palgrave Macmillan.
Estrada-Garcia, T., Lopez-Saucedo, C., Zamarripa-Ayala, B., Thompson, M. R., Gutierrez-Cogco, L., Mancera-Martinez, A. and Escobar-Gutierrez, A. (2004), 'Prevalence of Escherichia coli and Salmonella spp. in street-vended food of open markets (tianguis) and general hygienic and trading practices in Mexico City'. *Epidemiology and infection* 132.06, pp. 1181–4.
Evans, D., Campbell, H. and Murcott, A. (2012), 'A brief pre-history of food waste and the social sciences'. *The Sociological Review* 60.S2, pp. 5–26.
Evans, Stephen (2012), 'Berlin and its "democratic" canteen culture'. http://www.bbc.co.uk/news/magazine-19703080, 6 October 2012. Retrieved 22 January 2013.
Everett, Sally and Aitchison, Cara (2008), 'The role of food tourism in sustaining regional identity: A case study of Cornwall, South West England'. *Journal of Sustainable Tourism* 16.2, pp. 150–67.

Falzon, Mark-Anthony (2004), 'Paragons of lifestyle: Gated communities and the politics of space in Bombay'. *City & Society* 16.2, pp. 145–67.

Farhangmehr, Minoo and Veiga, Paula (1995), 'The changing consumer in Portugal'. *International Journal of Research in Marketing* 12.5, December, pp. 485–502.

Farhangmehr, Minoo, Marques, Susana and Silva, Joaquim (2000), 'Consumer and retailer perceptions of hypermarkets and traditional retail stores in Portugal'. *Journal of Retailing and Consumer Services* 7.4, pp. 197–206.

Farnham, S. D., McCarthy, J. F., Patel, Y., Ahuja, S., Norman, D., Hazlewood, W. R. and Lind, J. (2009), 'Measuring the impact of third place attachment on the adoption of a place-based community technology'. In *Proceedings of the SIGCHI Conference on Human Factors in Computing Systems*, pp. 2153–6.

Farr, Douglas (2008), *Sustainable Urbanism Design with Nature*. New Jersey: Wiley.

Faure, Alain (ed.) (1991), *Les Premiers banlieusards: Aux origines des banlieues de Paris 1860–1940*. Paris: Créaphis.

Feagan, Robert (2007), 'The place of food: Mapping out the "local" in local food systems'. *Progress in Human Geography* 31.1, pp. 23–42.

Feeley-Harnik, Gillian (1994), *The Lord's Table: The Meaning of Food in Early Judaism and Christianity*. Washington, DC: Smithsonian Institution Press.

Feeley-Harnik, Gillian (1995), 'Religion and food: An anthropological perspective'. *Journal of the American Academy of Religion* 63.3, pp. 565–82.

Feenstra, G. W. (1997), 'Local food systems and sustainable communities'. *American Journal of Alternative Agriculture* 12.01, pp. 28–36.

Feinberg, Richard A. and Meoli, Jennifer (1991), 'A brief history of the mall'. *Advances in Consumer Research* 18.1, pp. 426–7.

Fenton, Alexander (2003), 'Feeding the shearers: Endogenous developments in Scottish harvest foods', in Marc Jacobs and Peter Scholliers (eds), *Eating Out in Europe: Picnics, Gourmet Dining and Snacks since the Late Eighteenth Century*. Oxford: Berg, pp. 39–52.

Fernández-Armesto, Felipe (2002), *Near a Thousand Tables: A History of Food*. New York: The Free Press.

Fernando, Nisha (2005), 'Taste, smell and sound on the street in Chinatown and little Italy'. *Architectural Design* 75.3, pp. 20–5.

Fernie, John and Arnold, Stephen J. (2002), 'Wal-Mart in Europe: Prospects for Germany, the UK and France'. *International Journal of Retail & Distribution Management* 30.2, pp. 92–102.

Fiedler, Rob and Addie, Jean-Paul (2008), 'Canadian cities on the edge: Reassessing the Canadian suburb'. *Occasional Paper Series* 1.1, pp. 1–32.

Field, Carol (1990), *Celebrating Italy*. New York: William Morrow and Company.

Fife, John (2012), 'Bringing supermarkets into food deserts: An analysis of retail intervention policies'. Available at SSRN 2197864.

Finkelstein, Joanne (1989), *Dining Out: A Sociology of Modern Manners*. Oxford: The Polity Press.

Finkelstein, Joanne (1999), 'Foodatainment'. *Performance Research* 4, pp. 130–6.

Fischler, Claude (1988), 'Food, self and identity'. *Social Science Information* 27.2, pp. 275–92.

Fishman, Charles (2006), *The Wal-Mart Effect: How the World's Most Powerful Company Really Works–and How it's Transforming the American Economy*. London: Penguin.

Fishman, Robert (1987), *Bourgeois Utopias: The Rise and Fall of Suburbia*. New York: Basic Books.

Fishman, Robert (1991), 'The garden city tradition in the post-suburban age'. *Built Environment* 17.3/4, pp. 232–41.

Fishman, Robert (1996), *Beyond Suburbia: The Rise of the Technoburb*. London: Routledge.

Fishman, Robert (1998), 'Beyond utopia: Urbanism after the end of cities'. *Ciutat Real, Ciutat Idea*, pp. 29–36.

Fishman, Robert (2002), 'The bounded city', in Parsons and Schuyler (eds), *From Garden City to Green City*. Baltimore and London: The Johns Hopkins Press.

Flannery, Tim (1994), *The Future Eaters*. New York: George Braziller.

Flores, Heather C. (2006), *Food not Lawns: How to Turn your Yard into a Garden and your Neighborhood into a Community*. White River Junction, VT: Chelsea Green.

Floyd, J. (2004), 'Coming out of the kitchen: Texts, contexts and debates'. *Cultural Geographies* 11.1, pp. 61–73.

Fogelson, Robert M. (2005), *Bourgeois Nightmares: Suburbia, 1870-1930*. New Haven and London: Yale University Press.

Fonseca, Vanessa (2009), 'Targeting hispanics/latinos beyond locality: Food, social networks and Nostalgia in online shopping', in A. Lindgreen (ed.), *The New Cultures of Food: Marketing Opportunities from Ethnic, Religious and Cultural Diversity*. Farnham, Surrey: Gower Publishing, Ltd., pp. 163–80.

Fonte, Maria (2008), 'Knowledge, food and place: A way of producing, a way of knowing'. *Sociologia Ruralis* 48.3, pp. 200–22.

Food, London (2006), *Healthy and Sustainable Food for London: The Mayor's Food Strategy*. London: London Development Agency.

Food Standards Agency (2004), 'Do food deserts exist? A multi-level, geographical analysis of the relationship between retail food access, socio-economic position and dietary intake (N09010)'. Project Report, Monday 17 May, London.

Ford, J. (1945), Design of modern interiors. New York: Architectural Book Publishing Co

Forsberg, Håkan (1995), 'Out of town shopping centres in Sweden – a political struggle for purchasing power'. *Scandinavian Housing and Planning Research* 12.2, pp. 109–13.

Fort, F., Ferras, N. and Amilien, V. (2007), 'Hyper-real territories and urban markets: Changing conventions for local food–case studies from France and Norway'. *Anthropology of Food*, S2. Special issue on local food products and systems.

Foucault, Michel (1967), 'Of other spaces', in Lieven De Cauter and Michiel Dehaene (eds and trans.) (2008), *Heterotopia and The City Public Space in a Post Civil Society*. London; New York: Routledge.

Fowler, Alys (2011), *The Thrifty Forager*. London: Kyle.

Fox, Renata (2007), 'Reinventing the gastronomic identity of Croatian tourist destinations'. *International Journal of Hospitality Management* 26.3, pp. 546–59.

Franck, Karen A. (2005a), 'Food for the city, food in the city'. *Architectural Design* 75.3, pp. 35–42.

Franck, Karen A. (2005b), 'The city as dining room, market and farm'. *Architectural Design* 75.3, pp. 5–10.

Francis, Mark and Griffith, Lucas (2011), 'The meaning and design of farmers' markets as public space an issue-based case study'. *Landscape Journal* 30.2, pp. 261–79.
Freeman, Andrea (2007), 'Fast food: Oppression through poor nutrition'. *California Law Review* 95, p. 2221.
Freeman, Donald B. (1991), *City of Farmers: Informal Urban Agriculture in the Open Spaces of Nairobi, Kenya*. Canada: Magill-Queen's University Press.
Freeman, June (2003), *The Making of the Modern Kitchen: A Cultural History*. Oxford: Berg.
Frederick, Christine (1918), *The New Housekeeping: Efficiency Studies in Home Management*. Garden City, NY: Doubleday, Page.
Freidberg, Susanne (2004), *French Beans and Food Scares: Culture and Commerce in an Anxious Age*. Oxford: Oxford University Press.
Freund, Peter and Martin, George (2007), 'Hyperautomobility, the social organization of space, and health'. *Mobilities* 2.1, pp. 37–49.
Friedland, John (2005), 'The French paradox', *The Morning News*. http://www.themorningnews.org/article/roundtable-the-french-paradox.
Friedmann, Avi (2002), *Planning the New Suburbia: Flexibility by Design*. Vancouver: UBC Press.
Frochot, Isabelle (2003), 'An analysis of regional positioning and its associated food images in French tourism regional brochures'. *Journal of Travel & Tourism Marketing* 14.3–4, pp. 77–96.
Frost, L. (1991), *The New Urban Frontier: Urbanisation and City-building in Australasia and the American West*. Kensington, NSW: UNSW Press.
Frumkin, Howard (2002), 'Urban sprawl and public health'. *Public Health Reports* 117.3, p. 201.
Frumkin, Howard, Frank, Lawrence and Jackson, Richard J. (2004), *Urban Sprawl and Public Health: Designing, Planning, and Building for Healthy Communities*. Washington, DC: Island Press.
Fulfrost, B. and Howard, P. (2006), 'Mapping the markets: The relative density of retail food stores in densely populated census blocks in the central coast region of California'. Center for Agroecology & Sustainable Food Systems. UC Santa Cruz: Center for Agroecology and Sustainable Food Systems. Retrieved from https://escholarship.org/uc/item/34j371tf.
Funes, F., García, L., Bourque, M., Pérez, N. and Rosset, P. (2002), *Sustainable Agriculture and Resistance: Transforming Food Production in Cuba*. Food First Books.
Furey, Sinead, Farley, Heather and Strugnell, Christopher (2002), 'An investigation into the availability and economic accessibility of food items in rural and urban areas of Northern Ireland'. *International Journal of Consumer Studies* 26.4, pp. 313–21.
Galantay, E. Y. (ed.) (1987), *The Metropolis in Transition*. New York: Icus Books.
Gallanter, Eden (2012), 'Ciudad Jardín Lomas del Palomar: Deriving ecocity design lessons from a garden city'. *Planning Perspectives* 27.2, pp. 297–307.
Gallegos, Danielle (2005), 'Cookbooks as manuals of taste', in David Bell and Joanne Hollows (eds), *Ordinary Lifestyles: Popular Media, Consumption and Taste*. Maidenhead: Open University Press, pp. 99–110.
Gallent, Nick and Andersson, John (2007), 'Representing England's rural-urban fringe'. *Landscape Research* 32.1, pp. 1–21.

Galli, Mariassunta et al. (2010), 'Agricultural management in peri-urban areas'. The experience of an international workshop. Land Lab–Scuola Superiore Sant'Anna (Italy), INRA et AgroParisTech-ENGREF, UMR Métafort Clermont Ferrand (France), Felici Editore, Ghezzano, Italy.

Gans, Herbert J. (1962), 'Urbanism and suburbanism as ways of life: A re-evaluation of definitions', in Arnold Rose (ed.), *Human Behaviour and Social Process: An Interactionist Approach*. London: Routledge and Keegan Paul, pp. 625–48.

Garnaut, Christine and Hutchings, Alan (2003), 'The colonel light gardens garden suburb commission: Building a planned community'. *Planning Perspectives* 18.3, pp. 277–93.

Garnett, Daisy (2012), http://www.standard.co.uk/news/london/take-me-to-the-river-caf-hammersmith-restaurant-favoured-by-heston-blumenthal-celebrates-25-years-8144099.html, 17 September 2012. Retrieved 22 January 2013.

Garnett, Tara (1996), 'Growing food in cities: A report to highlight and promote the benefits of urban agriculture in the UK'. London: National Food Alliance; SAFE Alliance.

Garreau, Joel (1991), *Edge City: Life on the New Urban Frontier*. New York: Doubleday.

Gastil, Raymond W. and Ryan, Zoë (2004), *Open: New Designs for Public Space* (vol. 16). New York: Princeton Architectural Press.

Gaynor, Andrea (2004), 'Animal husbandry and house wifery? Gender and suburban household food production in Perth and Melbourne 1890–1950 1'. *Australian Historical Studies* 36.124, pp. 238–54.

Gaynor, Andrea (2006), *Harvest of the Suburbs: An Environmental History of Growing Food in Australian Cities*. Crawley, WA: University of Western Australia Press.

Gazillo, Stephen (1981), 'The evolution of restaurants and bars in Vieux-Quebec since 1900'. *Cahiers de géographie du Québec* 25.64, pp. 101–18.

Geddes-Brown, L. (2007), *The Walled Garden*. London; New York: Merrill Publishing.

Gehl, Jan (2010), *Cities for People*. Washington, DC: Island Press.

Geist, Johann Friedrich (1983), *Arcades: The History of a Building Type*. Cambridge, MA: MIT Press.

Geller, Matthew (2010), http://www.sustainablecitynews.com/foodtrucksla.html. Accessed 16 January 2013.

Geniş, Şerife (2007), 'Producing elite localities: The rise of gated communities in Istanbul'. *Urban Studies* 44.4, pp. 771–98.

Germov, John and Williams, Lauren (eds) (2008), *A Sociology of Food and Nutrition: The Social Appetite*. Oxford: Oxford University Press.

Getz, Arthur (1991), *Urban Foodsheds*. Permaculture Activist 24.26.

Giard, Luce (1998), 'Doing cooking', in Michel de Certeau, Luce Giard and Pierre Mayol (eds), *The Practice of Everyday Life. Vol. 2, Living and Cooking*, trans. Timothy J. Tomasik. Minneapolis; London: University of Minnesota Press.

Gieryn, T. F. (2000), 'A space for place in sociology'. *Annual Review of Sociology* 26, pp. 463–96.

Gilchrist, Hannah (2012), 'Leon founders take over school dinner fight'. http://www.redonline.co.uk/news/in-the-news/leon-founders-take-over-school-dinner-fight, 9 July 2012. Retrieved 22 January 2013.

Gillespie, G., Hilchey, D. L., Hinrichs, C. C. and Feenstra, G. (2007), 'Farmers' markets as keystones in rebuilding local and regional food systems', in C. C. Hinrichs and T. A. Lyson (eds), *Remaking the North American Food System: Strategies for Sustainability*. Lincoln, NE: University of Nebraska Press, pp. 65–83.

Gilman, Nicholas (2011), *Good Food in Mexico City: Food Stalls, Fondas and Fine Dining*. iUniverse.

Gilman, Sander (2008), *Fat: A Cultural History of Obesity*. Cambridge, MA: Polity Press.

Gilroy, Paul (2004), *After Empire: Melancholia or Convivial Culture?* London: Routledge.

Ginn, F. (2012), 'Dig for victory! New histories of wartime gardening in Britain'. *Journal of Historical Geography* 38.3, pp. 294–305.

Giorda, Erica (2012), 'Farming in Motown: Competing narratives for urban development and urban agriculture in Detroit', in Andre Viljoen and Johannes S. C. Wiskerke (eds), *Sustainable Food Planning: Evolving Theory and Practice*. The Netherlands: Wageningen Academic Publishers, pp. 271–81.

Giovannucci, D., Barham, E. and Pirog, R. (2010), 'Defining and marketing "local" foods: geographical indications for US products'. *The Journal of World Intellectual Property* 13.2, pp. 94–120.

Girardet, Herbert (1999), *Creating Sustainable Cities* (No. 2). White River Junction, VT: Chelsea Green Publishing.

Girling, Cynthia and Kellett, Ronald (2005), *Skinny Streets and Green Neighbourhoods Design for Environment and Community*. Washington, DC; Covelo; London: Island Press.

Giroir, Guillame (2007), 'Spaces of leisure: Gated golf communities in China', in Fulong Wu (ed.), *China's Emerging Cities*. London: Routledge, pp. 235–55.

Girouard, Mark (1985), *Cities and People: A Social and Architectural History*. New Haven; New York: Yale University Press.

Giuliano, Genevieve and Narayan, Dhiraj (2003), 'Another look at travel patterns and urban form: The US and Great Britain'. *Urban Studies* 40.11, pp. 2295–312.

Glasze, Georg (2003), 'Segmented governance patterns – fragmented urbanism: The development of guarded housing estates in Lebanon'. *The Arab World Geographer* 6.2, pp. 79–100.

Glasze, Georg (2006), 'Segregation and seclusion: The case of compounds for western expatriates in Saudi Arabia'. *GeoJournal* 66, pp. 83–8.

Glasze, Georg and Alkhayyal, A. (2002), 'Gated housing estates in the arab world: Case studies in Lebanon and Riyadh, Saudi Arabia'. *Environment and Planning B: Planning and Design* 29.3, pp. 321–36.

Glasze, Georg, Frantz, Klaus and Webster, Chris (2002), 'The global spread of gated communities'. *Environment and Planning B: Planning and Design* 29.3, pp. 315–20.

Glasze, Georg, Webster, Chris and Frantz, Klaus (2006), *Private Cities: Global and Local Perspectives*. London: Routledge.

Gliessman, S. R. (1998), *Agroecology: Ecological Processes in Sustainable Agriculture*. Boca Raton, FL: CRC Press.

Gnad, Martin (2002), 'Gated communities in South Africa – experiences from Johannesburg'. *Environment and Planning B: Planning and Design* 29.3, pp. 337–53.

Godina-Golija, Maja (2003), 'Food culture in Slovene urban inns and restaurants between the end of the nineteenth century and workd war II', in Marc Jacobs and Peter Scholliers (eds), *Eating Out in Europe: Picnics, Gourmet Dining and Snacks since the Late Eighteenth Century*. Oxford: Berg, pp. 125–35.

Goffman, E. (1959), *The Presentation of Self in Everyday Life*. Edinburgh: Edinburgh University Press.

Goheen, Peter G. (1998), 'Public space and the geography of the modern city'. *Progress in Human Geography* 22.4, pp. 479–96.

Goldman, Arieh and Hino, Hayiel (2005), 'Supermarkets vs. traditional retail stores: Diagnosing the barriers to supermarkets' market share growth in an ethnic minority community'. *Journal of Retailing and Consumer Services* 12.4, pp. 273–84.

Goldman, Arieh, Krider, Robert and Ramaswami, Seshan (1999), 'The persistent competitive advantage of traditional food retailers in Asia: Wet markets' continued dominance in Hong Kong'. *Journal of Macromarketing* 19.2, pp. 126–39.

Goldshleger, Naftaly, Amit-Cohen, Irit and Shoshany, Maxim (2006), 'A step ahead of time: Design, allocation and preservation of private open space in the 1920s – the case of a garden suburb in Israel'. *GeoJournal* 67.1, pp. 57–69.

Goode, J. J. (2012), 'World's best street food'. Concierge.com. http://www.concierge.com/ideas/spa/tours/2274?page=4.

Goodland, R. (1997), 'Environmental sustainability in agriculture: Diet matters'. *Ecological Economics* 23.3, pp. 189–200.

Goodman, D. (2002), 'Rethinking food production–consumption: Integrative perspectives'. *Sociologia Ruralis*, 42.4, pp. 271–7.

Goodman, D. (2004), 'Rural Europe redux? Reflections on alternative agro-food networks and paradigm change'. *Sociologia Ruralis* 44.1, pp. 3–16.

Goodwin, H. and Francis, J. (2003), 'Ethical and responsible tourism: Consumer trends in the UK'. *Journal of Vacation Marketing* 9.3, pp. 271–84.

Gorton, Delvina, Bullen, Chris R. and Ni Mhurchu, Cliona (2010), 'Environmental influences on food security in high-income countries'. *Nutrition Reviews* 68.1, pp. 1–29.

Gospodini, Aspa (2006), 'Portraying, classifying and understanding the emerging landscapes in the post-industrial city'. *Cities* 23.5, pp. 311–30.

Goswami, Paromita and Mishra, Mridula S. (2009), 'Would Indian consumers move from kirana stores to organized retailers when shopping for groceries?' *Asia Pacific Journal of Marketing and Logistics* 21.1, pp. 127–43.

Gottdiener, Mark (1994), *The New Urban Sociology*. New York; London: McGraw-Hill.

Gottmann, Jean (1964), *Megalopolis: The Urbanized Northeastern Seaboard of the United States*. Cambridge, MA: MIT Press.

Graham, Emma-Jayne (2005), 'Dining al fresco with the living and the dead in Roman Italy', in Maureen Carroll, D. M. Hadley and Hugh Willmott (eds), *Consuming Passions: Dining from Antiquity to the Eighteenth Century*. Revealing history. Stroud: Tempus, pp. 49–65.

Graham, Sonia and Connell, John (2006), 'Nurturing relationships: The gardens of Greek and Vietnamese migrants in Marrickville, Sydney'. *Australian Geographer* 37.3, pp. 375–93.

Graham, S. and Marvin, S. (2001), *Splintering Urbanism: Networked Infrastructures, Technological Mobilities and the Urban Condition*. Psychology Press.

Grant, Jill (2005), 'Planning responses to gated communities in Canada'. *Housing Studies* 20.2, pp. 273–85.

Grant, Jill and Mittelsteadt, Lindsey (2004), 'Types of gated communities'. *Environment and Planning B* 31, pp. 913–30.

Grant, Richard (2005), 'The emergence of gated communities in a West African context: Evidence from Greater Accra, Ghana'. *Urban Geography* 26.8, pp. 661–83.

Gray, Patience (1987), *Honey From a Weed Fasting and Feasting in Tuscany, Catalonia, The Cyclades and Apulia*. London: Papermac.

Greene, J. M. (2009), 'Localization: Implementing the Right to Food'. *Drake Journal of Agricultural Law* 14, pp. 377–99.

Greenwood, J. and Standford, J. (2008), 'Preventing or improving obesity by addressing specific eating patterns'. *Journal of the American Board of Family Medicine* 21, pp. 135–40.

Gregory, D. P. (2008), *Cliff May and the Modern Ranch House*. New York: Rizzoli.

Grignon, Claude (2001) 'Commensality and social morphology: An essay of typology', in Peter Scholliers (ed.), *Food, Drink and Identity: Cooking, Eating and Drinking in Europe Since the Middle Ages*. Oxford: Berg, pp. 23–33.

Grigson, Jane (1982), *Jane Grigson's Fruit Book*, illustrated by Yvonne Skargon. Harmonsworth, Middlesex: Penguin Books.

Grimm, N. B., Faeth, S. H., Golubiewski, N. E., Redman, C. L., Wu, J., Bai, X. and Briggs, J. M. (2008), 'Global change and the ecology of cities'. *Science* 319.5864, pp. 756–60.

Groening, G. (2005), 'The world of small urban gardens'. *Chronica Horticulturae* 45.2, pp. 22–5.

Gruen, Victor (1964), *The Heart of Our Cities. The Urban Crisis: Diagnosis and Cure*. New York: Simon and Schuster.

Gulati, Ashok (2008), 'The rise of supermarkets and their development implications: International experience relevant for India (vol. 752)'. *International Food Policy Research Institute*, Washington, DC.

Guthman, Julie (2011), *Weighing in: Obesity, Food Justice, and the Limits of Capitalism*. Berkeley, CA; London: University of California Press.

Guthman, Julie (2013), 'Too much food and too little sidewalk? Problematizing the obesogenic environment thesis'. *Environment and Planning A* 45.1, pp. 142–58.

Guthrie, John, Guthrie, Anna and Lawson, Rob (2006), 'Farmers' markets: The small business counter-revolution in food production and retailing'. *British Food Journal* 108.7, pp. 560–73.

Guzik, H. (2004), 'Lazy Susan': The Avant-Garde, Women and Taylorism in the Czechoslovak Household'. *Umění (Art)* 52.3, pp. 247–60 (Abstract only).

Gyimóthy, S. and Mykletun, R. J. (2009), 'Scary food: Commodifying culinary heritage as meal adventures in tourism'. *Journal of Vacation Marketing* 15.3, pp. 259–73.

Habermas, J. (1989), *The Structural Transformation of the Public Sphere, An Inquiry into a Category of Bourgeois Society*, trans. T. Burger and F. Lawrence. Oxford: Polity Press.

Haeg, Fritz et al. (2008), *Edible Estates*. New York: Metropolis Books.

Haine, W. Scott (1992), '"Café Friend": Friendship and fraternity in parisian working-class cafés, 1850–1914'. *Journal of Contemporary History* 27.4, pp. 607–26.

Haine, W. Scott (1996), *The World of the Paris Cafe, Sociability among the French Working Class, 1789-1914*. London: Johns Hopkins University Press.

Halepete, Jaya, Seshadri Iyer, K. V. and Park, Soo Chul (2008), 'Wal-Mart in India: A success or failure?' *International Journal of Retail & Distribution Management* 36.9, pp. 701–13.

Hall, Anthony Clive (2006), 'A plea for front gardens'. *Urban Design* 97, pp. 14–16.

Hall, C. M. and Mitchell, Richard (2002), 'Tourism as a force for gastronomic globalization and localization', in Anne-Mette Hjalager and Greg Richards (eds), *Tourism and Gastronomy*. London: Routledge, pp. 71–87.

Hall, C. M. and Sharples, Liz (2003), 'The consumption of experiences or the experience of consumption? An introduction to the tourism of taste'. *Food Tourism Around the World: Development, Management and Markets*, pp. 1–24.

Hall, C. M., Fullagar, S., Markwell, K. and Wilson, E. (2012), 'The contradictions and paradoxes of slow food: Environmental change, sustainability and the conservation of taste', in Simone Fullagar, Kevin Markwell and Erica Wilson (eds), *Slow Tourism: Experiences and Mobilities*. Bristol; Buffalo; Toronto: Channel View Publications, pp. 53–68.

Hall, Peter (1988), *Cities of Tomorrow: An Intellectual History of Urban Planning and Design in the Twentieth Century*. Oxford: Basil Blackwell.

Hall, Peter (1992), *Urban and Regional Planning*. London: Routledge.

Hall, Peter (1993), 'Forces shaping urban Europe'. *Urban Studies* 30.6, pp. 883–98.

Hall, Peter and Ward, Colin (1998), *Sociable Cities: The Legacy of Ebenezer Howard* Chichester: J. Wiley.

Hall, S. (2012), *City, Street and Citizen: The Measure of the Ordinary* (vol. 9). London: Routledge.

Hall, Tony (2010), *The Life and Death of the Australian Backyard*. Collingwood, VC: CSIRO Publishing.

Hallberg, B., Richardson, J. and Leonard, B. (2009), 'Using community gardens to augment food security efforts in low-income communities'. *Virginia Tech Final Paper*. Retrieved online: www.ipg.vt.edu/Papers/Hallberg%20Major%20Paper.pdf.

Halseth, Greg (1998), *Cottage Country in Transition A Social Geography of Change and Contention in the Rural-Recreational Countryside*. Montreal and Kingston; London; Ithaca: McGill-Queen's University Press.

Haltiwanger, John, Jarmin, Ron and Krizan, Cornell John (2010), 'Mom-and-Pop Meet Big-Box: Complements or Substitutes?' *Journal of Urban Economics* 67.1, pp. 116–34.

Halweil, Brian (2002), *Home Grown: The Case for Local Food in a Global Market* (vol. 163). Danvers, MA: Worldwatch Institute.

Hamlett, Jane, Bailey, Adrian R., Alexander, Andrew and Shaw, Gareth (2008), 'Ethnicity and consumption south asian food shopping patterns in Britain, 1947–75'. *Journal of Consumer Culture March* 8.1, pp. 91–116.

Hamilton, Neil D. (1998), 'Right-to-farm laws reconsidered: Ten reasons why legislative efforts to resolve agricultural nuisances may be ineffective'. *Drake Journal of Agric L*. 3, p. 103.

Hampton, K. N. and Gupta, N. (2008), 'Community and social interaction in the wireless city: Wi-fi use in public and semi-public spaces'. *New Media & Society* 10.6, pp. 831–50.
Handy, Susan L. and Kelly J. Clifton (2001), 'Local shopping as a strategy for reducing automobile travel'. *Transportation* 28.4, pp. 317–46.
Hansen, Torben (2003), 'Intertype competition: Specialty food stores competing with supermarkets'. *Journal of Retailing and Consumer Services* 10.1, pp. 35–49.
Hardy, Matthew (1993), 'The gastronomic landscape: Food production and the cultural value of the countryside'. *Gastronomic Symposiette Series*, University of Adelaide.
Hardy, Matthew (1994), 'The Future of Food & Dining in Post-modern France'. *Gastronomic Symposiette Series*, University of Adelaide.
Hardy, Matthew (2004), 'The Renaissance of the Traditional City', in *Axess* 1.10. Stockholm.
Hardy, Matthew (ed.) (2009), *The Venice Charter Revisited: Modernism, Conservation and Tradition in the 21st Century*. Newcastle-upon-Tyne: Cambridge Scholars Publishing, pp. 187–97.
Harrington, R. J. and Ottenbacher, M. C. (2010), 'Culinary tourism – A case study of the gastronomic capital'. *Journal of Culinary Science & Technology* 8.1, pp. 14–32.
Harris, Marvin (1985), *Good to Eat: Riddles of Food and Culture*. New York: Simon and Schuster.
Harris, Neil (1987), 'Spaced out at the shopping center', in Nathan Glazer and Mark Lilla (eds), *The Public Face of Architecture: Civic Culture and Public Spaces*. New York: Free Press; London: Collier Macmillan, pp. 320–8.
Harris, Richard and Larkham, Peter (eds) (1999), *Changing Suburbs: Foundation, Form and Function*. London: Spon.
Harris, Richard and Lewis, Robert (2001), 'The Geography of North American Cities and Suburbs, 1900–1950 A New Synthesis'. *Journal of Urban History* 27.3, pp. 262–92.
Harvey, M., McMeekin, A. and Warde, A. (eds) (2004), *Qualities of Food*. Manchester: Manchester University Press.
Harvey, Thomas and Works, Martha A. (2002), 'Urban sprawl and rural landscapes: Perceptions of landscape as amenity in Portland, Oregon'. *Local Environment* 7.4, pp. 381–96.
Hashimoto, A. and Telfer, D. J. (2006), 'Selling Canadian culinary tourism: Branding the global and the regional product'. *Tourism Geographies* 8.1, pp. 31–55.
Hauck-Lawson, A. and Deutsch, J. (eds) (2009), *Gastropolis: Food and New York City*. New York: Columbia University Press.
Haughton, G. and Hunter, C. (2004), *Sustainable Cities*. London: Routledge.
Haworth-Hoeppner, Susan (2004), 'The critical shapes of body image: The role of culture and family in the production of eating disorders'. *Journal of Marriage and Family* 62.1, pp. 212–27.
Hayden, Dolores (1981), *The Grand Domestic Revolution: A History of Feminist Designs for American Homes, Neighbourhoods and Cities*. Cambridge, MA: MIT Press.
Hayden, Dolores (2009), *Building Suburbia: Green Fields and Urban Growth, 1820–2000*. New York: Random House Digital.
Hayes-Conroy, A. and Martin, D. G. (2010), 'Mobilising bodies: Visceral identification in the slow food movement'. *Transactions of the Institute of British Geographers* 35.2, pp. 269–81.

Hayes-Conroy, J. and Hayes-Conroy, A. (2010), 'Visceral geographies: Mattering, relating, and defying'. *Geography Compass* 4.9, pp. 1273–83.
Hayley F. (2011), 'Rochelle Canteen'. http://www.yelp.co.uk/biz/rochelle-canteen-london Dated 10.1.2011, Retrieved 22 January 2013.
Hayward, R. and McGlynn, S. (eds) (2002), Editorial. *Urban Design International* 7, pp. 127–9.
Head, Lesley M. and Muir, Patricia A. (2007), *Backyard: Nature and Culture in Suburban Australia*. Wollongong: University of Wollongong Press.
Hedden, W. P. (1929), *How Great Cities are Fed*. Boston, MA: D.C. Heath and Company.
Helbich, Marco and Leitner, Michael (2009), 'Spatial analysis of the urban-to-rural migration determinants in the Viennese metropolitan area: A transition from suburbia to postsuburbia?' *Applied Spatial Analysis and Policy* 2.3, pp. 237–60.
Hemingway, Wayne (2012), 'The British high street is dead – let's celebrate: Most town centres are boring clones, and the closure of large retailers will open up creative space for quirky start-ups'. http://www.guardian.co.uk/commentisfree/2012/mar/22/british-high-street-dead-lets-celebrate Accessed 11 January 2013.
Henderson, John (2004), *The Roman Book of Gardening*. London; New York: Routledge.
Hendon, Julia (2003), 'Feasting at home: Community and house solidarity among the Maya', in Tamara L. Bray (ed.), *The Archaeology and Politics of Food and Feasting in early States and Empires*. London: Springer, pp. 203–33.
Henisch, Bridget Ann (1986), *Fast and Feast: Food in Medieval Society*. London: Penn State University Press.
Hernandez-Lopez, Ernesto (2011), 'LA's taco truck war: How law cooks food culture contests'. *The University of Miami Inter-American Law Review* 43, pp. 233–68.
Hesse, Markus (2004), 'Land for logistics: Locational dynamics, real estate markets and political regulation of regional distribution complexes'. *Tijdschrift voor economische en sociale geografie* 95.2, pp. 162–73.
Hesse, Markus and Rodrigue, J. P. (2004), 'The transport geography of logistics and freight distribution'. *Journal of Transport Geography* 12.3, pp. 171–84.
Hessler, M. (2009), 'The Frankfurt Kitchen: The model of modernity and the "Madness" of traditional users, 1926 to 1933', in R. Oldenziel and K. Zachmann (eds), *Cold War Kitchen: Americanization, Technology, and European Users*. Cambridge, MA: MIT Press, pp. 163–84.
Hester, R. T. (2006), *Design for Ecological Democracy*. Cambridge, MA: MIT Press.
Heuzenroeder, Angela (1999), *Barossa Food*. Kent Town; South Australia: Wakefield Press.
Hewer, P. A. (2003), 'Consuming gardens: Representations of paradise, nostalgia and postmodernism'. *European Advances in Consumer Research* 6, pp. 327–31.
Hidding, Marjan, Needham, Barrie and Wisserhof, Johan (2000), 'Discourses of town and country'. *Landscape and Urban Planning* 48.3, pp. 121–30.
Hinchcliffe, S. and Whatmore, S. (2006), 'Living cities: Towards a politics of conviviality'. *Science as Culture* 15.2, pp. 123–38.
Hinrichs, C. Clare (2003), 'The practice and politics of food system localization'. *Journal of Rural Studies* 19.1, pp. 33–45.

Hinrichs, C. and Lyson, Thomas A. (2007), *Remaking the North American Food System: Strategies for Sustainability*. Lincoln and London: University of Nebraska Press.

Hird, Alison (2010), http://www.english.rfi.fr/node/44476/3330. Retrieved 21 January 2013.

Hirvi, Laura (2011), 'Sikhs in Finland: Migration histories and work in the restaurant sector', in K. A. Jacobsen and K. Myrvold (eds), *Sikhs in Europe: Migration, Identities and Representations*. Farnham: Ashgate, pp. 95–114.

Hise, Greg (1997), *Magnetic Los Angeles: Planning the Twentieth Century Metropolis*. Baltimore, MD: JHU Press.

Hitchings, Russell (2003), 'People, plants and performance: On actor network theory and the material pleasures of the private garden'. *Social & Cultural Geography* 4.1, pp. 99–114.

Hjalager, Anne-Mette (2002), 'A typology of gastronomy tourism', in Anne-Mette Hjalager and Greg Richards (eds), *Tourism and Gastronomy*. London: Routledge, pp. 21–35.

Hjalager, Anne-Mette and Corigliano, M. A. (2000), 'Food for tourists – determinants of an image'. *International Journal of Tourism Research* 2.4, pp. 281–93.

Hjalager, Anne-Mette and Richards, Greg (eds) (2004), *Tourism and Gastronomy*. London: Routledge.

HM Government (2010), *Food 2030*. London: Department for Environment, Food and Rural Affairs.

Hobsbawm, Eric and Ranger, T. (eds) (1983), *The Invention of Tradition*. Cambridge: Cambridge University Press.

Hoek, Hans Wijbrand and Van Hoeken, Daphne (2003), 'Review of the prevalence and incidence of eating disorders'. *International Journal of Eating Disorders* 34.4, pp. 383–96.

Hoehner, C. M., Brennan, L. K., Brownson, R. C., Handy, S. L. and Killingsworth, R. (2003), 'Opportunities for integrating public health and urban planning approaches to promote active community environments'. *American Journal of Health Promotion* 18.1, pp. 14–20.

Hollister, C. Warren (1974), *Medieval Europe. A Short History*, 3rd edn. London; New York; Toronto: John Wiley and Sons.

Holt, Georgina and Amilien, Virginie (2007), 'Introduction: From local food to localised food'. *Anthropology of Food* S2. Special issue on local food products and systems.

Hook, Derek and Vrdoljak, Michele (2002), 'Gated communities, heterotopia and a "rights" of privilege: A "heterotopology" of the South African security-park'. *Geoforum* 33.2, pp. 195–219.

Horrigan, Leo, Lawrence, Robert S. and Walker, Polly (2002), 'How sustainable agriculture can address the environmental and human health harms of industrial agriculture'. *Environmental Health Perspectives* 110.5, p. 445.

Hough, Michael (1984), *City Form and Natural Process: Towards a New Urban Vernacular*. London; New York: Routledge.

Hough, Michael (1990), *Out of Place, Restoring Identity to the Regional Landscape*. New Haven; London: Yale University Press.

Houston, Peter (2005), 'Re-valuing the Fringe: Some findings on the value of agricultural production in Australia's peri-urban regions'. *Geographical Research* 43.2, pp. 209–23.
Houy, Yvonne (2009), 'From Taylor to just-in-time sustainability', in Sharon Kleinman (ed.), *The Culture of Efficiency: Technology in Everyday Life* (vol. 55). New York; Washington, DC: Peter Lang, pp. 20–38.
Howard, Ebenezer (1902), *Garden Cities of Tomorrow*. London: Dodo Press (Facsimile of 2nd edn).
Howard, P. H., Fitzpatrick, M. and Fulfrost, B. (2011), 'Proximity of food retailers to schools and rates of overweight ninth grade students: An ecological study in California'. *BMC Public Health* 11.1, p. 68.
Howe, Joe and Wheeler, Paul (1999), 'Urban food growing: The experience of two UK cities'. *Sustainable Development* 7.1, pp. 13–24.
Huang, Shu-Li, Wang, Szu-Hua and Budd, William W. (2009), 'Sprawl in Taipei's peri-urban zone: Responses to spatial planning and implications for adapting global environmental change'. *Landscape and Urban Planning* 90.1, pp. 20–32.
Huat, Chua Beng and Rajah, Ananda (2001), 'Hybridity, ethnicity and food in Singapore', in David Y. H. Yu and Tan Chee-beng (eds), *Changing Chinese Foodways in Asia*. Hong Kong: Chinese University of Hong Kong, pp. 161–97.
Huerta, A. (2011), *Examining the Perils and Promises of an Informal Niche in a Global City: A Case Study of Mexican Immigrant Gardeners in Los Angeles*. Berkeley: City & Regional Planning. Retrieved from http://escholarship.org/uc/item/32q197vh.
Hughes, John, O'Brien, Jon, Rodden, Tom, Rouncefield, Mark and Viller, Stephen (2000), 'Patterns of home life: Informing design for domestic environments'. *Personal Technologies* 4, pp. 25–38.
Humphery, Kim (1998), *Shelf Life: Supermarkets and the Changing Cultures of Consumption*. Cambridge: Cambridge University Press.
Hutchings, Alan (2011), 'Garden suburbs in Latin America: A new field of international research?' *Planning Perspectives: An International Journal of History, Planning and the Environment* 26.2, pp. 313–17.
Hynes, H. Patricia and Howe, Genevieve (2002), 'Urban horticulture in the contemporary United States: Personal and community benefits'. *Acta Hort (ISHS)* 643, pp. 171–81. http://www.actahort.org/books/643/643_21.htm.
Ignatieva, Maria, Stewart, Glenn H. and Meurk, Colin (2011), 'Planning and design of ecological networks in urban areas'. *Landscape and Ecological Engineering* 7.1, pp. 17–25.
Ignatov, E. and Smith, S. (2006), 'Segmenting Canadian culinary tourists'. *Current Issues in Tourism* 9.3, p. 235.
Ikpe, Eno Blankson (1994), *Food and Society in Nigeria: A History of Food Customers, Food Economy and Cultural Change 1900-1989*. Stuttgart: Franz Steiner Verlag.
Ilbery, B. W. and Bowler, I. R. (1998), 'From agricultural productivism to post-productivism', in B. W. Ilbery (ed.), *The Geography of Rural Change*. Essex: Longman, pp. 57–84.
Ilbery, B. and Kneafsey, M. (2000), 'Producer constructions of quality in regional speciality food production: A case study from south west England'. *Journal of Rural Studies* 16.2, pp. 217–30.

Ilbery, B. and Kneafsey, M. (2000), 'Registering regional speciality food and drink products in the United Kingdom: The case of PDOs and PGIs'. *Area* 32.3, pp. 317–25.

Ilbery, Brian and Maye, Damian (2005), 'Food supply chains and sustainability: Evidence from specialist food producers in the Scottish/English borders'. *Land Use Policy* 22.4, pp. 331–44.

Ilbery, Brian, Chiotti, Quentin and Rickard, Timothy (eds) (1997), *Agricultural Restructuring and Sustainability: A Geographical Perspective*. Wallingford: CAB International.

Ilbery, B. W., Healey, M. and Higginbottom, J. (1997), 'On and off-farm business diversification by farm house- holds in England', in B. W. Ilbery, Q. Chiotti and T. Rickard (eds), *Agricultural Restructuring and Sustainability – A Geographical Perspective*. Wallingford: CAB International, pp. 135–51.

Illich, Ivan (1973), *Tools for Conviviality*. New York: Harper & Row.

Irvine, Seana, Johnson, Lorraine and Peters, Kim (1999), 'Community gardens and sustainable land use planning: A case-study of the Alex Wilson community garden'. *Local Environment* 4.1, pp. 33–46.

Jabareen, Y. R. (2006), 'Sustainable urban forms their typologies, models, and concepts'. *Journal of Planning Education and Research* 26.1, pp. 38–52.

Jack, I., Holmes, K. and Kerr, R. (1984), 'Ah toy's garden: A Chinese market-garden on the Palmer river Goldfield, north Queensland'. *The Australian Journal of Historical Archaeology* 2, October, pp. 51–8.

Jackson, Frank (1985), *Sir Raymond Unwin: Architect, Planner and Visionary*. London: Zwemmer.

Jackson, Kenneth T. (1985), *Crabgrass Frontier the Suburbanization of the United States*. New York; Oxford: Oxford University Press.

Jackson, Kenneth T. (1996), 'All the world's a mall: Reflections on the social and economic consequences of the American shopping center'. *The American Historical Review* 101.4, pp. 1111–21.

Jackson, Laura E. (2003), 'The relationship of urban design to human health and condition'. *Landscape and Urban Planning* 64.4, pp. 191–200.

Jacobs, J. (1961), *The Death and Life of Great American Cities*. New York: Random House.

Jacobs, J. (1972), *The Economy of Cities*. Harmondsworth: Penguin Books.

Jakle, John A. and Keith A. Sculle (1999), *Fast Food: Roadside Restaurants in the Automobile Age*. New York: John Hopkins.

Jansma, J. E. and Visser, A. J. (2011), 'Agromere: Integrating urban agriculture in the development of the city of Almere'. *UA Magazine* 25, pp. 28–31.

Jarosz, L. (2008), 'The city in the country: Growing alternative food networks in Metropolitan areas'. *Journal of Rural Studies* 24.3, pp. 231–44.

Jarvis, Helen, Pratt, Andy and Cheng-Chong Wu, Peter (2001), *The Secret Life of Cities: The Social Reproduction of Everyday Life*. London: Prentice Hall.

Jeffery, R. W., Baxter J., McGuire, M. and Linde, J. (2006), 'Are fast food restaurants an environmental risk factor for obesity?' *International Journal of Behavioral Nutrition & Physical Activity* 3, p. 2.

Jeffres, L. W., Bracken, C. C., Jian, G. and Casey, M. F. (2009), 'The impact of third places on community quality of life'. *Applied Research in Quality of Life* 4.4, pp. 333–45.
Jencks, Charles (1993), *Heteropolis: Los Angeles, the Riots and the Strange Beauty of Hetero-Architecture*. London: Academy Editions.
Jenks, M. and Dempsey, N. (eds) (2005), *Future Forms and Design for Sustainable Cities*. London: Routledge.
Jepson Jr., E. J. and Edwards, M. M. (2010), 'How possible is sustainable urban development? An analysis of planners' perceptions about new urbanism, smart growth and the ecological city'. *Planning, Practice & Research* 25.4, pp. 417–37.
Jerram, Leif (October 2006), 'Kitchen sink dramas: Women, modernity and space in Weimar Germany'. *Cultural Geographies* 13.4, pp. 538–56.
Johnson, Lorraine (2009), 'Revisting victory: Gardens past, gardens future', in Christina Palassio and Alana Wilcox (eds), *The Edible City Toronto's Food from Farm to Fork*. Toronto: Coach House Books.
Johnson, Louise (2000a), *Placebound: Australian Feminist Geographies*. South Melbourne: Oxford University Press.
Johnson, Louise (2000b), 'Powerlines: A cultural geography of domestic open space', in Elaine Stratford (ed.), *Australian Cultural Geographies*. Melbourne: Oxford University Press (hereafter OUP), pp. 87–108.
Johnson, L. C. (2006), 'Browsing the modern kitchen – A feast of gender, place and culture'. (Part 1). *Gender, Place & Culture* 13.2, pp. 123–32.
Johnston, J. (2008), 'The citizen-consumer hybrid: Ideological tensions and the case of Whole Foods Market'. *Theory and Society* 37.3, pp. 229–70.
Johnston, J., Biro, J. and MacKendrick, N. (2009), 'Lost in the supermarket: The corporate-organic foodscape and the struggle for food democracy'. *Antipode: A Radical Journal of Geography* 41, pp. 509–32.
Jonas, A. E. (2006), 'Pro scale: Further reflections on the "scale debate" in human geography'. *Transactions of the Institute of British Geographers* 31.3, pp. 399–406.
Jones, Louisa (1997), *Kitchen Gardens of France*. London: Thames and Hudson.
Jones, Martin (2007), *Feast: Why Humans Share Food*. Oxford: Oxford University Press.
Josling, Tim (2006), 'The war on terroir: Geographical indications as a transatlantic trade conflict'. *Journal of Agricultural Economics* 57.3, pp. 337–63.
Jun, Myung-Jin (2004), 'The effects of Portland's urban growth boundary on urban development patterns and commuting'. *Urban Studies* 41.7, pp. 1333–48.
Karrholm, Mattias (2012), *Retailising Space: Architecture, Retail and the Territorialisation of Public Space*. Farnham, Surrey: Ashgate Publishing.
Kasarda, John (2001), 'From airport city to aerotropolis'. *Airport World* 6.4 (August/September), pp. 42–5.
Kasarda, John and Dogan, Mattei (1989), *The Metropolis Era*. London: Sage.
Kayden, Jerold S. (2000), *Privately Owned Public Space: The New York City Experience*. Wiley.
Kazuko, Emi and Fukuokoa, Yasuko (2001), *Japanese Food and Cooking: A Timeless Cuisine: Its Traditions, Techniques, Ingredients and Recipes*. London: Lorenz.
Keister, D. (2005), *Courtyards: Intimate Outdoor Spaces*. Layton, Utah: Gibbs Smith.

Kelbaugh, Douglas (1997), *Common Place: Toward Neighborhood and Regional Design*. Seattle: University of Washington Press.

Kelbaugh, Douglas (2002), *Repairing the American Metropolis: Common Place Revisited*. Washington, DC: University of Washington Press.

Kesteloot, Christian and Pascale Mistiaen (1997), 'From ethnic minority niche to assimilation: Turkish restaurants in Brussels'. *Area* 29.4, pp. 325–34.

Kiang, Heng Chye (1994), 'Kaifeng and Yangzhou', in Zeynep Çelik, Diane Favro and Richard Ingersoll (eds), *Streets: Critical Perspectives on Public Space*. Berkeley, CA: University of California Press.

Kimes, Sheryl E. and Robson, Stephani K. A. (2004), 'The impact of restaurant table characteristics on meal duration and spending'. *Cornell Hotel and Restaurant Administration Quarterly* 45.4, pp. 333–46.

Kirby, Martha (2012), 'Too much of a good thing? Society, affluence and obesity in Britain, 1940-1970'. *eSharp* Issue 18: Challenges of Development, pp. 44–63.

Kirkpatrick, J. B., Daniels, G. D. and Zagorski, T. (2007), 'Explaining variation in front gardens between suburbs of Hobart, Tasmania, Australia'. *Landscape and Urban Planning* 79.3, pp. 314–22.

Kirwan, J. (2006), 'The interpersonal world of direct marketing: Examining conventions of quality at UK farmers' markets'. *Journal of Rural Studies* 22, pp. 301–12.

Kittel, G. and Snow, R. F. (1990), *Diners: People and Places*. London: Thames and Hudson.

Kivela, J. and Crotts, J. C. (2006a), 'Gastronomy tourism: A meaningful travel market segment'. *Journal of Culinary Science & Technology* 4.2-3, pp. 39–55.

Kivela, J. and Crotts, J. C. (2006b), 'Tourism and gastronomy: Gastronomy's influence on how tourists experience a destination'. *Journal of Hospitality & Tourism Research* 30.3, pp. 354–77.

Kjell, P. and Potts, R. (2005), *Clone Town Britain: The Survey Results on the Bland State of the Nation*, ed. M. Murphy. London: New Economics Foundation.

Klauser, Wilhelm (2008), 'From Pot au Feu to processed food: The restaurant as a modernist location', in Petra Hagen Hodgson and Rolf Toyka (eds), *The Architect, the Cook, and Good Taste*. Birkhauser Basel: Walter de Gruyter, pp. 110–19.

Klein, Naomi (2000), *No Logo: Taking Aim at the Brand Bullies*. London: Flamingo.

Kling, Rob, Olin, Spencer C. and Poster, Mark (eds) (1995), *Postsuburban California: The Transformation of Orange County since World War II*. Berkeley, CA: University of California Press.

Klopfer, L. (1993), 'Padang restaurants: Creating "ethnic" cuisine in Indonesia'. *Food and Foodways* 5.3, pp. 293–304.

Kloppenburg Jr, Jack, Hendrickson, John and Stevenson, George W. (1996), 'Coming in to the foodshed'. *Agriculture and Human Values* 13.3, pp. 33–42.

Kloppenburg, Jr., J., Lezberg, S., De Master, K., Stevenson, G. W. and Hendrickson, J. (2000), 'Tasting food, tasting sustainability: Defining the attributes of an alternative food system with competent, ordinary people'. *Human Organization* 59.2, pp. 177–86.

Kneafsey, M. (2010), 'The region in food – important or irrelevant?' *Cambridge Journal of Regions, Economy and Society* 3.2, pp. 177–90.

Knox, Paul L. (2005), 'Creating ordinary places: Slow cities in a fast world'. *Journal of Urban Design* 10.1, pp. 1–11.
Knox, Peter (1992), 'The packaged landscapes of post-suburban America', in J. Whitehand and P. Larkham (eds), *Urban landscapes: International Perspectives.* London: Routledge.
Koc, A. Ali, Boluk, Gulden and Kovaci, Sureyya (2009), 'Concentration of food retailing and anti-competitive practices in Turkey'. EAAE Seminar, September 3–6.
Koc, M. et al. (eds) (1999), *For Hunger-Proof Cities: Sustainable Urban Food Systems.* Ottawa and Toronto: IDRC.
Koc, M. and Dahlberg, K. A. (1999), 'The restructuring of food systems: Trends, research, and policy issues'. *Agriculture and Human Values* 16.2, pp. 109–16.
Kohn, Margaret (2004), *Brave New Neighborhoods: The Privatization of Public Space.* London: Routledge.
Kostof, Spiro (1991), *The City Shaped, Urban Patterns and Meanings Through History.* London: Thames and Hudson.
Kostof, Spiro (1992), *The City Assembled: The Elements of Urban Form Through History.* London: Thames and Hudson.
Kowinski, William Severini (1985), *The Malling of America: An Inside Look at the Great Consumer Paradise.* New York: W. Morrow.
Kumin, Beat (2003), 'Eating out before the restaurant', in Marc Jacobs and Peter Scholliers (eds), *Eating Out in Europe: Picnics, Gourmet Dining and Snacks since the Late Eighteenth Century.* Oxford: Berg, pp. 71–87.
Kunstler, James Howard (1994), *Geography of Nowhere: The Rise and Decline of America's Man-Made Landscape.* New York; London: Simon and Schuster.
Kynaston, David (2010), *Austerity Britain, 1945-1951.* USA: Bloomsbury Publishing.
Ladd, Melissa (2010), 'Market shopping: Marché des Enfants Rouges'. http://hipparis.com/2010/05/06/market-shopping-marche-des-enfants-rouges/.
Lai, A. E. (2009), 'A neighbourhood in Singapore: Ordinary people's lives "downstairs"'. Asia Research Institute Working Paper No. 113.
Lake, Amelia A. and Townshend, Tim G. (2006), 'Obesogenic environments: Exploring the built and food environments'. *The Journal of the Royal Society for the Promotion of Health* Issue 126, pp. 262–7.
Lake, Amelia A., Townshend, Tim G. and Alvanides, Seraphim (eds) (2010), *Obesogenic Environments: Complexities, Perceptions and Objective Measures.* Chichester, West Sussex: Wiley-Blackwell.
Lake, A. A., Burgoine, T., Greenhalgh, F., Stamp, E. and Tyrrell, R. (2010), 'The foodscape: classification and field validation of secondary data sources'. *Health & Place* 16.4, pp. 666–73.
Landman, Karina (2002), 'Gated communities in South Africa: Building bridges or barriers'. International Conference on Private Urban Governance, Mainz, Germany.
Lang, Robert (2003), *Edgeless Cities: Exploring the Elusive Metropolis.* Washington, DC: Brookings Institution Press.
Lang, R. E., Sanchez, T. and LeFurgy, J. (2006), *Beyond Edgeless Cities: Office Geography in the New Metropolis.* National Center for Real Estate Research. Washington: National Association of Realtors.

Lang, Tim (1997), *Food Policy for the 21st Century. Can it be both Radical and Reasonable?* Thames Valley University Centre for Food Policy Discussion Paper 4.

Lang, Tim (2009), 'Reshaping the food system for ecological public health'. *Journal of Hunger & Environmental Nutrition* 4.3–4, pp. 315–35.

Lang, Tim (2010), 'From "value-for-money" to "values-for-money"? Ethical food and policy in Europe'. *Environment and Planning A* 42.8, pp. 1814–32.

Lang, Tim (Undated), 'How new is the world food crisis? Thoughts on the long dynamic of food democracy, food control & food policy in the 21st century', Paper to Conference: *Visible Warnings: The World Food Crisis in Perspective*. April 3–4, Cornell University, Ithaca NY, USA.

Lang, Tim and Caraher, Martin (2001), 'Is there a culinary skills transition? Data and debate from the UK about changes in cooking culture'. *Journal of the HEIA* 8.2, pp. 2–14.

Lang, Tim and Michael Heasman (2004), *Food Wars: The Global Battle for Mouths, Minds and Markets*. London: Earthscan.

Lang, Tim and Rayner, Geof (2012), *Ecological Public Health: Reshaping the Conditions for Good Health*. London: Routledge.

Lang, Tim, Barling, David and Caraher, Martin (2009), *Food Policy: Integrating Health, Environment and Society*. Oxford: Oxford University Press.

Lang, Tim, Caraher, Martin, Dixon, P. and Carr-Hill, R. (1999), *Cooking Skills and Health*. London: Health Education Authority.

Lanjouw, J. O. and Lanjouw, P. (2001), 'The rural non-farm sector: Issues and evidence from developing countries'. *Agricultural Economics* 26.1, pp. 1–23.

Lapping, M. B. and Furuseth, O. J. (1999), 'Introduction and overview', in O. J. Furuseth and M. B. Lapping (eds), *Contested Countryside: The Rural Urban Fringe in North America*. Aldershot, UK: Ashgate, pp. 1–5.

Laquian, Aprodicio A. (2005), *Beyond Metropolis: The Planning and Governance of Asia's Mega-Urban Regions*. Baltimore, MD: Johns Hopkins University Press.

Larkcom, Joy (1996), 'The floating gardens of Amiens'. *The Garden* 121.9.

Larkcom, Joy (1998), 'An edible spectrum: The vibrant potager at Bosmelet'. *The Garden* 123.7

Larsen, Kristian and Jason Gilliland (2008), 'Mapping the evolution of "food deserts" in a Canadian city: Supermarket accessibility in London, Ontario, 1961–2005'. *International Journal of Health Geographics* 7.1, p. 16.

Latham, A. and McCormack, D. (2004), 'Moving cities: Rethinking the materialities of urban geographies'. *Progress in Human Geography* 28, pp. 701–24.

La Trobe, Helen (2001), 'Farmers' markets: Consuming local rural produce'. *International Journal of Consumer Studies* 25.3, pp. 181–92.

Laurier, Eric (2004), 'Busy Meeting Grounds: The café, the scene and the business'. *Paper for ICT: Mobilizing Persons, Places and Spaces an International Specialist Meeting on ICT, Everyday Life and Urban Change*, Utrecht.

Laurier, Eric (2008), 'How breakfast happens in the café'. *Time & Society* 17.1, pp. 119–34.

Laurier, Eric and Philo, Chris (2004), 'The Cappuccino Community: Café and civic life in the contemporary city' (draft) published by the Department of Geography & Topographic Science, University of Glasgow at: http://www.geog.gla.ac.uk/olpapers/elaurier002.pdf.

Laurier, Eric, Whyte, Angus and Buckner, Kathy (2001), 'An ethnography of a neighbourhood café: Informality, table arrangements and background noise'. *Journal of Mundane Behaviour* 2.2, pp. 195–232.

Lavialle, Claudine, André Lavialle and Béziat, Marc (1999), *Recettes paysannes en Lot-et-Garonne*. Rodez: Editions Subervie.

Lawrence, D. L. and Low, S. M. (1990), 'The built environment and spatial form'. *Annual Review of Anthropology* 19, pp. 453–505.

Lawson, K. (2004), 'Libraries in the USA as traditional and virtual "third places"'. *New Library World* 105.3/4, pp. 125–30.

Lawson, Laura J. (2005), *City Bountiful*. Berkeley, CA: University of California Press.

Lee, Jenny (2009), *The Market Hall Revisited: Cultures of Consumption in Urban Food Retail during the Long Twentieth Century*. Diss: Linköping.

Lee, I. and Arcodia, C. (2011), 'The role of regional food festivals for destination branding'. *International Journal of Tourism Research* 13.4, pp. 355–67.

Leenaert, Tobias (2012), 'Meat moderation as a challenge for government and civil society: The Thursday Veggie Day campaign in Ghent, Belgium', in Andre Viljoen and Johannes S. C. Wiskerke (eds), *Sustainable Food Planning: Evolving Theory and Practice*. The Netherlands: Wageningen Academic Publishers, pp. 189–98.

Lees-Maffei, G. (2007), 'Accommodating "Mrs. Three-in-One": Homemaking, home entertaining and domestic advice literature in post-war Britain'. *Women's History Review* 16.5, pp. 723–54.

Lefebvre, Henri (1991), *The Production of Space*. Oxford: Blackwell.

Le Goix, Renaud and Webster, Chris J. (2008), 'Gated communities'. *Geography Compass* 2.4, pp. 1189–214.

Leisch, Harald (2002), 'Gated communities in Indonesia'. *Cities* 19.5, pp. 341–50.

Leitch, Alison (2003), 'Slow food and the politics of pork fat: Italian food and European identity'. *Ethnos* 68.4, pp. 437–62.

Lento, Katrina (2006), 'The role of nature in the city: Green space in Helsinki, 1917-60', in P. Clark (ed.), *The European City and Green Space: London, Stockholm, Helsinki and St. Petersburg, 1850-2000*. Farnham, Surrey: Ashgate.

Leonardi, R. and Nanetti, R. (eds) (1990), *The Regions and European Integration: The Case of Emilia-Romagna*. London: Pinter Publishers.

Leonardi, R. and Nanetti, R. (1994), *Regional Development in a Modern European Economy: The Case of Tuscany*. London: Pinter Publishers.

Leontidou, L., Afouxenidis, A., Kourliouros, E. and Marmaras, E. (2007), 'Infrastructure related urban sprawl: Mega-events and hybrid peri-urban landscapes in southern Europe', in C. Couch, G. Petschel-Held and L. Leontidou (eds), *Urban Sprawl in Europe*. Oxford: Blackwell, pp. 71–98.

Levenstein, H. (1988), *Revolution at the Table (Vol. 7)*. Berkeley, CA: University of California Press.

Levenstein, H. (1993), *Paradox of Plenty: A Social History of Eating in Modern America*. New York; Oxford: Oxford University Press.

Lévêque, G. and Valéry, M. F. (1995), *French Garden Style*. London: Frances Lincoln Limited.

Levkoe, Charles Zalman (2011), 'Towards a transformative food politics'. *Local Environment* 16.7, pp. 687–705.

Liebs, Chester (1995), *Main Street to Miracle Mile: American Roadside Architecture*. Baltimore: JHU Press.
Lind, David and Barham, Elizabeth (2004), 'The social life of the tortilla: Food, cultural politics, and contested commodification'. *Agriculture and Human Values* 21.1, pp. 47–60.
Little, R., Maye, D. and Ilbery, B. (2010), 'Collective purchase: Moving local and organic foods beyond the niche market'. *Environment and Planning. A*, 42.8, pp. 1797–1813.
Llewellyn, Mark (2004), 'Designed by women and designing women: Gender, planning and the geographies of the kitchen in Britain 1917-1946'. *Cultural Geographies* 11.1, pp. 42–60.
Locher, Mira, Ando, Tadao and Shiratori, Yoshio (2006), *Super Potato Design: The Complete Works of Takashi Sugimoto: Japan's Leading Interior Designer*. Vermont, Singapore: Tuttle Publishing.
Loewen, J. W. (2005), *Sundown Towns: A Hidden Dimension of American Racism*. New York: The New Press.
Long, L. M. (2010), 'Culinary tourism and the emergence of an Appalachian cuisine: Exploring the foodscape of Asheville, NC'. *North Carolina Folklore Journal* 57.1, pp. 4–19.
Long, Lucy M. (2012), 'Culinary tourism', in M. Pilcher Jeffrey (ed.), *The Oxford Handbook of Food History*. Oxford: Oxford University Press, pp. 389–406.
Long, L. T. and Vargas, L. A. (2005), *Food Culture in Mexico*. Westport, CT: Greenwood Publishing Group.
Lopez, Russ (2004), 'Urban sprawl and risk for being overweight or obese'. *American Journal of Public Health* 94.9, p. 1574.
Losada, H., Martinez, H., Vieyra, J., Pealing, R., Zavala, R. and Cortés, J. (1998), 'Urban agriculture in the metropolitan zone of Mexico City: Changes over time in urban, suburban and peri-urban areas'. *Environment and Urbanization* 10.2, pp. 37–54.
Low, Setha M. (2000), *On the Plaza: The Politics of Public Space and Culture*. Austin: University of Texas Press.
Low, Setha M. (2003), *Behind the Gates: Life, Security, and the Pursuit of Happiness in Fortress America*. New York: Routledge.
Low, Setha M. (2003), 'The edge and the center: Gated communities and the discourse of urban fear', in Setha M. Low, Denise Lawrence-Zúñiga (eds), *The Anthropology of Space and Place: Locating Culture*. Malden, MA; Oxford : Blackwell, pp. 387–407.
Lozano, E. E. (1990), *Community Design and the Culture of Cities: The Crossroad and the Wall*. Cambridge: Cambridge University Press.
Luard, Elisabeth (1986), *European Peasant Cookery: The Rich Tradition*. Corgi Books.
Lucca, Alessandra and Elizabeth Aparecida Ferraz da Silva Torres (2006), 'Street-food: The hygiene conditions of hot-dogs sold in São Paulo, Brazil'. *Food Control* 17.4, pp. 312–16.
Ludwig, J., Sanbonmatsu, L., Gennetian, L., Adam, E., Duncan, G. J., Katz, L. F. and McDade, T. W. (2011), 'Neighborhoods, obesity, and diabetes—A randomized social experiment'. *New England Journal of Medicine* 365.16, pp. 1509–19.
Lupton, Deborah (1996), *Food, the Body and the Self*. London, Thousand Oaks; New Delhi: Sage.

Lupton, Ellen and Miller, J. A. (1992), *The Bathroom, the Kitchen and the Aesthetics of Waste: A Process of Elimination.* New York: Kiosk.
Lynch, Kevin (1960), *The Image of the City.* Cambridge, MA: MIT Press.
Lyon, Phil and Colquhoun, Anne (1999), 'Selectively living in the past: Nostalgia and lifestyle'. *Journal of Consumer Studies & Home Economics* 23.3, pp. 191–6.
Lyon, Phil, Anne Colquhoun and Emily Alexander (2003), 'Deskilling the domestic kitchen: National tragedy or the making of a modern myth?' *Food Service Technology* 3.3–4, pp. 167–75.
McBride, Jason (2009), 'Food fighters: The Stop Community Food Centre and the end of food banks', in Christina Palassio and Alana Wilcox(eds) (2005), *The Edible City Toronto's Food from Farm to Fork.* Toronto: Coach House Books.
MacCannell, D. (1976), *The Tourist: A New Theory of the Leisure Class.* Berkeley, CA: University of California Press.
McClintock, Nathan (2011), 'From industrial garden to food desert: Demarcated devaluation in the flatlands of Oakland, California', in Alison Hope Alkon and Julian Agyeman (eds), *Cultivating Food Justice: Race, Class, and Sustainability.* Cambridge, MA; London: MIT Press, pp. 89–120.
McCulloch, J. (2011), *The Modern Kitchen Garden: Design, Ideas and Practical Tips.* Mulgrave, Victoria: Images Publishing.
Macdiarmid, J. I., Kyle, J., Horgan, G. W., Loe, J., Fyfe, C., Johnstone, A. and McNeill, G. (2012), 'Sustainable diets for the future: Can we contribute to reducing greenhouse gas emissions by eating a healthy diet?' *The American Journal of Clinical Nutrition* 96.3, pp. 632–9.
McDouall, Robin (1974), *Clubland Cooking.* London: Phaidon Press.
McFeely, Mary Drake (2000), *Can She Bake a Cherry Pie? American Women and the Kitchen in the Twentieth Century.* Amherst, MA: University of Massachusetts Press.
McGowan, B. (2005), 'The Economics and organisation of Chinese mining in Colonial Australia'. *Australian Economic History Review* 45.2, pp. 119–38.
McHarg, Ian (1969), *Design with Nature.* Garden City, NY: Natural History Press; Doubleday.
McIntosh, A. and Zey, M. (1989), 'Women as gatekeepers of food consumption: A sociological critique'. *Food and Foodways* 3.4, pp. 317–32.
McKenzie, Evan (1994), *Privatopia: Homeowner Associations and the Rise of Residential Private Government.* New Haven; London: Yale University Press.
McKinnon, Alan (2009), 'The present and future land requirements of logistical activities'. *Land Use Policy* 26, pp. S293–301.
Mclaughlin, Katy (2009), 'Food truck nation'. *The Wall Street Journal Online.* June 5, 2009. http://online.wsj.com/articles/SB10001424052970204456604574201934018170554.
Mclaughlin, Katy (2012), http://www.online.wsj.com/article/SB10001424052970204276304577263813654892788.html Friday 23 March 2012, Accessed 21 January 2013.
McManus, Ruth and Philip J. Ethington (2007), 'Suburbs in transition: New approaches to suburban history'. *Urban History* 34.2, p. 317.
Macionis, John J. and Parrillo, Vincent N. (2006), *Cities and Urban Life.* Upper Saddle River, NJ: Pearson Education.

Madanipour, Ali (1999), 'Why are the design and development of public spaces significant for cities?' *Environment and Planning B* 26, pp. 879–92.

Madsen, M. F., Kristensen, S. B. P., Fertner, C., Busck, A. G. and Jørgensen, G. (2010), 'Urbanisation of rural areas: A case study from Jutland, Denmark'. *Geografisk Tidsskrift-Danish Journal of Geography* 110.1, pp. 47–63.

Mahfouz, Naguib (1994), *Palace Walk* (vol. 1). London: Random House.

Maitland, Robert (2007), 'Conviviality and everyday life: The appeal of new areas of London for visitors'. *International Journal of Tourism Research* 10, pp. 15–25.

Mand, H. N. and Cilliers, S. (Undated), *Jane Jacob and Designing Diversity: Investigating Gastronomic Quarters and Food Courts of Shopping Malls and Vitality of Public Spaces.*

Mandel, R. (1996), 'A Place of their own', in Barbara Daly Metcalf (ed.), *Making Muslim Space in North America and Europe*. Berkeley, CA: University of California Press, pp. 147–66.

Mandelblatt, Bertie (2012), 'Geography of food', in Jeffrey M. Pilcher (ed.), *The Oxford Handbook of Food History*. Oxford; New York: Oxford University Press.

Marcus, Clare Cooper and Carolyn Francis (eds) (1997), *People Places: Design Guidelines for Urban Open Space*. New York: Wiley.

Marcus, George H. (2013), 'Louis Kahn and the Architectural Barbecue'. *Journal of Design History* 26.2, pp. 168–81.

Markowitz, Lisa (2010), 'Expanding access and alternatives: Building farmers' markets in low-income communities'. *Food and Foodways* 18.1–2, pp. 66–80.

Marsden, Terry (1998), 'New rural territories: Regulating the differentiated rural spaces'. *Journal of Rural Studies* 14.1, pp. 107–17.

Marsden, Terry (1998), 'Agriculture beyond the treadmill? Issues for policy, theory and research practice'. *Progress in Human Geography* 22.2, pp. 265–75.

Marsden, Terry (2010), 'Mobilizing the regional eco-economy: Evolving webs of agri-food and rural development in the UK'. *Cambridge Journal of Regions, Economy and Society* 3.2, pp. 225–44.

Marsden, Terry and Smith, Everard (2005), 'Ecological entrepreneurship: Sustainable development in local communities through quality food production and local branding'. *Geoforum* 36.4, pp. 440–51.

Marsh, Robin (1998), 'Building on traditional gardening to improve household food security'. *Food Nutrition and Agriculture* 22, pp. 4–14.

Marshall, F., Waldman, L., MacGregor, H., Mehta, L. and Randhawa, P. (2009), 'On the edge of sustainability: Perspectives on peri-urban dynamics'. STEPS Working Paper 35, Brighton: STEPS Centre.

Marshall, Richard (2004), 'Asian megacities', in E. Robbins and R. El-Khoury (eds), *Shaping The City: Studies in History, Theory and Urban Design*. New York: Routledge; Taylor & Francis Group, pp. 194–211.

Marshall, S. (2009), *Cities Design & Evolution*. London; New York: Routledge.

Martínez, J. A., Moreno, B. and Martínez-González, M. A. (2004), 'Prevalence of obesity in Spain'. *Obesity Reviews* 5.3, pp. 171–2.

Maruyama, M. and Trung, L. V. (2007), 'Supermarkets in Vietnam: Opportunities and obstacles'. *Asian Economic Journal* 21.1, pp. 19–46.

Marvin, Simon (2001), *Splintering Urbanism: Networked Infrastructures, Technological Mobilities and the Urban Condition*. London: Routledge; Taylor & Francis.
Massey, Doreen and Wield, David (1992), *High-Tech Fantasies: Science Parks in Society, Science and Space*. London: Routledge.
Matthews, H., Taylor, M., Percy-Smith, B. and Limb, M. (2000), 'The unacceptable Flaneur: The shopping mall as a teenage hangout'. *Childhood* 7.3, pp. 279–94.
Maxwell, Daniel, Levin, Carol and Csete, Joanne (1998), 'Does urban agriculture help prevent malnutrition? Evidence from Kampala'. *Food Policy* 23.5, pp. 411–24.
Maye, D., Holloway, L. and Kneafsey, M. (eds) (2007), *Alternative Food Geographies: Representation and Practice*. Oxford: Elsevier.
Mayer, Heike and Knox, Paul L. (2006), 'Slow cities: Sustainable places in a fast world'. *Journal of Urban Affairs* 28.4, pp. 321–34.
Mayer, H. and Knox, P. L. (2009), 'Pace of life and quality of life: The slow city charter', in M. Joseph Sirgy, Rhonda Phillips and Don R. Rahtz (eds), *Community Quality-of-Life Indicators: Best Cases III*. Netherlands: Springer, pp. 21–40.
Meenar, Mahbubur R., Featherstone, Jeffrey P. and McCabe, Julia (2012), 'Urban agriculture in post-industrial landscape: A case for community-generated urban design'. 48th ISOCARP Congress.
Meiberger, E. and Weichbold, M. (2010), 'How can mountain quality food reduce the vulnerability of mountain farming systems', in 9th European IFSA Symposium. Vienna (Austria), pp. 1626–35.
Melching, Karen (2005), Frankfurt Kitchen: Patina follows function V and A, http://www.vam.ac.uk/content/journals/conservation-journal/issue-53/frankfurt-kitchen-patina-follows-function/. Accessed 2 December 2012.
Meller, H. (1994), *Patrick Geddes: Social Evolutionist and City Planner*. London: Routledge.
Mennell, Stephen (1985), *All Manners of Food: Eating and Taste in England and France from the Middle Ages to the Present*. Urbana: University of Illinois Press.
Mennell, Stephen, Murcott, Anne and Van Otterloo, Anneke H. (1992), *The Sociology of Food: Eating, Diet and Culture*. London: Sage.
Mensah, P., Yeboah-Manu, D., Owusu-Darko, K. and Ablordey, A. (2002), 'Street foods in Accra, Ghana: How safe are they?' *Bulletin of the World Health Organization* 80.7, pp. 546–54.
Metcalf, Barbara Daly (ed.) (1996), *Making Muslim Space in North America and Europe*. Berkeley, CA: University of California Press.
Metcalfe, A., Owen, J., Dryden, C. and Shipton, G. (2011), 'Concrete chips and soggy semolina: The contested spaces of the school dinner hall'. *Population Space and Place* 17.4, Special Issue, pp. 377–89.
Meyers, Tom (2008), '"Berlin tip": Cafeterias dish up cheap meals', http://www.eurocheapo.com/blog/berlin-cheap-eats-mensa-cafeterias-dish-up-square-meals.html. Retrieved 22 January, 2013.
Meyer-Ohle, Hendrik (2009), 'Two Asian malls: Urban shopping centre development in Singapore and Japan'. *Asia Pacific Business Review* 15.1, pp. 123–35.
Miao, Pu (2003), 'Deserted streets in a jammed town: The gated community in Chinese cities and its solution'. *Journal of Urban Design* 8.1, pp. 45–66.

Midmore, David J. and Jansen, Hans G. P. (2003), 'Supplying vegetables to Asian cities: is there a case for peri-urban production?' *Food Policy* 28.1, pp. 13–27.

Miele, Mara and Murdoch, Jonathan (2002), 'The practical aesthetics of traditional cuisines: Slow food in Tuscany'. *Sociologia Ruralis* 42.4, pp. 312–28.

Mikkelsen, Bent Egberg (2011), 'Images of foodscapes: Introduction to foodscape studies and their application in the study of healthy eating out-of-home environments'. *Perspectives in Public Health* 131.5, pp. 209–16.

Miles, Malcolm (1997), *Art, Space and the City*. London; New York: Routledge.

Miles, Malcolm (1998), 'Strategies for the convivial city: A new agenda for education for the built environment'. *Journal of Art & Design Education* 17.1, pp. 17–25.

Miles, Malcolm (2000), 'Café-Extra: Culture, representation and the everyday', in Nick Stanley and Ian Cole (eds), *Beyond the Museum: Art, Institutions, People*. Oxford: Museum of Modern Art, pp. 30–7.

Miles, Steven and Miles, Malcolm (2004), *Consuming Cities*. New York: Palgrave Macmillan.

Milewicz, A., Jędrzejuk, D., Lwow, F., Białynicka, A. S., Łopatynski, J., Mardarowicz, G. and Zahorska-Markiewicz, B. (2005), 'Prevalence of obesity in Poland'. *Obesity Reviews* 6.2, pp. 113–14.

Miller, Char (2003), 'In the sweat of our brow: Citizenship in American domestic practice during WWII – Victory gardens'. *The Journal of American Culture* 26.3, pp. 395–409.

Miller, C., Wakefield, M., Kriven, S. and Hyland, A. (2002), 'The air we breathe: Evaluation of smoke-free dining in South Australia: Support and compliance among the community and restaurateurs'. *Australian and New Zealand Journal of Public Health* 26.1, pp. 38–44.

Miller, Mervyn (2010), *English Garden Cities: An Introduction*. Swindon: English Heritage.

Miller, Tim (2010), 'The birth of the Patio Daddy-O: Outdoor grilling in postwar America'. *The Journal of American Culture* 33.1, pp. 5–11.

Millstone, Erik and Lang, Tim (2003), *The Atlas of Food: Who Eats What, Where and Why*. London: Earthscan.

Minton, Anna (2006), 'The privatisation of public space'. London: The Royal Institution of Chartered Surveyors.

Minton, Anna (2012), *Ground Control: Fear and happiness in the Twenty-First-Century City*. UK: Penguin.

Mintz, S. W. (1985), *Sweetness and Power*. New York: Viking.

Mintz, Sydney and Kellogg, Susan (1988), *Domestic Revolutions: A Social History of American Family Life*. New York: Free Press.

Mitchell, Christine M. (2010), 'The rhetoric of celebrity cookbooks'. *The Journal of Popular Culture* 43.3, pp. 524–39.

Mock, M. (2011), *The Modernization of the American Home Kitchen, 1900-1960*. Doctoral dissertation, Carnegie Mellon University.

Moginon, D. F., Toh, P. S. and Saad, M. (2012), 'Indigenous food and destination marketing', in Artinah Zainal, Salleh Mohd Radzi, Rahmat Hashim, Chemah Tamby Chik and Rozita Abu (eds), *Current Issues in Hospitality and Tourism Research and Innovations Organised by Faculty of Hotel and Tourism Management*. University Teknologi MARA, Shah Alam, Selangor, London, Malaysia: Taylor and Francis, pp. 355–8.

Molz, J. G. (2007), 'Eating difference the cosmopolitan mobilities of culinary tourism'. *Space and Culture* 10.1, pp. 77–93.
MOMA (2011), *Counter Space: Design and the Modern Kitchen*, http://www.moma.org/visit/calendar/exhibitions/1062.
MOMA *Counter Space: Design and the Modern Kitchen*, http://www.moma.org/interactives/exhibitions/2010/counter_space/the_frankfurt_kitchen. Accessed 2 December 2012.
Monfries, Alice (2012), '"Farming" in Adelaide Parklands could be a vision of future', *The Sunday Mail*, 07 January 2012, http://www.cityfarmer.info/2012/01/09/farming-in-adelaide-parklands-could-be-a-vision-of-future/. Accessed online 10 May 2013.
Montanari, A. and Staniscia, B. (2009), 'Culinary tourism as a tool for regional re-equilibrium'. *European Planning Studies* 17.10, pp. 1463–83.
Montanari, Massimo (2006), *Food is Culture*. New York: Columbia University Press.
Montgomery, John (1998), 'Making a city: Urbanity, vitality and urban design'. *Journal of Urban Design* 3.1, pp. 93–116.
Moran, Emilio F. (2006), *People and Nature: An Introduction to Human Ecological Relations*. Oxford: Blackwell.
Moran, Patricia (2003), '"A sinkside, stoveside, personal perspective": Female authority and kitchen space in contemporary women's writing'. *Scenes of the Apple: Food and the Female Body in Nineteenth-and Twentieth-Century Women's Writing*. Albany: State University of New York Press, p. 215.
Moravansky, Akos (2007), 'The Reproducibility of Taste', in Hodgson, Petra Hagen and Rolf Toyka (eds), *The Architect, the Cook, and Good Taste*. Birkhauser: Basel, pp. 146–51.
Morgan, Kevin (2008), 'Greening the realm: Sustainable food chains and the public plate', in Graham, Haughton and Kevin, Morgan (eds), Sustainable Regions. *Regional Studies* 42.9, pp. 1237–50.
Morgan, Kevin (2010), 'Local and green, global and fair: The ethical foodscape and the politics of care'. *Environment and Planning A* 42.8, pp. 1852–67.
Morgan, Kevin and Sonnino, R. (2010), 'The urban foodscape: World cities and the new food equation'. *Cambridge Journal of Regions, Economy and Society* 3.2, pp. 209–24.
Morgan, Kevin, Marsden, Terry and Murdoch, Jonathan (2006), *Worlds of Food: Place, Power, and Provenance in the Food Chain*. Oxford: Oxford University Press.
Morland, K., Diez Roux, A. V. and Wing, S. (2006), 'Supermarkets, other food stores, and obesity: The atherosclerosis risk in communities study'. *American Journal of Preventive Medicine* 30.4, pp. 333–9.
Morland, K., Wing, S., Diez Roux, A. V. and Poole, C. (2002), 'Neighborhood characteristics associated with the location of food stores and food service places'. *American Journal of Preventive Medicine* 22.1, pp. 23–9.
Morris, A. E. J. (1994), *A History of Urban Form Before the Industrial Revolutions*. Harlow, New York: Wiley, Longman.
Morrison, Kathryn A. (2003), *English Shops and Shopping: An Architectural History*. New Haven and London: Yale University Press; English Heritage.

Morrow, Daniel (2011), 'Ardeer: A postwar Ukrainian suburban village'. *Australian Historical Studies* 42.3, pp. 390–403.

Mosupye, Francina M. and Alexander von Holy (2000), 'Microbiological hazard identification and exposure assessment of street food vending in Johannesburg, South Africa'. *International Journal of Food Microbiology* 61.2, pp. 137–45.

Mougeot, Luc J. A. (1994), 'African city farming from a world perspective', in Egziabher, A. G., D. Lee-Smith, D. G. Maxwell, P. A. Memon, L. J. A. Mougeot, and C. J. Sawio (eds), *Cities Feeding People: An Examination of Urban Agriculture in East Africa*. Ottawa: International Development Research Centre, pp. 1–24.

Mougeot, Luc J. A. (2000), 'Urban agriculture: Definition, presence, potentials and risks', in N. Bakker, M. Dubbeling, S. Guendel, U. Sabel Koschella and H. D. Zeeuw (eds), *Growing Cities, Growing Food: Urban Agriculture on the Policy Agenda. A Reader on Urban Agriculture*. DSE.

Mougeot, L. J. (ed.) (2005), *Agropolis: The Social, Political and Environmental Dimensions of Urban Agriculture*. London; Sterling, VA: Earthscan, IDRC.

Moughtin, J. C. C. and Shirley, Peter (2005), *Urban Design: Green Dimensions*. London: Routledge.

Mouzon, Steve (2008), 'Being authentically green goes beyond "gizmo green"'. http://www.cooltownstudios.com/2008/04/30/being-authentically-green-goes-beyond-gizmo-green/.

Mr Foodie, http://www.mrfoodie.co.uk/164/my-favourites/best-in-north-west-london/pacific-plaza-wembley-oriental-city-reincarnated/. Accessed 11 January 2013.

Mullins, P. and Kynaston, C. (2000), 'Household production and subsistence goods: The "urban peasant" thesis reassessed', in Patrick Troy (ed.), *History of European Housing In Australia*. Cambridge: Cambridge University Press, pp. 142–63.

Mumford, Lewis (1938), *The Culture of Cities*. London: Routledge (Thoemmes Press, 1997, orig. 1938).

Mumford, Lewis (1961), 'The medieval town', in Nathan Glazer and Mark Lilla (eds) (1987), *The Public Face of Architecture: Civic Culture and Public Spaces*. New York: Free Press; London: Collier Macmillan, pp. 60–74.

Muñiz, I. and Galindo, A. (2005), 'Urban form and the ecological footprint of commuting: The case of Barcelona'. *Ecological Economics* 55.4, pp. 499–514.

Murayama, M., Parker, G., Fullagar, S., Markwell, K. and Wilson, E. (2012), 'Fast Japan, slow Japan': Shifting to slow tourism as a rural regeneration tool in Japan', in Simone Fullagar, Kevin Markwell and Erica Wilson (eds), *Slow Tourism: Experiences and Mobilities*. Bristol: Channel View, p. 170.

Murcott, Anne (1982), 'On the social significance of the "cooked" dinner in South Wales'. *Social Science Information/sur les sciences sociales* 21.4–5, pp. 677–96.

Murcott, Anne (1997), 'Family meals – A thing of the past?', in Pat, Caplan (ed.) (2002), *Food, Health and Identity*. London: Routledge.

Murdoch, J. and Marsden, T. (1994), *Reconstituting Rurality: Class, Community and Power in the Development Process*. London: UCL Press.

Murdoch, Jonathan and Miele, Mara (2004), 'A new aesthetic of food? Relational reflexivity in the "alternative" food movement', in Mark Harvey, Andrew McMeekin and Alan Warde (eds), *Qualities of Food*. Manchester: Manchester University Press, pp. 156–75.

Murrain, Paul (1993), 'Urban expansion: Look back and learn', in Richard Hayward and Sue McGlynn (eds), *Making Better Places: Urban Design Now*. Oxford: Butterworth Architecture, pp. 83–94.

Murrain, Paul (2002), 'Understand urbanism and get off its back'. *Urban Design International* 7.3, pp. 131–42.

Myhrvold, Nathan (2011), 'The art in gastronomy: A modernist perspective'. *Gastronomica: The Journal of Food and Culture* 11.1, pp. 13–23.

Mycoo, Michelle (2006), 'The retreat of the upper and middle classes to gated communities in the poststructural adjustment era: The case of Trinidad'. *Environment and Planning A* 38.1, p. 131.

Myrick, Richard (2012), *Running a Food Truck for Dummies*. Hoboken, NJ: John Wiley and Sons.

Naccarato, Peter and LeBasco, Kathleen (2012), *Culinary Capital*. London, New York: Berg.

Napton, Darrell (1992), 'Farm diversification in the United States', in I. R. Bowler, C. R. Bryant and M. D. Nellis (eds), *Contemporary Rural Systems in Transition 1*. Wallingford: CAB International, pp. 87–99.

Nasir, Kamaludeen Mohamed and Pereira, Alexius A. (2008), 'Defensive dining: Notes on the public dining experiences in Singapore'. *Contemporary Islam* 2.1, pp. 61–73.

Nasr, Joe L. and Komisar, June D. (2012), 'The integration of food and agriculture into urban planning and design practices', in André Viljoen and Johannes S. C. Wiskerke (eds), *Sustainable Food Planning: Evolving Theory and Practice*. The Netherlands: Wageningen Academic Publishers, p. 47.

Navarro, Manuelina Porto Nunes and Pimentel, Roberto Leal (2007), 'Speech interference in food courts of shopping centres'. *Applied acoustics* 68.3, pp. 364–75.

Nelson, Arthur C. (1992), 'Characterizing exurbia'. *Journal of Planning Literature* 6.4, pp. 350–68.

Nelson, Arthur C. (1999), 'The exurban battleground', in Owen J. Furuseth and Mark B. Lapping (eds), *Contested Countryside: The Rural Urban Fringe in North America*. Aldershot: Ashgate, pp. 137–50.

Nelson, Arthur C. and Terry Moore (1993), 'Assessing urban growth management: The case of Portland, Oregon, the USA's largest urban growth boundary'. *Land Use Policy* 10.4, pp. 293–302.

Nelson, George and Wright, Henry (1945), *Tomorrow's House: A Complete Guide for the Homebuilder*. London: The Architectural Press; New York: Simon and Schuster.

Nestlé, Marion (2007), 'Dinner for six billion', in Peter, Mendel and Faith D'Aluisio (eds), Photographed by Peter Menzel *Hungry Planet: What the World Eats*. Berkeley, CA: Ten Speed; Enfield: Airlift [distributor].

Neuner, Kailee, Sylvia Kelly and Samina Raja (2011) 'Planning to eat? Innovative local government plans and policies to build healthy food systems in the United States'. Food Systems Planning and Healthy Communities Lab University at Buffalo, The State University of New York.

Newman, Felicity and Gibson, Mark (2005), 'Monoculture versus multiculturalism: Trouble in the Aussie kitchen' in David, Bell and Joanne, Hollows (eds), *Ordinary Lifestyles: Popular Media, Consumption and Taste*. Maidenhead: Open University Press, pp. 82–98.

Newman, L. L. and Burnett, K. (2013), 'Street food and vibrant urban spaces: Lessons from Portland, Oregon'. *Local Environment* 18.2, pp. 233–48.

Newman, P. W. and Kenworthy, J. R. (1989), 'Gasoline consumption and cities: A comparison of US cities with a global survey'. *Journal of the American Planning Association* 55.1, pp. 24–37.

Newman, P. W. and Kenworthy, J. R. (2006), 'Urban design to reduce automobile dependence'. *Opolis* 2.1

Newman, P. W. and Kenworthy, J. R. (2011), '"Peak Car Use": Understanding the demise of automobile dependence'. *World Transport Policy and Practice* 17.2, pp. 31–42.

Ng, Cheuk Fan (2003), 'Satisfying shoppers' psychological needs: From public market to cyber-mall'. *Journal of Environmental Psychology* 23.4, pp. 439–55.

Nichol, Lucy (2003), 'Local food production: Some implications for planning'. *Planning Theory & Practice* 4.4, pp. 409–27.

Nicholson-Lord, David (1987), *The Greening of the Cities*. London: Routledge & Kegan Paul.

Nielsen, N. C., Nissen, K. A. and Just, F. (2010), 'Rural tourism–return to the farm perspective'. Proceedings, 19th Nordic Symposium on Tourism and Hospitality Research.

Noguchi, Paul H. (1994), 'Savor slowly. Ekiben: The fast food of high-speed Japan'. *Ethnology* 22.4, Autumn, pp. 317–30.

Nolin, Catharina (2006), 'Stockholm's urban parks: Meeting places and social contexts from 1860–1930', in P. Clark (ed.), *The European City and Green Space: London, Stockholm, Helsinki and St. Petersburg, 1850–2000*. Aldershot: Ashgate.

Norberg-Hodge, Helena, Merrifield, Todd and Gorelick, Steven (2002), *Bringing the Food Economy Home: Local Alternatives to Global Agribusiness*. London: Zed Books; Halifax, NS: Fernwood; Bloomfield, CT: Kumarian.

Nordahl, Darrin (2009), *Public Produce: The New Urban Agriculture*. Island Press.

Norris, Patricia, Taylor, Gary and Wyckoff, Mark (2011), 'When urban agriculture meets michigan's right to farm act: The pig's in the parlor', Michigan State Law Revue, p. 365.

Nystrom, M. (1994), 'Focus kitchen design: A study of housing in Hanoi'. Lund: Lund University, p. 203.

Oatley, Nick (1997), 'Lexicons of suburban and ex-urban development', Research Project, City Words Programme, UNESCO-CNRS.

Obosu-Mensah, Kwaku (1999), *Food Production in Urban Areas: A Study of urban agriculture in Accra, Ghana*. Aldershot: Ashgate.

Ochs, Elinor and Merav Shohet (2006), 'The cultural structuring of mealtime socialization'. *New Directions for Child and Adolescent Development* 111, pp. 35–49.

Oddy, Derek J., Peter J. Atkins and Amilien, Virginie (2009), *The Rise of Obesity in Europe: A Twentieth Century Food History*. Aldershot: Ashgate Publishing.

O'Gorman, K. D. (2009), 'Origins of the commercial hospitality industry: From the fanciful to factual'. *International Journal of Contemporary Hospitality Management* 21.7, pp. 777–90.

Oguntona, Clara R. B. and O. Kanye (1995), 'Contribution of street foods to nutrient intakes by Nigerian adolescents'. *Nutrition and Health* 10.2, pp. 165–71.

Ohnuki-Tierney, Emiko (1997), 'McDonald's in Japan: Changing manners and etiquette', in L. Watson (ed.), *Golden Arches East: McDonald's in East Asia*. Stanford, CA: Stanford University Press, pp. 161–82.
Okumus, Bendegul, Okumus, Fevzi and McKercher, Bob (2007), 'Incorporating local and international cuisines in the marketing of tourism destinations: The cases of Hong Kong and Turkey'. *Tourism Management* 28.1, pp. 253–61.
Oldenburg, Ray (1989), *The Great Good Place: Cafés, Coffee Shops, Community Centers, Beauty Parlors, General Stores, Bars, Hangouts, and How They Get You Through the Day*. New York: Paragon House.
Oldenburg, Ray (1997), 'Our vanishing third places'. *Planning Commissioners Journal* 25, pp. 8–10.
Oldenziel, Ruth and Zachmann, Karin (2009), *Cold War Kitchen: Americanization, Technology, and European Users*. Cambridge: MIT Press.
Olkowski, H. and Olkowski, W. (1975), *The City People's Book of Raising Food*. Emmaus, PA: Rodale Press.
O'Neil, David K. (Winter 2014), 'Transforming markets into market places'. *Urban Design* 129, pp. 25–7.
O'Neill, P. and Whatmore, S. (2000), 'The business of place: Networks of property, partnership and produce'. *Geoforum* 31.2, pp. 121–36.
Oram, J., Conisbee, M. and Simms, A. (2003), *Ghost Town Britain II: Death on the High Street*. London: New Economics Foundation.
Oram, J., MacGillivray, A. and Drury, J. (2002), *Ghost Town Britain: The Threat from Economic Globalisation to Livelihoods, Liberty and Local Economic Freedom*. London: New Economics Foundation.
Orsini, Stefano (2011), 'Deregulating "the rural" threatening land management regime. Experiences of space in the Tuscan countryside'. *arXiv preprint arXiv:1106.0803*
Orsini, Stefano (2013), 'Landscape polarisation, hobby farmers and a valuable hill in Tuscany: Understanding landscape dynamics in a peri-urban context'. *Geografisk Tidsskrift-Danish Journal of Geography* 113.1, pp. 53–64.
Otiman, P. I. (2008), 'Sustainable development strategy of agriculture and rural areas in romania on medium and long term-rural romania XXI–'. *Agricultural Economics and Rural Development* 5.1–2, pp. 4–18.
Ousselin, Edward (2006), 'Film and the Popular Front: "La Belle Equipe" and "Le Crime de M. Lange"'. *The French Review* 79.5, April, pp. 952–62.
Pacini, C., Lazzerini, G., Migliorini, P. and Vazzana, C. (2011), 'An indicator-based framework to evaluate sustainability of farming systems: Review of applications in Tuscany'. *Italian Journal of Agronomy* 4.1, pp. 23–40.
Pacciani, Andrea and Italiani, Sabina (2005), *L'Arte Del Prosciutto. Il prosciutto nelle sue migliori manifestazioni pittorichie*. Fidenza, Parma: Mattioli 1885.
Pamuk, Orhan (2006), *Istanbul: Memories and the City*. New York: Vintage.
Parasecoli, F. (2004), *Food Culture in Italy*. Westport, CT; London: Greenwood Publishing Group.
Parasecoli, Fabio and de Abreu e Lima, Paulo (2012), 'Eat your way through culture: Gastronomic tourism as performance and bodily experience', in Fullagar, Simone, Markwell, Kevin and Wilson, Erica (eds), *Slow Tourism: Experiences and Mobilities*. Bristol: Buffalo; Toronto: Channel View Publications.

Parasecoli, Fabio and Scholliers, Peter (2012), *A Cultural History of Food*. London: Berg.
Parham, Susan (1990), 'The table in space: A planning perspective', *Meanjin* 49.2, p. 213.
Parham, Susan (1992), *Gastronomic Strategies for Australian Cities*, Urban Futures, 2, 2, Canberra.
Parham, Susan (1993a), 'Gastronomy and urban form' South Australian Winter Planning Seminar, Adelaide: Planning Education Foundation Papers.
Parham, Susan (1993b), 'Convivial green space' Proceedings, Canberra: Seventh Australian Symposium of Gastronomy.
Parham, Susan (1995a), 'Megalopolis' *Arena* April/16 May, Melbourne.
Parham, Susan (1995b), 'Strategic issues for the ecological city', OECD/World Health Organisation Symposium, Conference Paper, Madrid.
Parham, Susan (1996a), 'Food and megalopolis', Proceedings 9th Symposium of Australian Gastronomy, March 1996, Sydney.
Parham, Susan (1996b), 'Food and Travel', Paper, The Oxford Symposium of Food and Cooking, Oxford.
Parham, Susan (1996c), 'Gastronomic architecture: The cafe and beyond'. *Architecture Bulletin* October, Sydney: RAIA.
Parham, Susan (1996d), *Productive Land-Use on the Urban Fringe: A Comparative Study in Planning for Regional Economic Development in Languedoc and Tuscany*, Report, Department of Housing and Urban Development, South Australia.
Parham, Susan (1998), 'Fat city: Why Bologna works', *Australian Financial Times* Saturday Review Section, February 1998.
Parham, Susan (2001), *Developing Market Typologies: An Urban Design Analysis* Master's Thesis, London: University of Westminster (Unpublished)
Parham, Susan (2005), 'Designing the gastronomic quarter', in Franck, K (ed.), *Food and the City*. In *Architectural Design*, 75, 3, May/June. Chichester: Wiley Academic.
Parham, Susan (2006a), 'The European city model and its critics', Discussion Paper, *Urban Age Programme*, LSE, January.
Parham, Susan (2006b), 'Marylebone high street assessment report', Urbanism Awards 2006, London: The Academy of Urbanism (unpublished).
Parham, Susan (2007), *Let's Rip it All Up*. Viewpoint. Journal of Urbanism.
Parham, Susan (2008), 'The relationship between approaches to conservation and the idea of nostalgia: Looking at food-centred spaces within cities' Chapter, in Matthew Hardy (ed.), *The Venice Charter Revisited: Modernism, Conservation and Tradition in the 21st Century World*. Newcastle upon Tyne: Cambridge Scholars Publishing.
Parham, Susan (2011), 'Theory and practice of agrarian urbanism' Review, *Journal of Urbanism: International Research on Placemaking and Urban Sustainability* 4.2, pp. 193–4.
Parham, Susan (2012a), *Market Place: Food Quarters, Design and Urban Renewal in London*. Newcastle upon Tyne: Cambridge Scholars Publishing.
Parham, Susan (2012b), 'Retrofitting for Food – Edible Urbanism in inmidtown'. Report. Unpublished.
Parham, Susan (2013), 'Shaping sustainable urbanism: Are garden cities the answer?' Conference paper. *Shaping Canberra: The Lived Experience of Place, Home and Capital*, Humanities Research Centre Conference. Australian National University.

Parham, Susan and Konvitz, Josef (1996), *Innovative Policies for Sustainable Urban Development*. Paris: OECD.
Parham, Susan and McCabe, Ben (2014), *Making Space for Food in Hatfield*. Hatfield: University of Hertfordshire.
Parlette, V. and Cowen, D. (2011), 'Dead Malls: Suburban activism, local spaces, global logistics'. *International Journal of Urban and Regional Research* 35.4, pp. 794–811.
Parrott, N., Wilson, N. and Murdoch, J. (2002), 'Spatializing quality: Regional protection and the alternative geography of food'. *European Urban and Regional Studies* 9.3, pp. 241–61.
Parsons, Kermit C. and Schuyler, David. (eds) (2002), *From Garden City to Green City the Legacy of Ebenezer Howard*. Baltimore and London: The Johns Hopkins University Press.
Pascali, Lara (2006), 'Two stoves, two refrigerators, due cucine: The Italian immigrant home with two kitchens'. *Gender, Place and Culture: A Journal of Feminist Geography* 13.6, pp. 685–95.
Patch, Diana Craig (1991), 'The origin and early development of urbanism in ancient Egypt: A regional study' Dissertation.
Patel, Raj (2007), *Stuffed and Starved: Markets, Power and the Hidden Battle for the World Food System*. London: Portobello.
Paül, Valerià and McKenzie, Fiona Haslam (2013), 'Peri-urban farmland conservation and development of alternative food networks: Insights from a case-study area in metropolitan Barcelona (Catalonia, Spain)'. *Land Use Policy* 30.1, pp. 94–105.
Paül, Valerià and Tonts, Matthew (2005), 'Containing urban sprawl: Trends in land use and spatial planning in the metropolitan region of Barcelona'. *Journal of Environmental Planning and Management* 48.1, pp. 7–35.
Paull, John (2011), 'Incredible edible todmorden: Eating the street'. *Farming Matters* 27.3, pp. 28–9.
Paxson, Heather (2005), 'Slow food in a fat society: Satisfying ethical appetites'. *Gastronomica* 5.1, pp. 14–18.
Paxson, Heather (2010), 'Locating value in artisan cheese: Reverse engineering terroir for new world landscapes'. *American Anthropologist* 112.3, pp. 444–57.
Peace, Adrian (2006), 'Barossa slow: The representation and rhetoric of slow food's regional cooking'. *Gastronomica* 6.1, pp. 51–9.
Pearce, Jamie and Witten, Karen (2010), 'Introduction: Bringing a geographical perspective to understanding the "Obesity Epidemic"', in Jamie Pearce and Karen Witten (eds), *Geographies of Obesity*. Farnham: Ashgate.
Peattie, Lisa (1998), 'Convivial cities', in M. Douglass and J. Friedmann (eds), *Cities for Citizens: Planning and the Rise of Civil Society in a Global Age*. Chichester, New York: John Wiley and Sons, pp. 247–53.
Pegler, M. M. (1991), *Visual Merchandising and Display*. Fairchild Fashion & Merchandising Group.
Perlman, Janice (1993), 'Mega-cities: Global urbanization and innovation'. The Mega-Cities Project. Hartford: Publication MCP-013.
Perlman, J. E. (2005), 'The myth of marginality revisited: The case of Favelas in Rio De Janeiro', in Lisa M. Hanley, Blair A. Rubie and Joseph S. Tulchin (eds), *Becoming Global and the New Poverty of Cities*. Washington, DC: Woodrow Wilson International Center for Scholars Comparative Urban Studies Project, pp. 9–54.

Peters, C. J., Bills, N. L., Wilkins, J. L. and Fick, G. W. (2009), 'Foodshed analysis and its relevance to sustainability'. *Renewable Agriculture and Food Systems* 24.01, pp. 1–7.

Peters, C. J., Lembo, Arthur J. and Fick, Gary W. (2005), A Tale of Two Foodsheds: Mapping Local Food Production Capacity Relative to Local Food Requirements. http://www.crops.confex.com/crops/viewHandout.cgi?uploadid=226.

Petherick, Tom and Eclare, Melanie (2006), *The Kitchen Gardens at Heligan: Lost Gardening Principles Rediscovered.* London: Weidenfeld & Nicolson.

Petrini, Carlo (2007), *Slow Food Nation: Why Our Food should be Good, Clean, and Fair.* New York: Rizzoli; Enfield: Publishers Group UK [distributor].

Phelps, N. A., Parsons, N., Ballas, D. and Dowling, N. (2006), *Post-Suburban Europe: Planning and Politics at the Margins of Europe's Capital Cities.* Basingstoke: Palgrave Macmillan.

Phelps, N. A. and Wu, Fulong (eds) (2011), *International Perspectives on Suburbanization.* Basingstoke: Palgrave Macmillan.

Philips, A. (2013), *Designing Urban Agriculture: A Complete Guide to the Planning, Design, Construction, Maintenance and Management of Edible Landscapes.* Hoboken, NJ: John Wiley & Sons.

Pickett, S. T. and Cadenasso, M. L. (2008), 'Linking ecological and built components of urban mosaics: An open cycle of ecological design'. *Journal of Ecology* 96.1, pp. 8–12.

Pietrykowski, Bruce (2004), 'You are what you eat: The social economy of the slow food movement'. *Review of Social Economy* 62.3, pp. 307–32.

Pilcher, Jeffrey M. (2008), 'Taco Bell, Maseca, and slow food: A postmodern apocalypse for Mexico's peasant cuisine', in Richard Wilk (ed.), *Fast Food/Slow Food: The Cultural Economy of the Global Food System.* Plymouth: Altamira Press.

Pillsbury, Richard (1998), *No foreign Food: The American Diet in Time and Place.* Boulder, CO; Oxford: Westview Press, Inc.

Pine, B. and Gilmore, J. (1999), *The Experience Economy: Work is Theater and every Business is a Stage.* Boston, MA: Harvard Business School Press.

Pink, Sarah (2008), 'Sense and sustainability: The case of the Slow City movement'. *Local Environment* 13.2, pp. 95–106.

Pink, Sarah (2012), *Situating Everyday Life: Practices and Places.* London: Sage Publications Limited.

Pincetl, S. and Gearin, E. (2005), 'The reinvention of public green space'. *Urban Geography* 26.5, pp. 365–84.

'Planning for food. Towards a prosperous, resilient and healthy food system through Victoria's Metropolitan Planning Strategy' (2012), Food Alliance and National Heart Foundation of Australia (Victorian Division).

Plantinga, Andrew J., Lubowski, Ruben N. and Stavins, Robert N. (2002), 'The effects of potential land development on agricultural land prices'. *Journal of Urban Economics* 52.3, pp. 561–81.

Policy Commission on the Future of Farming and Food (2002), *Farming and Food: A Sustainable Future.* Report of the Policy Commission on the Future of Farming and Food. Great Britain.

Pollan, M. (2006), *The Omnivore's Dilemma: A Natural History of Four Meals.* New York; London: Penguin.

Pongracz, Petra (2004), 'Attracting supermarkets to underserved, urban markets-a case study in a low-income neighborhood of Durham, North Carolina'. *NEURUS Papers.*

Ponsonby, Margaret (2003), 'Ideals, reality and meaning: Homemaking in England in the first half of the nineteenth century'. *Journal of Design History* 16.3, pp. 201–14.

Poole, Steven (2012), *You aren't What You Eat: Fed up with Gastroculture.* London: Union Books.

Popkin, Barry, M. (2010), 'The emerging obesity epidemic: An introduction', in Jamie, Pearce and Karen, Witten (eds), *Geographies of Obesity. Environmental Understandings of the Obesity Epidemic.* Farnham: Ashgate.

Portas, Mary (2011), *The Portas Review. An Independent Review into the Future of our High Streets.* London: Mary Portas.

Pothukuchi, Kameshwari (2005), 'Attracting supermarkets to inner-city neighborhoods: economic development outside the box'. *Economic Development Quarterly* 19.3, pp. 232–44.

Pothukuchi, Kameshwari (2009), 'Community and regional food planning: Building institutional support in the United States'. *International Planning Studies* 14.4, pp. 349–67.

Pothukuchi, Kameshwari and Kaufman, Jerome L. (1999), 'Placing the food system on the urban agenda: The role of municipal institutions in food systems planning'. *Agriculture and Human Values* 16.2, pp. 213–24.

Pothukuchi, Kameshwari and Kaufman, Jerome L. (2000), 'The Food System'. *Journal of the American Planning Association* 66.2, pp. 113–24.

Potteiger, Matthew (2013), 'Eating places food systems, narratives, networks, and spaces'. *Landscape Journal* 32.2, pp. 261–75.

Pottier, Johan (1999), *Anthropology of Food: The Social Dynamics of Food Security.* Malden, MA: Blackwell Publishers.

Poulain, J. P. (2002), 'The contemporary diet in France: "de-structuration" or from commensalism to "vagabond feeding"'. *Appetite* 39.1, pp. 43–55.

Pow, Choon-Piew (2009), *Gated Communities in China: The Politics of the Good Life.* London; New York: Routledge.

Power, Emma R. (2005), 'Human–nature relations in suburban gardens'. *Australian Geographer* 36.1, pp. 39–53.

Pozzi, Andrea (2011), 'Who is hurt by e-commerce? Crowding out and business stealing in online grocery'. Social Science Research Network (SSRN)

Pratt, Jeff (2007), 'Food values: The local and the authentic'. *Critique of Anthropology* 27.3, pp. 285–300.

Præstholm, Seren and Kristensen, Søren Pilgaard (2007), 'Farmers as initiators and farms as attractors for non-agricultural economic activities in peri-urban areas in Denmark'. *Geografisk Tidsskrift-Danish Journal of Geography* 107.2, pp. 13–27.

Presenza, A. and Del Chiappa, G. (2013), 'Entrepreneurial strategies in leveraging food as a tourist resource: A cross-regional analysis in Italy'. *Journal of Heritage Tourism* 8.2-3, pp. 182–92.

Press Association (2012), 'High street shops in "death spiral", warns veteran retailer', theguardian.com, Thursday 2 February, 2012, 11.30 GMT.

Pritchard, W. N. (2000), 'Beyond the modern supermarket: geographical approaches to the analysis of contemporary Australian retail restructuring'. *Australian Geographical Studies* 38.2, pp. 204–18.

Privitera, D. (2009), *Factors of Development of Competitiveness: The Case of Organic-Agritourism*. In presentation at the 113th EAAE Seminar, Belgrade, Republic of Serbia December, pp. 9–11.

Probyn, E. (2000), *Carnal Appetites: Foodsexidentities*. London: Routledge.

Prospersi, Patricia Silvia (2009), 'The Health impacts of farming on producers in Rosario, Argentina in Redwood', in Mark (ed.), *Agriculture in Urban Planning: Generating Livelihoods and Food Security*. Abingdon: Routledge, pp. 167–79.

Psomopoulos, Panayotis (1987), 'Toward Megalopolis', in Ervin Y. Galantay (ed.), *The Metropolis in Transition*. New York: Paragon.

Pumain, Denise (2004), 'Urban Sprawl: Is there a French Case?', in Harry W. Richardson and Chang-Hee Christine Bae (eds), *Urban Sprawl in Western Europe ad the United States*. Aldershot: Asghate, pp. 137–57.

Punakivi, M., Yrjölä, H. and HolmstroÈm, J. (2001), 'Solving the last mile issue: Reception box or delivery box?'. *International Journal of Physical Distribution & Logistics Management* 31.6, pp. 427–39.

Qviström, Mattias (2007), 'Landscapes out of order: Studying the inner urban fringe beyond the rural–urban divide'. *Geografiska Annaler: Series B, Human Geography* 89.3, pp. 269–82.

Rabikowska, Marta and Burrell, Kathy (2004), 'The material worlds of recent Polish migrants: Transnationalism, food, shops and home'. In *Polish Migration to the UK in the 'New' European Union After 2009*, pp. 211–32.

Rand, Gerrie E. D., Heath, Ernie and Alberts, Nic (2003), 'The role of local and regional food in destination marketing: A South African situation analysis'. *Journal of Travel & Tourism Marketing* 14.3–4, pp. 97–112.

Ravenscroft, Neil and van Westering, Jetske (2002), 'Gastronomy and intellectual property', in Anne-Mette Hjalager and Greg Richards (eds), *Tourism and Gastronomy*. London: Routledge, pp. 153–65.

Ravetz, Alison (1989), 'A view from the interior' *A View from the Interior: Feminism, Women and Design*. London: The Women's Press.

Rayner, Jay (2011), 'Farmers' markets won't change the world: Shopping for dinky, artisan products is fun – but don't think you're making a stand against The Man'. *The Observer* Sunday 17 July, 2011.

Reardon, Thomas and Berdegué, Julio A. (2008), 'The rapid rise of supermarkets in Latin America: Challenges and opportunities for development'. *Development Policy Review* 20.4, pp. 371–88.

Reardon, Thomas and Hopkins, Rose (2006), 'The supermarket revolution in developing countries: policies to address emerging tensions among supermarkets, suppliers and traditional retailers'. *The European Journal of Development Research* 18.4, pp. 522–45.

Reardon, T., Henson, S. and Berdegué, J. (2007), '"Proactive fast-tracking" diffusion of supermarkets in developing countries: Implications for market institutions and trade'. *Journal of Economic Geography* 7.4, pp. 399–431.

Reardon, T., Berdegué, J., Barrett, C. B. and Stamoulis, K. (2007), *Household Income Diversification into Rural Nonfarm Activities: Transforming the Rural Non-farm Economy*. Baltimore, MD: Johns Hopkins University Press.

Reardon, T., Timmer, C. P., Barrett, C. B. and Berdegué, J. (2003), 'The rise of supermarkets in Africa, Asia, and Latin America'. *American Journal of Agricultural Economics* 85.5, pp. 1140–6.

Rebora, Giovanni (2001), *Culture of the Fork: A Brief History of Food in Europe*. New York; Chichester: Columbia University Press.

Redwood, Mark (ed.) (2009), *Agriculture in Urban Planning; Generating Livelihoods and Food Security*. Abingdon: Routledge.

Reeder, David (2006), 'London and Green Space, 1850–2000', in P. Clark (ed.), *The European City and Green Space: London, Stockholm, Helsinki and St. Petersburg, 1850-2000*. Aldershot: Ashgate.

Rees, W. (1992), 'Ecological footprints and appropriate carrying capacity: What urban economics leaves out'. *Environment and Urbanization* 4.2, pp. 121–30.

Reeve, Alan (1993), 'The ontology of the built environment: The production of places and buildings in a culture of Historical Amnesia', in Richard Hayward and Sue McGlynn (eds), *Making Better Places: Urban Design Now*. Oxford: Butterworth-Heinemann, pp. 30–4.

Regoli, Francesca, Vittuari, Matteo and Segrè, Andrea (2011), 'Policy options for sustainability. A preliminary appraisal of rural tourism in Romania: The case of Maramureş'. *Food, Agri-Culture and Tourism*. Berlin, Heidelberg: Springer, pp. 41–55.

Reid, Neil, Gatrell, Jay D. and Ross, Paula S. (2012), *Local Food Systems in Old Industrial Regions: Concepts, Spatial Context, and Local Practices*. Farnham: Ashgate Publishing.

Relph, Edward (1987), *The Modern Urban Landscape*. Baltimore: Johns Hopkins University Press.

Renting, H., Marsden, T. K. and Banks, J. (2003), 'Understanding alternative food networks: Exploring the role of short food supply chains in rural development'. *Environment and Planning A* 35.3, pp. 393–411.

Rheinländer, T., Olsen, M., Bakang, J. A., Takyi, H., Konradsen, F. and Samuelsen, H. (2008), 'Keeping up appearances: Perceptions of street food safety in urban Kumasi, Ghana'. *Journal of Urban Health* 85.6, pp. 952–64.

Rich, Rachel (2011), *Bourgeois Consumption: Food, Space and Identity in London and Paris, 1850–1914*. Manchester: Manchester University Press.

Richards, Greg (2002), 'Gastronomy: An essential ingredient in tourism production and consumption', in Anne-Mette Hjalager and Greg Richards (eds) (2004), *Tourism and Gastronomy*. London: Routledge, pp. 2–20.

Richards, J. M. (1938), *High Street* with illustrations by Eric Ravilious. London: Country Life Ltd.

Richardson, Harry Ward and Bae, Chang-Hee Christine (eds) (2004), *Urban Sprawl in Western Europe and the United States*. Aldershot; Burlington, VT: Ashgate Publishing.

Riley, M. (1994), 'Marketing eating out: The influence of social culture and innovation'. *British Food Journal* 96.10, pp. 15–18.

Riordan, John (2006), *Restaurants by Design*. New York: Collins Design.

Rishbeth, Clare (2001), 'Ethnic minority groups and the design of public open space: An inclusive landscape?' *Landscape Research* 26.4, pp. 351–66.

Ritzer, George (1995), *The McDonaldization of Society: An Investigation into the Changing Character of Contemporary Social Life*. Thousand Oaks, CA, London: Pine Forge Press.

Ritzer, George (2008), *The McDonaldization of Society*. Los Angeles, London; Pine Forge Press.

Ritzer, G. and Liska, A. (1997), '"McDisneyization" and "post tourism": Complementary perspectives on contemporary tourism', in C. Rojek, and J. Urry (eds), *Touring Cultures: Transformations of Travel and Theory*. London: Routledge, pp. 96–109.

Robbins, Deborah (1994), 'Via Della Lungaretta: The making of a medieval street', in Zeynep, Çelik, Diane Favro and Richard Ingersoll (eds), *Streets: Critical Perspectives on Public Space*. Berkeley, CA; London: University of California Press.

Roberts, Brian H. (2004), 'The application of industrial ecology principles and planning guidelines for the development of eco-industrial parks: An Australian case study'. *Journal of Cleaner Production* 12.8, pp. 997–1010.

Roberts, Marion (1991), *Living in a Man Made World: Gender Assumptions in Modern Housing Design*. London: Routledge.

Roberts, Paul (2008), *The End of Food: The Coming Crisis in the World Food Industry*. Houghton: Mifflin Harcourt.

Robinson, Jennifer Meta and Hartenfeld, J. A. (2007), *The Farmers' Market Book: Growing Food, Cultivating Community*. Indiana University Press.

Robson, S. K. (2002), 'A review of psychological and cultural effects on seating behavior and their application to foodservice settings'. *Journal of Foodservice Business Research* 5.2, pp. 89–107.

Roddy, Rachel (2013), 'A foodie autumn break in Abruzzo, Italy', *The Guardian*, Saturday 30 November, 2013.

Roden, Claudia (1989), *The Food of Italy*. London, Melbourne: Arrow Books.

Rofe, Matthew W. (2006), 'New landscapes of gated communities: Australia's Sovereign Islands'. *Landscape Research* 31.3, pp. 309–17.

Roitman, Sonia (2005), 'Who segregates whom? The analysis of a gated community in Mendoza, Argentina'. *Housing Studies* 20.2, pp. 303–21.

Rome, Adam (2001), *The Bulldozer in the Countryside. Suburban Sprawl and the Rise of American Environmentalism*. Cambridge: Cambridge University Press.

Romig, Kevin (2005), 'The upper sonoran lifestyle: Gated communities in Scottsdale, Arizona'. *City & Community* 4.1, pp. 67–86.

Ronald, Richard and Hirayama Yosuke, (2009), 'Home alone: the individualization of young, urban Japanese singles'. *Environment and Planning*. A 41.12, pp. 2836–54.

Root, Waverley (1971), *The Food of Italy*. New York: Vintage.

Rose, D., Bodor, J. N. and Swalm, C. M. (2009), Deserts in New Orleans? Illustrations of urban food access and implications for policy. University of Michigan National Poverty Center/USDA Economic Research Service Research, 'Understanding the Economic Concepts and Characteristics of Food Access'. Retrieved from http://www.npc.umich.edu/news/events/food-access/rose_et_al.pdf.

Roseberry, William (1996), 'The rise of yuppie coffees and the reimagination of class in the United States'. *American Anthropologist* 98.4, pp. 762–75.
Rosenbaum, Mark S. (2006), 'Exploring the social supportive role of third places in consumers' lives'. *Journal of Service Research* 9.1, pp. 59–72.
Rosenbaum, M. S., Ward, J., Walker, B. A. and Ostrom, A. L. (2007), 'A cup of coffee with a dash of love an investigation of commercial social support and third-place attachment'. *Journal of Service Research* 10.1, pp. 43–59.
Rousseau, Signe (2012), *Food Media: Celebrity Chefs and the Politics of Everyday Interference*. Oxford: Berg.
Roux, Albert and Roux, Michel (1989), *The Roux Brothers. French Country Cooking*. London, Basingstoke: Papaermac.
Rowe, Peter (1991), *Making a Middle Landscape*. Cambridge, MA; London: MIT Press.
Rowe, Peter (1997), *Civic Realism*. Cambridge, MA; London: MIT Press.
Royle, Tony and Towers, Brian (eds) (2002), *Labour Relations in the Global Fast-Food Industry*. London: Routledge.
Rudlin, David and Falk, Nicholas (2001), *Building the 21st Century Home: The Sustainable Urban Neighbourhood*. Oxford: Architectural Press.
Rudofsky, Bernard (1964), *Architecture without Architects: A Short Introduction to Non-Pedigreed Architecture*. Albuquerque: University of New Mexico Press.
Rudofsky, Bernard (1980), *Now I Lay Me Down to Eat: Notes and Footnotes on the Lost Art of Living*. Garden City, NY: Anchor Press and Doubleday.
Ruggles, D. Fairchild (2008), *Islamic Gardens and Landscapes*. Philadelphia, Bristol: University of Pennsylvania Press.
Rutherford, J. W. (2003), *Selling Mrs. Consumer: Christine Frederick and the Rise of Household Efficiency*. Athens, GA; London: University of Georgia Press.
Ryan, I., Cowan, C., McCarthy, M. and O'sullivan, C. (2004), 'Segmenting Irish food consumers using the food-related lifestyle instrument'. *Journal of International Food & Agribusiness Marketing* 16.1, pp. 89–114.
Ryder, Bethan (2007), *New Restaurant Design*. London: Laurence King.
Rykwert, J. (1988), *The Idea of a Town: The Anthropology of Urban Form in Rome, Italy and the Ancient World*. MIT Press Published by arrangement with Princeton University Press, Princeton, New Jersey.
Saarikangasa, Kirsi (2006), 'Displays of the Everyday: Relations between gender and the visibility of domestic work in the modern Finnish kitchen from the 1930s to the 1950s'. *Gender, Place & Culture: A Journal of Feminist Geography* 13.2, pp. 161–72.
Sage, Colin (2003), 'Social embeddedness and relations of regard: Alternative 'good food'networks in south-west Ireland'. *Journal of Rural Studies* 19.1, pp. 47–60.
Sage, Colin (2010), 'Re-imagining the Irish foodscape'. *Irish Geography* 43.2, pp. 93–104.
Salcedo, Rodrigo and Torres, Alvaro (2004), 'Gated communities in Santiago: Wall or frontier?' *International Journal of Urban and Regional Research* 28.1, pp. 27–44.
Salvalaggio, Nantas (1984), *The Carnival of Venice*. Trans. Howard Roger MacLean. Milan: Amilcare Pizzicato.
Sandgren, Fredrik (2009), 'From "peculiar stores" to "a new way of thinking": Discussions on self-service in Swedish trade journals, 1935–1955'. *Business History* 51.5, pp. 734–53.

Santibanez, Roberto (2012), *Tacos, Tortas and Tamales: Flavours from the Griddles, Pots, and Street Side Kitchens of Mexico*. Hoboken, NJ: John Wiley.

Santich, Barbara (2004), 'The study of gastronomy and its relevance to hospitality education and training'. *International Journal of Hospitality Management* 23.1, pp. 15–24.

Sarasúa, Carmen and Scholliers, Peter (2005), 'The rise of a food market in European history', in Carmen, Sarasúa, Peter, Scholliers and Leen, Van Molle (eds), *Land, Shops and Kitchens. Technology and the Food Chain in Twentieth-Century Europe*. Turnhout: Brepols, pp. 13–29.

Sassatelli, Roberta and Davolio, Federica (2010), 'Consumption, pleasure and politics: Slow food and the politico-aesthetic problematization of food'. *Journal of Consumer Culture* 10.2, pp. 202–32.

Sayers, Dorothy (1935), *Gaudy Night*. London: Victor Gollancz.

Sbicca, Joshua (2012), 'Growing food justice by planting an anti-oppression foundation: opportunities and obstacles for a budding social movement'. *Agriculture and Human Values* 29.4, pp. 455–66.

Scarpato, Rosario (2002), 'Sustainable gastronomy as a tourist product', in Anne-Mette Hjalager and Greg Richard (eds), *Tourism and Gastronomy*. London: Routledge, pp. 132–52.

Scarpello, T., Poland, F., Lambert, N. and Wakeman, T. (2009), 'A qualitative study of the food-related experiences of rural village shop customers'. *Journal of Human Nutrition and Dietetics* 22.2, pp. 108–15.

Scharoun, L. (2012), *America at the Mall: The Cultural Role of a Retail Utopia*. Jefferson, NC: McFarland.

Schiller, Russell (1994), 'Vitality and viability: Challenge to the town centre'. *International Journal of Retail & Distribution Management* 22.6, pp. 46–50.

Schlosser, Eric (2002), *Fast Food Nation: What the All-American Meal is doing to the World*. UK: Penguin.

Schmiechen, James and Carls, Kenneth (1999), *The British Market Hall: A Social and Architectural History*. New Haven: Yale University Press.

Schofield, John (1987), *London*. London: Collins.

Scholderer, Joachim and Grunert, Klaus G. (2005), 'Consumers, food and convenience: The long way from resource constraints to actual consumption patterns'. *Journal of Economic Psychology* 26.1, pp. 105–28.

Schroeder, F. E. (1993), *Front Yard America: The Evolution and Meanings of a Vernacular Domestic Landscape*. Bowling Green State University: Popular Press.

Schuetz, Jenny (2013), 'Why are big box stores moving downtown?' USC price school of public policy. Federal Reserve System Community Development Research Conference. 11 April.

Schultz, Avalon and Sichley, Sara (Undated), 'Urban agriculture policy in San José'.

Schwentesius, Rita and Gómez, Manuel Ángel (2002), 'Supermarkets in Mexico: Impacts on horticulture systems'. *Development Policy Review* 20.4, pp. 487–502.

Scola, Roger (1975), 'Food markets and shops in Manchester 1770–1870'. *Journal of Historical Geography* 1.2, pp. 153–67.

Scola, Roger and Scola, Pauline (1992), *Feeding the Victorian City: The Food Supply of Manchester, 1770-1870*. Manchester: Manchester University Press.

Scruton, Roger (1987), 'Public space and the classical vernacular', in Nathan Glazer and Mark Lilla (eds), *The Public Face of Architecture*. London: The Free Press, pp. 13–25.
Seddon, George (December 1990), 'The Suburban Garden in Australia'. *Westerly* 4, pp. 5–13.
Segreto, L., Manera, C. and Pohl, M. (eds) (2009), *Europe at the Seaside: The Economic History of Mass Tourism in the Mediterranean*. Oxford: Berghahn Books.
Seiders, Kathleen., Simonides, Constantine and Tigert, Douglas J. (2000), 'The impact of supercenters on traditional food retailers in four markets'. *International Journal of Retail & Distribution Management* 28.4/5, pp. 181–93.
Selman, Paul (2002), 'Multi-function landscape plans: A missing link in sustainability planning?' *Local Environment* 7.3, pp. 283–94.
Senauer, Ben and Seltzer, Jon (2010), 'The changing face of food retailing'. *Choices* 25.4
Sennett, Richard (1974), *The Fall of Public Man*. Cambridge; London; Melbourne: Cambridge University Press.
Sennett, Richard (1994), *Flesh and Stone: The Body and the City in Western Civilization*. London: Faber.
Sennett, Richard (1987), 'The Public Domain', in Nathan Glazer and Mark Lilla (eds), *The Public Face of Architecture: Civic Culture and Public Spaces*. New York: The Free Press; London: Collier Macmillan, pp. 26–47.
Sered, Susan Starr (1988), 'Food and holiness: Cooking as a sacred act among Middle-Eastern Jewish women'. *Anthropological Quarterly* 61.3, pp. 129–39.
Serlio, Sebastiano (2001), *Sebastiano Serlio on Architecture*. Volume Two. Trans. Vaughan Hart and Peter Hicks. New Haven; London: Yale University Press.
Severson, Rebecca (1990), 'United we Sprout: A chicago community garden story', in Mark Francis and Randolph Hester (eds), *The Meaning of Gardens: Idea, Place and Action* Cambridge, MA: MIT Press, pp. 80–5.
Shaftoe, Henry (2012), *Convivial Urban Spaces: Creating Effective Public Places*. London: Routledge.
Shane, Grahame (2006), 'The emergence of landscape urbanism', in Charles, Waldheim (ed.), *The Landscape Urbanism Reader*. Princeton: Princeton Architectural Press, pp. 55–67.
Shapiro, Laura (2009), *Perfection Salad: Women and Cooking at the Turn of the Century* Berkeley, CA: University of California Press.
Shaw, Gareth, Curth, Louise and Alexander, Andrew (2004), 'Selling self-service and the supermarket: The Americanisation of food retailing in Britain, 1945–60'. *Business History* 46.4, pp. 568–82.
Shaw, Hillary J. (2006), 'Food deserts: Towards the development of a classification'. *Geografiska Annaler: Series B, Human Geography* 88.2, pp. 231–47.
Shaw, S., Bagwell, S. and Karmowska, J. (2004), 'Ethnoscapes as spectacle: Reimaging multicultural districts as new destinations for leisure and tourism consumption'. *Urban Studies* 41.10, pp. 1983–2000.
Sherwin, Chris, Tracy, Bhamra and Stephen, Evans (October 1998), 'The "eco-kitchen" project – using eco-design to innovate'. *The Journal of Sustainable Product Design*, The Centre for Sustainable Design, Issue 7, pp. 51–7.

Shields, R. (1989), 'Social spatialization and the built environment: The West Edmonton Mall'. *Environment and Planning D: Society and Space* 7, pp. 147–64.
Shirazi, Faegheh (2005) 'The sofreh: Comfort and community among women in Iran 1'. *Iranian Studies* 38.2, pp. 293–309.
Short, Anne, Guthman, Julie and Raskin, Samuel (2007), 'Food deserts, oases, or mirages? Small markets and community food security in the San Francisco bay area'. *Journal of Planning Education and Research* 26.3, pp. 352–64.
Short, Frances (2003), 'Domestic cooking skills – What are they?' *Journal of the HEIA* 10.3, pp. 13–22.
Short, Frances (2007), 'Cooking, convenience and dis-connection'. *Inter: A European Cultural Studies: Conference in Sweden 11–13 June 2007*.
Shorthose, Jim (2004), 'Nottingham's de facto Cultural Quarter: The lace market, independents and a convivial ecology', in David Bell and Mark Jayne (eds), *City of Quarters: Urban Villages in the Contemporary City*. Aldershot: Ashgate, pp. 149–62.
Shouse, Heather (2011), Food Trucks: Stories and Recipes from America's Best Kitchens on Wheels Ten Speed Press.
Shove, Elizabeth (2003), *Comfort, Cleanliness and Convenience: The Social Organization of Normality*. Oxford, New York: Berg.
Showell, D., Sundkvist, T., Reacher, M. and Gray, J. (2007), 'Norovirus outbreak associated with canteen salad in Suffolk, United Kingdom'. *Eurosurveillance Weekly Release* 12.48.
Shucksmith, Mark and Rønningen, Katrina (2011), 'The uplands after neoliberalism?–The role of the small farm in rural sustainability'. *Journal of Rural Studies* 27.3, pp. 275–87.
Sidali, K. L. (2011), 'A sideways look at farm tourism in Germany and in Italy', in *Food, Agriculture and Tourism*. Berlin, Heidelberg: Springer, pp. 2–24.
Sidali, K. L., Spiller, A. and Schulze, B. (eds) (2011), *Food, Agriculture and Tourism: Linking Local Gastronomy and Rural Tourism: Interdisciplinary Perspectives*. Heidelberg; Dordrecht; London; New York: Springer.
Sima, R., Micu, I., Maniutiu, D., Sima, N. and Lazar, V. (2010), 'Edible landscaping– integration of vegetable garden in the landscape of a private property'. *Bulletin UASVM Horticulture* 67.
Simeti, Mary Taylor (1991), *Pomp and Sustenance. Twenty-five Centuries of Sicilian Food*. New York: An Owl Book Henry Holt and Company.
Simmonds, I. G. (1993), *Cultural Constructs of the Environment*. London: Routledge.
Simmons, Dean and Chapman, Gwen E. (2012), 'The significance of home cooking within families'. *British Food Journal* 114.8, pp. 1184–95.
Simon, David, McGregor, Duncan and Thompson, Donald (2006), 'Contemporary perspectives on the peri-urban zones of cities in developing areas', in Duncan F. M. McGregor (ed.), *The Peri-urban Interface: Approaches to Sustainable Natural and Human Resource Use*. London: Earthscan, pp. 3–17.
Sims, Rebecca (2009), 'Food, place and authenticity: Local food and the sustainable tourism experience'. *Journal of Sustainable Tourism* 17.3, pp. 321–36.
Sitte, Camillo (1965), *City Planning According to Artistic Principles*, trans. George R. Collins and Christiane Crasemann Collins. London: Phaidon Press.
Smiley, David J. (2002), *Sprawl and Public Space: Redressing the Mall*. Washington, DC: National Endowment for the Arts; Princeton: Princeton Architectural Press.

Smith, Andrew and Sparks, Leigh (2000), 'The role and function of the independent small shop: The situation in Scotland'. *The International Review of Retail, Distribution and Consumer Research* 10.2, pp. 205–26.
Smith, Christopher M. and Kurtz, Hilda E. (2003), 'Community gardens and politics of scale in New York City'. *Geographical Review* 93.2, pp. 193–212.
Smith, C. and Morton, L. W. (2009), 'Rural food deserts: Low-income perspectives on food access in Minnesota and Iowa'. *Journal of Nutrition Education and Behavior* 41.3, pp. 176–87.
Smith, N. (2002), 'New globalism, new urbanism: Gentrification as global urban strategy'. *Antipode* 34.3, pp. 427–50.
Smoyer-Tomic, K. E., Spence, J. C. and Amrhein, C. (2006), 'Food deserts in the prairies? Supermarket accessibility and neighborhood need in Edmonton, Canada'. *The Professional Geographer* 58.3, pp. 307–26.
Smoyer-Tomic, K. E., Spence, J. C., Raine, K. D., Amrhein, C., Cameron, N., Yasenovskiy, V., ... and Healy, J. (2008), 'The association between neighborhood socioeconomic status and exposure to supermarkets and fast food outlets'. *Health and Place* 14.4, pp. 740–54.
Sobal, J. and Wansink, B. (2007), 'Kitchenscapes, tablescapes, platescapes, and foodscapes influences of microscale built environments on Food Intake'. *Environment and Behavior* 39.1, pp. 124–42.
Soja, Edward (1989), *Post Modern Geographies: The Reassertion of Space on Critical Social Theory*. London, New York: Verso.
Soja, Edward (2000), *Postmetropolis: Critical Studies of Cities and Regions*. Oxford, Malden, MA: Blackwell.
Soler, Ivette (2011), *The Edible Front Yard: The Mow-less, Grow-more Plan for a Beautiful, Bountiful Garden*. Portland, Oregon, London: Timber.
Sonnino, R. (2009), 'Quality food, public procurement, and sustainable development: The school meal revolution in Rome'. *Environment and Planning A* 41.2, pp. 425–40.
Sonnino, R. and Marsden, T. (2006), 'Beyond the divide: Rethinking relationships between alternative and conventional food networks in Europe'. *Journal of Economic Geography* 6.2, pp. 181–99.
Soriano, Domingo Ribeiro (2002), 'Customers' expectations factors in restaurants: The situation in Spain'. *International Journal of Quality & Reliability Management* 19.8/9, pp. 1055–67.
Sorkin, M. (ed.) (1992), *Variations on a Theme Park: The New American City and the End of Public Space*. New York: Hill and Wang.
Southworth, Michael (2005), 'Reinventing main street: From mall to townscape mall'. *Journal of Urban Design* 10.2, pp. 151–70.
Southworth, Michael and Ben-Joseph, Eran (2003), *Streets and the Shaping of Towns and Cities*. Washington, DC; London: Island Press.
Southworth, Michael and Owens, Peter M. (1993), 'The evolving metropolis: Studies of community, neighborhood, and street form at the urban edge'. *Journal of the American Planning Association* 59.3, pp. 271–87.
Spang, Rebecca L. (2000), *The Invention of the Restaurant: Paris and Modern Gastronomic Culture* (vol. 135). Cambridge, MA: Harvard University Press.

Spencer, Colin (1996), *The Heretic's Feast: A History of Vegetarianism*. London: Fourth Estate.
Spencer, Colin (2002), *British Food: An Extraordinary Thousand Years of History*. London: Columbia University Press.
Spens, Erin and Gilland, Christine (2012), 'Top 10 London street foods', http://www.guardian.co.uk/travel/2012/apr/24/top-10-london-street-food-stalls. Accessed 12 January 2013.
Spilková, J. and Fialová, D. (2013), 'Culinary tourism packages and pegional brands in Czechia'. *Tourism Geographies* 15.2, pp. 177–97.
Spilková, Jana and Šefrna, Luděk (2010), 'Uncoordinated new retail development and its impact on land use and soils: A pilot study on the urban fringe of Prague, Czech Republic'. *Landscape and Urban Planning* 94.2, pp. 141–8.
Spirn, Anne Whiston (1984), *The Granite Garden: Urban Nature and Human Design*. New York: Basic Books.
Square Meal (2013), 'London's 10 best pop ups', http://www.squaremeal.co.uk/feature/londons-10-best-pop-up-restauran/7084. Retrieved 21 January, 2013.
Srivastava, R. K. (2008), 'Changing retail scene in India'. *International Journal of Retail & Distribution Management* 36.9, pp. 714–21.
Staeheli, Lynn A., Mitchell, Don and Gibson, Kristina (2002), 'Conflicting rights to the city in New York's community gardens'. *GeoJournal* 58.2, pp. 197–205.
Stafford, M., Cummins, S., Ellaway, A., Sacker, A., Wiggins, R. D. and Macintyre, S. (2007) 'Pathways to obesity: Identifying local, modifiable determinants of physical activity and diet'. *Social Science & Medicine* 65.9, pp. 1882–97.
Stanhill, Gerald (1977), 'An Urban Agro-Ecosystem: The example of nineteenth century Paris'. *Agro-Ecosystems* 3, pp. 269–84.
Stanilov, Kiril (2004), 'Planning for sprawl – The evolution of a regional shopping center', in Brenda Case Scheer and Kiril Stanilov (eds), *Suburban Form: An International Perspective*. London: Routledge.
Steiner, Frederick (2011,) 'Landscape ecological urbanism: Origins and trajectories'. *Landscape and Urban Planning* 100.4, pp. 333–7.
Stevenson, Deborah (2003), *Cities and Urban Cultures*. Maidenhead: Open University Press.
Stevenson, G. and Born, H. (2007), 'The "red label" poultry system in France: Lessons for renewing an agriculture-of-the-middle in the United States', in Hinrichs, C. C. and Lyson, T. A. (eds), *Remaking the North American Food System: Strategies for Sustainability*, University of Nebraska Press, pp. 144–62.
Stilgoe, John (1988), *Borderlands: Origins of the American Suburb*. Borderland; New Haven, CT; London: Yale University.
Stobart, Jon, Hamm, Andrew and Morgan, Victoria (2007), *Spaces of Consumption. Leisure and Shopping in the English Town, c1689–1830*. London, New York: Routledge.
Stone, K. E. (1997), 'Impact of the Wal-Mart phenomenon on rural communities: Increasing understanding of public problems and policies', Published in Proceedings 'Increasing Understanding of Public Problems and Policies – 1997' By Farm Foundation, Chicago, Illinois, pp. 1–22.
Storper, Michael (1997a), *The Regional World: Territorial Development in a Global Economy*. New York, London: Guilford.

Storper, Michael (1997b), 'Regional economies as relational assets', in R. Lee and J. Wills (eds), *Geographies of Economies*. London: Edward Arnold, pp. 248–58.
Storper, Michael and Walker, R. (1989), *The Capitalist Imperative Territory, Technology and Industrial Growth*. New York/Oxford: Basil: Blackwell.
Stren, Richard, White, Rodney and Whitney, Joseph (eds) (1992), *Sustainable Cities, Urbanisation and the Environment in International Perspective*. Boulder, CO: Westview Press.
Suárez Carrasquillo, Carlos A. (2011), 'Gated communities and city marketing: Recent trends in Guaynabo, Puerto Rico'. *Cities* 28.5, pp. 444–51.
Sudjic, Deyan (1991), *The 100 Mile City*. London: Andre Deutsch.
Suen, Wong Hong (2007), 'A taste of the past: Historically themed restaurants and social memory in Singapore', in Cheung, Sydney and Che-Beng Tan (eds), *Food and Foodways in Asia: Resource, Tradition and Cooking*. London: Routledge, pp. 115–28.
Sun, J. and Chen, X. (2007), 'When local meets global: Residential differentiation, global connections, and consumption in Shanghai', in F. Wu (ed.), *China's Emerging Cities: The Making of New Urbanism*. London: Routledge, pp. 284–302.
Super, John C. (2002), 'Food and history'. *Journal of Social History* 36.1, pp. 165–78.
Supski, S. (2006), '"It Was Another Skin": The kitchen as home for Australian post-war immigrant women'. *Gender, Place & Culture* 13.2, pp. 133–41.
Sustain (2012), *Good Food for London 2012 – London Borough Maps of Progress on Healthy and Sustainable Food*. http://www.sustainweb.org/publications/?id=249.
Sustainable Development Commission (2008), *Green, Healthy and Fair: A Review of Government's Role in Supporting Sustainable Supermarket Food*. London: SDC.
Sustainable Development Commission (2011), *Looking Back, Looking Forward: Sustainability and UK Food Policy 2000–2011*. London: SDC.
Sutton, David and Vournelis, Leonidas (2009), 'Vefa or Mamalakis: Cooking up Nostalgia in contemporary Greece'. *South European Society and Politics* 14.2, pp. 147–66.
Swinburn, B. and Egger, G. (2002), 'Preventive strategies against weight gain and obesity'. *Obesity Reviews* 3.4, pp. 289–301.
Swinburn, B., Egger, G. and Raza, F. (1999), 'Dissecting obesogenic environments: The development and application of a framework for identifying and prioritizing environmental interventions for obesity'. *Preventive Medicine* 29, pp. 563–70.
Swyngedouw, E. and Heynen, N. C. (2003), 'Urban political ecology, justice and the politics of scale'. *Antipode* 35.5, pp. 898–918.
Symons, Michael (1982), *One Continuous Picnic: A History of Eating in Australia*. Adelaide: Duck Press.
Tachieva, Galina (2010), *Sprawl Repair Manual*. Washington, DC: Island Press.
Tacoli, C. (1998), 'Rural-urban Interactions: A Guide to the Literature'. *Environment and Urbanization* 10, pp. 147–66.
Talen, Emily (2002), 'Help for urban planning: The transect strategy'. *Journal of Urban Design* 7.3, pp. 293–312.
Talen, Emily (2005), *New Urbanism and American Planning: The Conflict of Cultures*. New York; London: Routledge.
Talen, Emily (2008), *Design for Diversity: Exploring Socially Mixed Neighbourhoods*. Architectural Press, Amsterdam: London: Elsevier.

Talen, Emily, Bohl, Charles and Hardy, Matthew (2008), 'Statement of Journal aims'. *Journal of Urbanism: International Research on Placemaking and Urban Sustainability* 1.1

Talukdar, Debabrata (2008), 'Cost of being poor: Retail price and consumer price search differences across inner-city and suburban neighborhoods'. *Journal of Consumer Research* 35.3, pp. 457–71.

Tanenbaum, Jason (2012), *Regulating Mobile Food Vending in Greenville, SC*. Diss. Clemson University.

Tansey, Geoff and Worsley, Tony (1995), *The Food System: A Guide*. London: Earthscan.

Tanskanen, Kari, Yrjölä, Hannu and Holmström, Jan (2002), 'The way to profitable internet grocery retailing–six lessons learned'. *International Journal of Retail & Distribution Management* 30.4, pp. 169–78.

Taskforce on Regional Development (1994), 'Developing Australia: A regional Perspective'. A Report to the Federal Government by the Taskforce on Regional Development, vol. 1, Canberra, Australia.

Tauger, Mark (2011), *Agriculture in World History*. London: Routledge.

Teaford, Jon C. (1997), *Post-suburbia: Government and Politics in the Edge Cities*. Baltimore, MD: Johns Hopkins University Press.

Teaford, Jon C. (2011), 'Suburbia and Post-Suburbia: A Brief History', in Nicholas, A. Phelps and Fulong Wu (eds), *International Perspectives on Suburbanization*. Basingstoke: Palgrave Macmillan.

Tegegne, Azage, Mekasha, Yoseph, Tadesse, Million and Yami, Alemu (2006), 'Market-oriented urban and peri-urban dairy systems', in René van Veenhuizen (ed.), *Cities Farming for the Future*. The Philippines: International Institute of Rural Reconstruction and ETC Urban Agriculture.

te Lintelo, Dolf J. H. (2009), 'The spatial politics of food hygiene: Regulating small-scale retail in Delhi'. *European Journal of Development Research* 21, pp. 63–80.

Thailand Food Forums (2013), http://thailand-uk.com/forums/showthread.php?8039-Asian-food-stalls-in-Wembley&s=08717147037309723a7fda81f3e7ecc3.

The gentle author (2013), 'So Long, Tubby Isaac's Jellied Eel Stall'. http://spitalfieldslife.com/2013/06/13/so-long-tubby-isaacs-jellied-eel-stall/.

Thierolf, Andrew (2012), 'The Nebraska sandhills food desert: Causes, identification, and actions towards a resolution'. Thesis. University of Nebraska.

Thomas, B. (2010), 'Food deserts and the sociology of space: Distance to food retailers and food insecurity in an urban American neighborhood'. *International Journal of Human and Social Sciences* 5.6, pp. 400–9.

Thompson, C. J. and Arsel, Z. (2004a), 'The starbucks brandscape and consumers' (anticorporate) experiences of glocalization'. *Journal of Consumer Research* 31.3, pp. 631–42.

Thompson, C. J. and Arsel, Z. (2004b), 'The starbucks brandscape and the discursive mapping of local coffee shop cultures'. *Journal of Consumer Research* 31, pp. 631–42.

Thoms, Ulrike (2003), 'Industrial canteens in Germany', in Marc Jacobs and Peter Scholliers (eds), *Eating Out in Europe: Picnics, Gourmet Dining and Snacks since the Late Eighteenth Century*. Oxford: Berg, pp. 351–72.

Thornton, Alexander (2008), 'Beyond the metropolis: Small town case studies of urban and peri-urban agriculture in South Africa'. *Urban forum* 19. 3, Netherlands: Springer, pp. 243–62.
Thornton, L. and Kavanagh, A. (2010), 'The local food environment and obesity', in J. Pearce and K. Witten (eds), *Geographies of Obesity: Environmental Understandings of the Obesity Epidemic*. Surrey: Ashgate.
Thrift, N. (2004), 'Driving in the city'. *Theory, Culture & Society* 21.4–5, pp. 41–59.
Thrift, N. and Glennie, P. (1993), 'Historical geographies of urban life and modern consumption', in G. Kearns and C. Philo (eds), *Selling places: The City as Cultural Capital. Past and Present*. Oxford: Pergamon, pp. 33–48.
Tibbalds, Francis (ed.) (2012), *Making People-friendly Towns: Improving the Public Environment in Towns and Cities*. Harlow: Taylor & Francis.
Tierney, R. Kenji and Ohnuki-Tierney, Emiko (2012), 'Anthropology of food', in Jeffrey M. Pilcher (ed.), *The Oxford Handbook of Food History*. Oxford, New York: Oxford University Press.
Tilleray, Brigitte (1995), *Recipes from the French Kitchen Garden*. London: Cassell.
Timothy, D. J. and Ron, A. S. (2013), 'Heritage cuisines, regional identity and sustainable tourism', in C. Michael Hall and Stefan Gossling (eds), *Sustainable Culinary Systems: Local Foods, Innovation, Tourism and Hospitality*. London, New York: Routledge.
Tinker, Irene (1997), *Street foods: Urban Food and Employment in Developing Countries*. New York; Oxford: Oxford University Press.
Tjallingii, Sybrand P. (2000), 'Ecology on the edge: Landscape and ecology between town and country'. *Landscape and Urban Planning* 48.3, pp. 103–19.
Tokatli, Nebahat and Boyaci, Yonca (1999), 'The changing morphology of commercial activity in Istanbul'. *Cities* 16.3, pp. 181–93.
Tomasik, T. J. (2001), 'Certeau a la Carte – Translating discursive terroir in the practice of everyday life: Living and cooking'. *The South Atlantic Quarterly* 100.2, pp. 519–42.
Tomlinson, Mark (1998), 'Changes in Taste in Britain, 1985–1992'. *British Food Journal* 100.6, pp. 295–301.
Tornaghi, Chiara (2012), 'Public space, urban agriculture and the grassroots creation of new commons: Lessons and challenges for policy makers', in Andre Viljoen and Johannes S. C. Wiskerke (eds), *Sustainable Food Planning: Evolving Theory and Practice*. The Netherlands: Wageningen Academic Publishers, pp. 349–63.
Torres, Haroldo., Alves, Humberto and Maria Aparecida De Oliveira (2007), 'São Paulo peri-urban dynamics: Some social causes and environmental consequences'. *Environment and Urbanization* 19.1, pp. 207–23.
Townshend, T. and Lake, A. A. (2009), 'Obesogenic urban form: Theory, policy and practice'. *Health & Place* 15.4, pp. 909–16.
Tracey, David (2007), *Guerrilla Gardening: A Manifesto*. Gabriola Island, British Columbia, Canada: New Society Publishers.
Trancik, Roger (1986), *Finding Lost Space: Theories of Urban Design*. New York: Van Nostrand Reinhold.
Trefon, Theodore (2009), 'Hinges and fringes: Conceptualising the peri-urban', in Francesca Locatelli and Paul Nugent (eds), *Central Africa in African Cities: Competing Claims on Urban Spaces* Leiden, Konicklijke: Brill.

Treib, Marc (1994), 'Underground in umeda', in Zeynep Çelik, Diane Favro and Richard Ingersoll (eds), *Streets: Critical Perspectives on Public Space*. Berkeley, CA; London: University of California Press.

Trevelyan, G. M. (2007), *English Social History: A Survey of Six Centuries-Chaucer to Queen Victoria*. Ghose Press.

Trubek, A. B. and Bowen, S. (2008), 'Creating the taste of place in the United States: Can we learn from the French?'. *GeoJournal* 73.1, pp. 23–30.

Trubek, Amy, Kolleen M. Guy and Bowen, Sarah (2010), 'Terroir: A French conversation with a transnational future'. *Contemporary French and Francophone Studies* 14.2, pp. 139–48.

Tudge, Colin (2002), *Food for the Future*. London: Dorling Kindersley.

Tulloh, Jojo (2011), *East End Paradise: Kitchen Garden Cooking in the City*. Photographs by Jason Lowe, engravings by Andy English. London: Vintage Books.

Turner, Bethany (2011), 'Embodied connections: Sustainability, food systems and community gardens'. *Local Environment* 16.6, pp. 509–22.

Turner, Tom (2005), *Garden History: Philosophy and Design 2000 BC-2000 AD*. London: Spon Press.

Tway, Timothea Larisa (2011), 'Roving restaurants: Mobile food vendors at the intersection of public space and policy'. Master's Thesis. Cal Poly. San Luis Obispo.

Urban Taskforce (1999), *Towards an Urban Renaissance*. London: DETR.

Urry, J. and Larsen, J. (2011), *The Tourist Gaze 3.0*. Los Angeles: Sage Publications Limited.

Valentine, Gill (1998), 'Food and the production of the civilised street', in Nicholas R. Fyfe (ed.), *Images of the Street: Planning, Identity, and Control in Public Space*. London, New York: Routledge.

Valentine, Gill (2001), 'Eating in: home, consumption and identity'. *The Sociological Review* 47.3, pp. 491–524.

Vall-Casas, P., Koschinsky, J. and Mendoza, C. (2011), 'Retrofitting suburbia through pre-urban patterns: Introducing a European perspective'. *Urban Design International* 16.3, pp. 171–87.

Vallianatos, M., Gottlieb, R. and Haase, M. A. (2004), 'Farm-to-school strategies for urban health, combating sprawl, and establishing a community food systems approach'. *Journal of Planning Education and Research* 23.4, pp. 414–23.

van Caudenberg, Anke and Heynen, Hilde (2004), 'The rational kitchen in the interwar period in Belgium: Discourses and realities'. *Home Cultures* 1.1, 1 March, pp. 23–50(28).

Van Esterik, Penny (1982), 'Celebrating ethnicity: Ethnic flavour in an urban festival'. *Ethnic, Groups* 4.4, pp. 207–27.

Van Leeuwen, C. and Seguin, G. (2006), 'The concept of terroir in viticulture'. *Journal of Wine Research* 17.1, pp. 1–10.

van Odijk, Tom (2014), 'Casablanca – a new sustainable market square'. *Urban Design* 129, Winter 2014, pp. 28–30.

van Otterloo, A. H. (2000), 'The rationalization of kitchen and cooking 1920–1970'. *Journal for the Study of Food and Society* 4.1, Spring 2000, pp. 19–26(8).

van Otterloo, A. H. (2005), 'Fast food and slow food: The fastening food chain and recurrent countertrends in Europe and the Netherlands (1890–1990)', in Carmen Sarasúa, Peter Scholliers and Leen van Molle (eds), *Land, Shops and Kitchens: Technology and the Food Chain in Twentieth-century Europe*. Turnhout: Brepols, pp. 255–78.

Vecchio, Riccardo (2009), 'European and United States farmers' markets: Similarities, differences and potential developments', Paper prepared for presentation at the 113th EAAE Seminar *A Resilient European Food Industry and Food Chain in a Challenging World*, Chania, Crete, Greece, 3–6 September 2009.

Viljoen, A., Bohn, K. and Howe, J. (eds) (2005), *Continuous Productive Urban Landscapes (CPULs): Designing Urban Agriculture for Sustainable Cities*. Oxford: Architectural Press.

Viljoen, Andre and Wiskerke, Johannes S. C. (eds) (2012), *Sustainable Food Planning: Evolving Theory and Practice*. The Netherlands: Wageningen Academic Publishers.

Villavicencio, Luis Valdonado (2009), 'Urban agriculture as a livelihood strategy in Lima, Peru', in Mark, Redwood (ed.) *Agriculture in Urban Planning; Generating Livelihoods and Food Security*. Abingdon: Routledge.

Visser, Margaret (1987), *Much Depends on Dinner: The Extraordinary History and Mythology, Allure and Obsessions, Perils and Taboos, of an Ordinary Meal*. Toronto: McClelland and Stewart.

Visser, Margaret (1993), *The Rituals of Dinner: The Origins, Evolution, Eccentricities, and Meaning of Table Manners*. London: Penguin.

Vitiello, Domenic (2008), 'Growing edible cities', in E. Birch and S. Wachter (eds) *Growing Greener Cities: Urban Sustainability in the Twenty-first Century*. Philadelphia: University of Pennsylvania Press, pp. 259–78.

Vix (2008), http://www.alotofgaul.blogspot.co.uk/2008/12/decline-of-french-cafe.html. Retrieved 21 January, 2013.

Vlach, Anna (2013), 'Monaco Matters' *Sunday Mail Home Magazine*. November 10: 4.

von der Dunk, Andreas et al. (2011), 'Defining a typology of peri-urban land-use conflicts–A case study from Switzerland'. *Landscape and Urban Planning* 101.2, pp. 149–56.

Von Henneberg, Krystyna (1994) 'Piazza Castello and the making of a Fascist colonial capital', in Zeynep, Çelik, Diane Favro and Richard Ingersoll (eds), *Streets: Critical Perspectives on Public Space*. Berkeley, CA; London: University of California Press.

Von Holy, A. and Makhoane, F. M. (2006), 'Improving street food vending in South Africa: Achievements and lessons learned'. *International Journal of Food Microbiology* 111.2, pp. 89–92.

Von Stackelberg, K. T. (2009), *The Roman Garden: Space, Sense, and Society*. London: Routledge.

Voyce, M. (2006), 'Shopping malls in Australia: The end of public space and the rise of "consumerist citizenship"?'. *Journal of Sociology* 42.3, pp. 269–86.

Voyce, M. (2007), 'Shopping malls in India: New social "Dividing practices"'. *Economic and Political Weekly* 42.22, pp. 2055–62.

Wackernagel, M. and Rees, W. (1996), *Our Ecological Footprint: Reducing Human Impact on the Earth*. Gabriola: New Society Publishers.

Wakefield, Sarah, Yeudall, Fiona, Taron, Carolin, Reynolds, Jennifer and Skinner, Ana (2007), 'Growing urban health: Community gardening in South-East Toronto'. *Health Promotion International* 22.2, pp. 92–101.
Waldheim, Charles (ed.) (2006), *The Landscape Urbanism Reader*. New York: Princeton Architectural Press.
Waldheim, Charles (2010), 'On landscape, ecology and other modifiers to urbanism'. *Topos* 71, pp. 20–24.
Waley, Daniel (1969), *The Italian City Republics*. London: Weidenfeld and Nicholson.
Walker, Richard and Lewis, Robert D. (2001), 'Beyond the crabgrass frontier: Industry and the spread of North American cities, 1850–1950'. *Journal of Historical Geography* 27.1, pp. 3–19.
Walser, Robin (2005), *Taylorism and Social Class: The Modernist Kitchen Designs of Margarete Schütte-Lihotzky, Eileen Gray, and Charlotte Perriand*. Saint Paul, Minnesota: University of St. Thomas.
Walsh, E., Babakina, O., Pennock, A., Shi, H., Chi, Y., Wang, T. and Graedel, T. E. (2006), 'Quantitative guidelines for urban sustainability'. *Technology in Society* 28.1, pp. 45–61.
Walsh, R. A., Paul, C. L., Tzelepis, F., Stojanovski, E. and Tang, A. (2008), 'Is government action out-of-step with public opinion on tobacco control? Results of a New South Wales population survey'. *Australian and New Zealand Journal of Public Health* 32.5, pp. 482–8.
Wang, S. and Guo, C. (2007), 'A tale of two cities: Restructuring of retail capital', in Wu Fulong (ed.) *China's Emerging Cities*. London: Routledge, pp. 256–83.
Wang, M. C., Cubbin, C., Ahn, D. and Winkleby, M. A. (2008), 'Changes in neighbourhood food store environment, food behaviour and body mass index, 1981–1990'. *Public Health Nutrition* 11.09, pp. 963–70.
Warde, Alan (1997), *Consumption, Food and Taste: Culinary Antinomies and Commodity Culture*. London: Sage.
Warde, Alan (1999), 'Convenience food: Space and timing'. *British Food Journal* 101.7, pp. 518–27.
Warde, Alan and Martens, L. (2000), *Eating Out: Social Differentiation, Consumption and Pleasure*. Cambridge: Cambridge University Press.
Warner, J., Talbot, D. and Bennison, G. (2013), 'The cafe as affective community space: Reconceptualizing care and emotional labour in everyday life'. *Critical Social Policy* 33.2, pp. 305–24.
Warwick, Hugh and Doig, Alison (2004), *Smoke – The Killer in the Kitchen. Indoor Air Pollution in Developing Countries*. London: Report, ITDG Publishing.
Watson, Sophie (2009), 'The magic of the marketplace: Sociality in a neglected public space'. *Urban Studies* 46, pp. 1577–91.
Watson, S. and Wells, K. (2005), 'Spaces of nostalgia: The hollowing out of a London market'. *Journal of Social and Cultural Geography* 6.1, pp. 17–30.
Watts, David C. H., Ilbery, Brian and Maye, Damian (2005), 'Making reconnections in agro-food geography: Alternative systems of food provision'. *Progress in Human Geography* 29.1, pp. 22–40.
Weaver, D. (2001), 'Ecotourism as mass tourism: Contradiction or reality?' *The Cornell Hotel and Restaurant Administration Quarterly* 42.2, pp. 104–12.

Webber, Melvin M. (1964), 'The urban place and the non-place urban realm', in M. M. Webber, et al. (eds), *Exploration into Urban Structure*. Philadelphia: University of Pennsylvania Press, pp. 79–153.

Weber, David (2012), *The Food Truck Handbook: Start, Grow, and Succeed in the Mobile Food Business*. Hoboken New Jersey: John Wiley and Sons.

Webster, R. (ed.) (2002), *Expanding Suburbia: Reviewing Suburban Narratives* (vol. 6). New York; Oxford: Berghahn Books.

Welch, R. M. and Graham, R. D. (1999), 'A new paradigm for world agriculture: Meeting human needs: Productive, sustainable, nutritious'. *Field Crops Research* 60.1, pp. 1–10.

Weller, Richard (2008), 'Landscape (sub) urbanism in theory and practice'. *Landscape Journal* 27.2, pp. 247–67.

Weisskoppel, Cordula (2004), 'A Sudanese snack bar in Berlin: Vitalization and presence in the diaspora'. *Between Resistance and Expansion: Explorations of Local Vitality in Africa* 18, p. 91.

Wharton, Edith (1908), *A Motor-flight Through France*. London: Picador, 1995.

Whatmore, S. and Thorne, L. (1997), 'Nourishing networks: Alternative geographies of food', in D. Goodman and M. Watts (eds), *Globalising Food: Agrarian Questions and Global Restructuring*. London: Routledge, pp. 287–304.

Wheeler, S. M. (2002) 'The new regionalism: Key characteristics of an emerging movement'. *Journal of the American Planning Association* 68.3, pp. 267–78.

Whelan, Amanda, Wrigley, Neil, Warm, Daniel and Cannings, Elizabeth (2002), 'Life in a "Food Desert"'. *Urban Studies* 39.11, pp. 2083–2100.

Whitaker, Ellen, Mahoney, Colleen and Jordan, Wendy Adler (2001), *Great Kitchens: Design Ideas from America's Top Chefs*. Newtown, CT: The Taunton Press.

White, M. (2007), 'Food access and obesity'. *Obesity Reviews* 8.s1, pp. 99–107.

Whitehand, J. W. and Carr, C. M. (2001). *Twentieth-century Suburbs: A Morphological Approach* (vol. 1). London: Routledge.

Whitehand, Jeremy and Larkham, Peter (1992), *Urban Landscapes: International Perspectives*. London, New York: Routledge.

Whitehead, Margaret (1998), 'Food deserts: What's in a name?' *Health Education Journal* 57.3, pp. 189–90.

Whitelegg, Drew (2002), 'From market stalls to restaurant row: The recent transformation of Exmouth Market'. *London Journal* 27.2, pp. 1–11.

Whyte, William H. (1980), *The Social Life of Small Urban Spaces*. Washington DC: The Conservation Foundation.

Whyte, William H. (1988), *City: Rediscovering the Centre*. New York: Doubleday.

Willan, Anne (1991), *La France Gastronomique*. UK: Pavilion.

Williams, Jacqueline, Martin, Paul and Stone, Christopher (2010), 'Using ecosystem services as a means to diffuse political land use decisions in peri-urban regions'. *CRC for Irrigation Futures*, Australia.

Wills, Wendy (2011), 'Introduction to food: Representations and meanings'. *Sociological Research Online* 16.2, p. 16.

Willson, Jane (2010), 'Market power', *The Sydney Morning Herald*, 29 June 2010. Accessed 23 December 2013, http://www.smh.com.au/small-business/managing/market-power-20100629-zgi1.h.

Wilkins, John M. and Hill, Shaun (2006), *Food in the Ancient World*. Maiden: Blackwell.
Wilson, Elizabeth (1991), *The Sphinx in the City*. London: Virago Press.
Winter, Michael (2003) 'Embeddedness, the new food economy and defensive localism'. *Journal of Rural Studies* 19.1, pp. 23–32.
Wirth, L. (1938), 'Urbanism as a Way of Life'. *American Journal of Sociology*, pp. 1–24.
Wolfert, Paula (1983), *The Cooking of South West France*. London: Papermac Macmillan.
Wood, M. E. (2002), *Ecotourism: Principles, Practices and Polices for Sustainability*. UNEP.
Woolf, Virginia (1984), *A Room of One's Own London, and, Three Guineas*, with an introduction by Hermione Lee. Chatto & Windus; London: Hogarth Press.
Worden, Suzette (1989), 'Powerful women: Electricity in the home', in Judy Attfield and Pat Kirkham (eds), *A View from the Interior: Feminism, Women and Design*. London: The Women's Press.
'Working at half-potential: Constructive analysis of home garden programmes in the Lima slums with suggestions for an alternative approach'. (1985), *Food and Nutrition Bulletin* 7.3. Vera Ni International Potato Centre, Lima, Peru.
Wright, Frank Lloyd (1935), *Broadacre City: A New Community Plan*. Architectural Record Publishing Company.
Wright, Gwendolyn (1980), *Moralism and the Model Home: Domestic Architecture and Cultural Conflict in Chicago 1873–1913*. Chicago: University of Chicago Press.
Wright, Gwendolyn (1981), *Building the Dream: A Social History of Housing in America*. New York: Pantheon Books.
Wright, L., Hickson, M. and Frost, G. (2006), 'Eating together is important: Using a dining room in an acute elderly medical ward increases energy intake'. *Journal of Human Nutrition and Dietetics* 19.1, pp. 23–6.
Wrigley, N. and Dolega, L. (2011), 'Resilience, fragility, and adaptation: New evidence on the performance of UK high streets during global economic crisis and its policy implications'. *Environment and Planning-Part A* 43.10, p. 2337.
Wrigley, N., Branson, J., Murdock, A. and Clarke, G. (2009), 'Extending the Competition Commission's findings on entry and exit of small stores in British high streets: Implications for competition and planning policy'. *Environment and Planning A* 41.9, p. 2063.
Wrigley, N., Warm, D. and Margetts, B. (2003), 'Deprivation, diet, and food-retail access: Findings from the Leeds "food deserts" study'. *Environment and Planning A* 35.1, pp. 151–88.
Wu, Fulong (2010), 'Gated and packaged suburbia: Packaging and branding Chinese suburban residential development'. *Cities: The International Quarterly on Urban Policy* 27.5, pp. 385–96.
Yan, Yunxiang (2000), 'Of hamburger and social space: Consuming McDonald's in Beijing', in Carole Counihan and Penny Van Esterik (eds), *Food and Culture: A Reader*. London: Routledge, pp. 449–71.
Yasmeen, Gisèle (1996), '"Plastic-bag housewives" and postmodern restaurants? public and private in Bangkok's foodscape'. *Urban Geography* 17.6, pp. 526–44.

Yasmeen, Gisèle (2000), 'Not "From Scratch": Thai food systems and "public eating"'. *Journal of Intercultural Studies* 21.3, pp. 341–52.

Yasmeen, Gisèle (2006), *Bangkok's Foodscape: Public Eating, Gender Relations, and Urban Change*. Bangkok: White Lotus Press.

Yegul, Fikret (1994), 'The street experience of ancient Ephesus', in Zeynep, Çelik, Diane Favro and Richard Ingersoll (eds), *Streets: Critical Perspectives on Public Space*. Berkeley, CA: University of California Press.

Yerbury, F. R. (ed.) (1947), *Modern Homes Illustrated*. London: Odhams Press.

Yrjo, Hannu (2001), 'Physical distribution considerations for electronic grocery shopping'. *International Journal of Physical Distribution & Logistics Management* 31.10, pp. 746–61.

Yurtseven, H. R. and Kaya, O. (2011), 'Slow tourists: A comparative research based on Cittàslow principles'. *American International Journal of Contemporary Research* 1.2, pp. 91–8.

Yurtseven, H. R. and Karakas, N. (2013), 'Creating a sustainable gastronomic destination: The case of Cittàslow Gokceada-Turkey'. *American International Journal of Contemporary Research* 3.3, March, pp. 91–100.

Zandstra, David L. (2004), 'In the city, but not of the city: Dutch truck farmers in the Calumet region', in Robert P. Swierenga, Donald Sinnema and Hans Krabbendam (eds), *The Dutch in America*, The association for the advancement of Dutch American studies. Fourtenth Biennial Conference Papers.

Zeiderman, Austin (2008), 'Cities of the future? megacities and the space/time of urban modernity'. *Critical Planning*, Summer, pp. 23–39.

Zhao, P. (2010), 'Sustainable urban expansion and transportation in a growing megacity: Consequences of urban sprawl for mobility on the urban fringe of Beijing'. *Habitat International* 34.2, pp. 236–43.

Zucker, Paul (1959), *Town and Square: From the Agora to the Village Green*. New York: Columbia University Press.

Zukin, Sharon (1982), *Loft Living: Culture and Capital in Urban Change*. New Brunswick, NJ: Rutgers University Press.

Zukin, Sharon (1991), *Landscapes of Power: From Detroit to Disney World*. Berkeley, CA: University of California Press.

Zukin, Sharon (1992), 'Postmodern urban landscapes: Mapping culture and power', in S. Lash and J. Friedman (eds), *Modernity and Identity*. London: Blackwell.

Zukin, Sharon (1995), *The Cultures of Cities*. Cambridge, MA: Oxford: Blackwell.

Zukin, Sharon (1998), 'Urban lifestyles: Diversity and standardisation in spaces of consumption'. *Urban studies* 35.5–6, pp. 825–39.

Zukin, Sharon (2004), *Point of Purchase: How Shopping Changed American Culture*. New York, London: Routledge.

Zukin, Sharon (2008), 'Consuming authenticity'. *Cultural Studies* 22.5, pp. 724–48.

Zukin, Sharon (2010), *Naked City: The Death and Life of Authentic Urban Places*. New York; Oxford: Oxford University Press.

Zukin, Sharon and Maguire, Jennifer S. (2004), 'Consumers and consumption'. *Annual Review of Sociology* 30, pp. 173–310.

Zweiniger-Bargielowska, I. (2000), *Austerity in Britain: Rationing, Controls, and Consumption, 1939-1955: Rationing, Controls, and Consumption, 1939-1955*. Oxford: University Press.

http://www.antiquehomestyle.com/plans/montgomery-ward/1930/30mw-hillcrest.htm. Accessed 2 December 2012.

http://www.antiquehomestyle.com/plans/sterling/1916-sterling/16sterling-browningb.htm. Accessed 2 December 2012.

http://www.archinect.com/blog/article/54676637/10-in-copenhagen-the-hottest-thing-since-sliced-rugbr-d. Accessed 27 August 2013.

http:// www. bettercities.net/news-opinion/blogs/kaid-benfield/20650/welcome-age-anti-mall-maybe. Accessed 17 January 2014.

http://www.burrensmokehouse.ie/blog/blog/food-tourism-and-eco-tourism. Accessed 5 January 2014.

http://www.capitalgrowth.org/. Accessed 10 May 2013.

http://www.capitalgrowth.org/spaces/?id=594&postcode=N5%20 2UH&borough=&limit_start=0&#info. Accessed 10 May 2013.

http://www.citiesmcr.wordpress.com/2013/03/11/feeding-the-city-the-politics-promise-of-urban-food/). Accessed 9 February 2014.

http://www.concierge.com/ideas/foodwine/tours/2274?page=4. Accessed 12 January 2013.

http://www.demotix.com/news/1428198/demolition-traditional-testaccios-food-market-rome#media-14281800. Accessed 21 August 2013.

http://www.edinburghgardenpartners.org.uk/benefits/. Accessed 21 August 2013.

http://www.citiesmcr.wordpress.com/2013/03/11/feeding-the-city-the-politics-promise-of-urban-food/. Accessed 20 April 2013.

http://www.finedininglovers.com/stories/paris-street-food-restaurants/. Accessed 7 January 2013.

http://www.foodcartsportland.com/maps/. Accessed 12 January 2013.

http://www.guardian.co.uk/business/2012/feb/02/high-street-shops-death-spiral?newsfeed=true. Accessed 11 January 2013.

http://www.standard.co.uk/lifestyle/esmagazine/wheels-on-meals-londons-best-gourmet-food-trucks-7959487.html. Accessed 12 January 2013.

http://www.hipparis.com/2010/05/06/market-shopping-marche-des-enfants-rouges/men/. Accessed 7 January 2013.

http://www.bbc.co.uk/homes/design/space_kitchens.shtml. Accessed 5 December 2012.

http://www.ipreferparis.typepad.com/i_prefer_paris/2006/10/le_marche_des_e.html. Accessed 7 January 2013.

http://www.londonist.com/2013/02/smithfield-quarter-plans-revealed.php. Accessed 11 August 2013.

http://www.midcenturyhomestyle.com/inside/kitchen/1950s/gallery/page15.htm. Accessed 3 December 2012.

http://www.ny.eater.com/archives/2010/11/new_yorks_10_most_ridiculous_gimmick_restaurants.php. Accessed 15 January 2013.

http://www.myhomefoodthatsamore.wordpress.com/2012/09/13/new-testaccio-market-in-rome/. Accessed 12 May 2013.

http://www.larattedutouquet.com/. Accessed 26 January 2014.
http://www.romaurbanism.wordpress.com/market-day/. Accessed 9 February 2014.
http://www.spectrumdurham.com/index.html. Accessed 7 August 2013.
http://www.sustainweb.org/londonfoodlink/. Accessed 5 January 2014.
http://www.switchboard.nrdc.org/blogs/kbenfield/malls_and_big_boxes_continue_t.html. Accessed 16 August 2013.
http://www.thailand-uk.com/forums/showthread.php?8039-Asian-food-stalls-in-Wembley. Accessed 12 May 2013.
http://www.trusselltrust.org/foodbank-projects. Accessed 11 March 2014.
http://www.umamimart.com/2011/09/skankynavia-torvehallerne-a-new-farmers-market-in-copenhagen/. Accessed 9 February 2014.
http://www.visithelsinki.fi/en/stay-and-enjoy/eat/hakaniemi-market-hall-food-lovers-paradise. Accessed 9 February 2014.
The greening of Gavin (website) http://www.thegreeningofgavin.com. Accessed 19 January 2014.

INDEX

Abel, Mary Hinman 33
Adams, Thomas 137
Adelaide Central Market 94–95
agora, Greece 97–98
agrarian urbanism 180
Agricultural Marketing Service, United States 88
Alexander, C.
 Pattern Language, A 179, 195
allotments, decline and revival of 170–173
Alternative Agro-Food Networks (AAFNs) 253
alternative food networks (AFNs) 247–250
American Planning Association 246
ancient cities and towns, food markets in 74–75
Antica Norcineria Viola 91
anti-urbanism 1–2
Appellation d'Origine Contrôlée (AOC) system 259
Arabian Peninsula, gated communities in 220
Argentina
 one-stop shopping formats in 149
 regional food tourism and connections to spatiality and design in 264
 urban edge food cultivation in 191–192
arrondissements, urban 86, 102
Atkinson, Edward 33
Austen, Jane
 Persuasion 113
Australia
 business parks in 228
 café cultures and spaces in 115
 community garden in 169
 edge city inflected spatiality in 219
 farmer's markets in 88
 food festivals in 127, 128
 food markets in 82, 90, 93
 food shops and food streets in 100, 101
 food truck in 106
 front yard farming in 64
 gardening in 51, 53–55, 60, 63, 65
 gastronomic regionality in 245
 kitchen design in 32, 35, 37, 38
 outdoor restaurant dining in 183
 population and economic activities in 216–217
 pop-up restaurants in 123
 regional food policy in 246, 266
 regional food tourism and connections to spatiality and design in 264
 regional meat products stall in 258
 shopping malls with food courts in 148
 Slow Food movement in 110, 255
 snack bars in 117
 street food in 104
 suburban malls in 151
 supermarkets in 141, 142, 143
 transformations 150
 urban edge food cultivation in 196, 197, 198, 203, 204, 205
 urban fingers pattern in 213
 urban food growing in 161, 162, 164, 174, 178–179
  public policy initiatives 184
 urban stores in 152
Australian Bureau of Statistics Home Production Survey 61
Australian Home Beautiful, The (magazine) 37
Austria
 café cultures and spaces in 113
 regional food tourism in 259–260

Barnett, Henrietta 135
bastides 77, 92
Bauman, Zygmunt 157
Belgium
 kitchen design in 32, 35
 snack bars in 118

big box stores 226
  transformation, and traditional food shops and spaces 149–153
biophilic city 212
bistro mondain 121, 122
Bluewater shopping centre, London 149, 153–154
Bologna, designed space of food streets in 109
Boni, Ada
  Italian Regional Cooking 243
boundedness, of city 134
Bové, Jose 259
Boyd, William
  Restless 128
Brazil 254
  one-stop shopping formats in 149
  sustainable gastronomy in 262
Breuer, Marcel 38
British Isles, food shops and food streets in 100
Broadacre City 212
Burren Eco-Tourism Network (BEN) 264
business parks and distribution spaces, and food realm 226–229

café cultures and spaces 111–115
Campo dei Fiori 91
Canada
  community gardens in 174, 177
  farmer's markets in 88
  food banks in 235
  food buildings and spaces in 265
  food festivals in 127
  food within mall setting in 224
  front yard farming in 64
  gardening in 54, 62–63
  gastronomic regionality in 245
  gated communities in 220
  kitchen design in 32, 35
  outdoor barbecues in 58
  regional food policy in 266
  restaurant space in 122
  slow food spaces and courts in 110–111
  suburbia 137
  urban agriculture in 165, 176
  urban edge food cultivation in 197, 204
  urbanization in 204
Capital Growth 176

Caribbean, gated communities in 220
Carrefour 221, 225
cars and suburban food space 143–144
Central Africa, urban edge in 197
Central America, gated communities in 220
chain store urban food shop 237
Chambers, Fiona 89
Charlotte Observer 64–65
China 199, 225, 241
  food-centred streets in ancient 98
  food markets in ancient 75
  gated communities in 220
  one-stop shopping formats in 149
  peasant farming in 240
  street food in 105
  urban edge food cultivation in 191, 195, 200–201
Chipperfield, David 125
citizen-consumer hybrid 9
city farms 174
civic agriculture 174
Clare Farm Heritage Tours Co-Op 264
classical Rome 240
  food markets in 75–76, 77
  inns, taverns, and modern restaurant spaces in 118
Cliffs of Moher 264
Coates, Wells 32
commensality 11, 20, 115
communal kitchen designs 34
community gardens 169, 174, 175–177
Congrés Internationaux d'Architecture Moderne (CIAM) 32
Consorzio del Formaggio Parmigiano Reggiano 260
consumption junction 35
continuous kitchen 32
conurbation development 216–218, 233
convenience food 164
convenience restaurant 121
conviviality 6, 7, 20–21, 47, 86, 90, 97, 103, 109, 113, 115, 122, 126, 130, 211, 222–223, 230, 231, 237, 238
  domestic 58
  food market and 72
  as framing element 10–12
  home food production and 61
  and kopitiam 116–117

INDEX 355

lack of 147
leisurely 23–24
outdoor dining spaces and 56
private garden's design relationship to 52–53
women and 23
see also urban food growing
Cowan, Bronwyn 89
Crystal Palace, notion of 135
Cuba, urban agriculture in 163, 169
culinary tourism 262
cultural capital 10, 41, 121, 171
Czech Republic, food buildings and spaces in 265

dark stores 227
David, Elizabeth
 French Provincial Cooking 243
 Italian Food 244
De Konnick, Louis-Herman 36
de'Medici, Lorenza
 Heritage of Italian Cooking, The 243
Denby, Elizabeth 32
Denmark
 big box stores transformation in 149
 food buildings and spaces in 265
 food markets in 83–84
 green fingers' plans in 212
 regional food tourism and connections to spatiality and design in 264
 urban edge in 196
dining space
 schools 126
 urban 124–125
domestic food space, in review 45–46
domestic garden 61
 see also under garden
Dom Narkomfin (Moscow) 34
dot.com fulfillment centres 227
Drew, Jane
 Kitchen Planning 32
Dudley Report into Design of Dwellings 37
Dufour, Francois 259

ecological footprints 201
Ecuador, urban agriculture in 163
edgeless city 228
edge-of-town cabbage field 198
Edinburgh Garden Partners 66

egalitarian sharing 24
Egypt 158
 café cultures and spaces in 113
 food markets in ancient 75
electronic grocery shopping (EGS) 227
embodiment, visceral geographies, and food 3–4
enotecas 91
Ersoch, Gioacchino 93
ethical foodscape 8
Ethiopia, urban edge food cultivation in 195
European City Model 8
euro-sprawl 218

farmer's markets and hybrid markets 88–90
fast food culture 155, 156, 229
fete special 121
fin de millennium metropolis 220
Finland
 café cultures and spaces in 115
 electronic grocery shopping (EGS) in 227
 food markets in 85
 green fingers' plans in 212
 kitchen design in 32, 35–36
 pop-up restaurants in 123
Five Points Community Food Center 235
floating vegetable gardens 192
Food 2020 (United Kingdom) 63
food at home 41–45
food banks 235
food courts, fast food, and suburban space shaping 153–156
food deserts 234–236
food festivals 127
food market
 as designed spaces, and lessons from primary research 90–96
 farmer's markets and hybrid markets and 88–90
 London, as example for diversity and adaptation 86–88
 maintenance and revival of 83–86
 modern city development and 80–83
 situated as outdoor rooms 71–74
 in review 96
 urban space and 74–80
Food Sensitive Planning and Design Guide 246

foodshed approach, to understand peri-urban food space   195
food shops and food streets   97–102
food swamps   235
food vans and pods   105–107
food webs   212
foraging, urban   178–179
Ford, James and Ford, Katherine Morrow
   Design of Modern Interiors   57
Fordist technique   31
Fordyce, Allmon   36, 43
France
   AOC system in   259
   café culture and spaces in   112, 113
   edge city inflected spatiality in   218
   farmer's markets in   88
   food buildings and spaces   264–265
   food markets in   77, 81, 82, 83, 86, 90, 92
   food shops and food streets in   100, 101
   gardening in   52, 60
   global connections and local food movements in   247
   inns, taverns, and modern restaurant spaces and   119, 120
   productive landscape design in   211
   regional cheese display in   244
   regional food and convivial urbanism in   251–252
   regional food policy in   266
   regionality and food in   243–245
   slow food spaces and courts in   109–110
   sustainable gastronomy in   262–263
   terroir in   257–259
   urban agriculture in   164, 169–170, 176, 182
   urban edge food cultivation in   192, 193
Frankfurt Kitchen   31–32
front garden   53, 54, 64
front yard farming   64

garden   47
   connection with house, through food   56–59
   contemporary practices in domestic food production and   52–55
   design requirements for vegetable   61–62
   implications for gastronomic   62–67
   productive   47–52
   reanimation of   59–61
Garden City   180, 191, 212, 217, 218
   diagram   133
   movement, in United Kingdom   33, 133–137
Gardener's Magazine   51
gastronomic cartographies   7
gastronomic marginalization   231–236
gastronomic townscape   97
   café cultures and spaces and   111–115
   conviviality and kopitiam and   116–117
   as designed space   107–109
   food shops and food streets and   97–102
   food vans and pods and   105–107
   inns, taverns, and modern restaurant spaces and   118–121
   restaurant spaces and   121–126
   slow food spaces and courts and   109–111
   snack bars and   117–118
   streets for dining and   126–128
   transformed street food and   102–105
gastronomy tourism   262
gated communities   220–221, 232
Geddes, Patrick   217
George, Henry   134
Germany
   café cultures and spaces in   113
   edge city inflected spatiality in   218
   food markets in   87
   gardening in   52
   hypermarkets as food-retailing form in   222
   inns, taverns, and modern restaurant spaces and   121
   kitchen design in   32, 35, 36
   mensa cafeterias in   125
   one-stop shopping formats in   149
   Slow Cities movement in   255
   snack bars in   117–18
   street food in   104
Ghana
   gated communities in   220
   street food in   105
Ginsberg, Moisei   34
global connections and local food movements   247–250

globesity 232, 238
Glover, John 50
golden triangle principle, of kitchen design 39
Goldfinger, Ernö 37
Good Planning for Good Food (United Kingdom) 63
Granada, food markets in 85
Greece 97–98, 244
  agora 97–98
  urban food growing in ancient 159
green urbanism 213
Grow Your Neighbour's Own scheme (United Kingdom) 66
Gruen, Victor 147
Guide to Food Sensitive Planning and Urban Design (Australia) 63
guinguette 193
Gummer, John 150

habitus 3–4, 121, 220, 254
Hagens, Hans Peter 83
Havana, urban agriculture in 163, 176
Healthy Community Food System Plan 266
Heart of Burren Walks 264
Hiroshige, Andō
  Night Scene at Sarawakacho 108
Hong Kong
  food buildings and spaces in 266
  food markets in 84
  McDonaldization in 224
hortillons 192
hortus 48
hortus conclusus (contained domestic garden) 47
house place 29
Howard, Ebenezer 136, 137, 212, 217
  Tomorrow 133
Hungary, edge city inflected spatiality in 218
hybrid malls 225
hyperautomobility 229

Illich, Ivan 157
immigrant gardeners 54–55
impermanence syndrome 203
Incredible Edible Todmorden 176
India 199, 222, 241, 256
  business parks in 228
  gated communities in 220–221
  malls in 223, 225
  street food in 103, 104, 105
  suburban malls in 151
  urban edge in 196
  urban food growing in ancient 158
Indonesia
  gated communities in 220
  one-stop shopping formats in 149
  street food in 104
inns, taverns, and modern restaurant spaces 118–121
institutional forms of dining 124–125
Ireland, regional food tourism and connections to spatiality and design in 264
Islamic Spain 48
Italy
  branding systems for food in 260–261
  café cultures and spaces in 113
  farmer's markets in 88
  food buildings and spaces in 265
  food festivals in 127, 28
  food markets in 81, 90–93
  food shops and food streets in 99, 100
  gastronomic regionality in 244–245
  regional food and convivial urbanism in 251–252
  regional food tourism and connections to spatiality and design in 264
  school dining rooms in 126
  Slow Cities movement in 255
  Slow Food movement in 253–254
  sustainable gastronomy in 262
  Tuscan *mezzadria* system in 241–243
  urban edge food cultivation in 196, 205
    case study 209–211
  urban food growing in mediaeval 159

Jacobsen, Howard
  Mighty Waltzer, The 87
Japan 241, 254
  kitchen design in 32
  McDonaldization in 224
  slow food spaces and courts in 110
  street food in 104
Japsenne, Angeline 36
Johnson, Howard 154

Kahn, Louis   57
Kerr, Robert
    English Gentleman's House, The   27
killer large stores   150
Kingsolver
    Animal, vegetable, miracle   59
kitchen design   25–39
    experts on   39–41
kitchen gardens   47, 59
    see also under garden
Kollektivus (Stockholm)   34
kopitiam and conviviality   116–117

Landshare (United Kingdom)   65
Latin America
    business parks in   228
    gated communities in   220
Lawrence, D. H.   180
Lebanon, gated communities in   220
leisure gardens   171
L'Enfant Rouge market   83, 102, 110
Lindhagen, Anna   52
Liverpool Food and Health Strategy (United Kingdom)   63
living-kitchen   43, 44
local food movements   9
locavores   9
London Food Link   61, 64
Loudon, John Claudius   51
Louis, Herbert   211
Lutyens, Edward   135

McAslan, John   83
McDonaldization   223–224, 228, 238
McHarg, Ian
    Design with Nature   195
Mahfouz, Naguib
    Palace Walk   113
Malaysia, street food in   104
March é L'Enfant Rouge food court markets   109
Marshall, Alfred   134
May, Cliff   57–58
    Sunset Western Ranch Houses   58
mediaeval garden   49
megalopolitan food realm   215
    business parks and distribution spaces and   226–229
    car and   229–230

gastronomic marginalization and   231–236
post-urban context for food space and   215–219
privatopia emergence and   219–221
retreat from public realm   230–231
supermarkets, hypermarkets, and malls and   221–226
sustainability and urban design and   236–238
mensa cafeterias   125
Mercado Central San Agustin   85
Mesopotamia, urban food growing in   158
Mexico   241
    one-stop shopping formats in   149
    street food in   103–105
    supermarkets in   140
    Taco Bell chain in   224
    urban edge food cultivation in   194, 200
mezzardria (share cropping) system   241–243
micro-food deserts   235
Milnis, Ignaty   34
Mirabeau, Marquis de   193
Modern Homes Illustrated   37, 38, 57
'mom and pop' food shops   149
Morocco, food markets in   85
Morris, William   42

National Farmers Market Directory, United States   88
Nelson, George
    Tomorrow's House   36, 57
Netherlands   33, 183
New Towns movement   137
New Urbanist theory and practice   195–196
New Zealand
    farmer's markets in   88
    urban edge food cultivation in   198
Norway
    farmers' market in   206
    global connections and local food movements in   247
    Slow Cities movement in   255

obesity and post-urban spatiality   232–233
obesogenity   232
office dining rooms   125

one-stop shopping formats 149
on farm spaces 265
open source food 176
Orwell, George
    *Down and Out in Paris and
        London* 22
osterias 91
Östermalm food hall 85
outdoor dining and garden
    spaces 56–57
outdoor restaurant dining 183
out-of-home foodscapes 72
out-of-town shopping centres 150

paradise gardens 47, 48
Parc de Villette 93
Parker, Barry 33, 133, 135
parodic restaurant 121
peri-urban zone 195, 197
    agriculture and 199
    sustainability and 199–202
Perry, Clarence 137, 143
Persia, garden spaces in 48
Peru, urban agriculture in 163
Piazza san Cosimato 91
Planning Our New Homes 37
platescape 25
plating food 24
pleasure gardens 193
pluriactivity affecting farming
    practice 199
*Policy Guide on Community and
    Regional Food Planning*
    (United States) 63, 246, 266
pop ups and night markets 105–107
Portugal
    food buildings and spaces in 266
    hypermarkets as food-retailing
        form in 222
    urban edge food cultivation in 192
postmodern aesthetics 231
post-urban spaces *see* megalopolitan
    food realm
post-war suburbia, and food 136–139
potagers 49, 59, 60, 169–170
Prague, hypermarkets as food-retailing
    form in 222
presumption of primacy 203
private garden 48

privatopia, emergence of 219–221
propaganda gardens 176
Protected Denomination of Origin
    (PDO)/Protected Geographical
    Indication (PGI) food
    products 260
public kitchens 33
public space 1, 7, 11, 13, 19, 80, 147
    centrality to urbanism and 73
    definition of 72
    see also individual entries

Queen Victoria Market, Australia 93–94

regional food 239–240
    alternative food networks (AFNs)
        and 247–250
    appellation and place-specific branding
        systems for 259–361
    buildings and spaces 264–266
    case studies, and convivial
        urbanism 250–252
    current food regions and 243–245
    historical situation, in spatial
        practice 240–241
    place-based approaches to, and food
        policy 245–247
    policy and design guidance, rise
        of 266–267
    Slow Food as contested expression
        of 252–256
    terroir and 256–259
    tourism 261–263
        connections to spatiality and
            design 263–264
    Tuscan *mezzadria* and 241–243
Regional Planning Association of
    America 245
regional shopping centres 147
restaurant spaces 121–126
roadside restaurants 154–155
Roman *convivium* 21–22
Roman houses, garden spaces in 48
Root, Waverley
    Food of Italy 244
Rouse, James 82
Roux brothers
    French Country Cooking 243
Russia, gated communities in 220

Saudi Arabia, malls in   225
Sayers, Dorothy
   Gaudy Night   23
Schütte-Lihotzky, Margarete   31, 32
Scotland, vegetable gardening in   51
self-indulgent street   103
self-service shops   141, 142
shared meal   20
Shaw, Howard Van Doren   132
shopping centres, mall, and food   144–146
   and civic management   146–149
short food supply chains (SFSC)   249
Singapore
   designed space of food streets in   109
   kopitiam food in   116–117
   restaurant space in   121
   slow food spaces and courts in   111
   street food in   104
Slow Cities   11, 252, 253, 255, 267
slow food spaces and courts   11, 109–111, 264, 267
   as contested expression, and regionalism   252–256
slow tourism   261
snack bars   117–118
social fortresses, malls as   151
social isolation   20
Sokolov, Richard   225
solidarity   21
South Africa, urban edge food cultivation in   198
South Korea, hypermarkets as food-retailing form in   222
Soviet Union, Former
   urban food growing in   162
Spain   244
   café cultures and spaces in   113
   edge city inflected spatiality in   218
   food buildings and spaces in   265
   food shops and food streets in   100
   food within mall setting in   224
   market-related outdoor room in   72
   streets for dining in   127
   sustainable gastronomy in   262
spatial design matter, food as   5–6
spatialized food relationships and political economy/ecology   4–5
Stein, Clarence   136
sub-Saharan cities, urban edge in   197

suburbia   130
   big box stores transformation and traditional food shops and spaces in   149–153
   cars and food space in   143–144
   food and burgeoning of   130–132
   food courts, fast food and shaping of   153–156
   post-war, and food   136–139
   shopping centres, mall, and food in   144–146
   and civic management   146–149
   supermarkets and shopping strips in   139–143
   town, country, and town-country and   133–136
supercentres   149, 151, 236
supermarkets and shopping strips, rise and decline of   139–143
super-regional malls   153–154
sustainability   6, 10, 20, 40, 90, 97, 138, 155, 183, 211, 223, 230, 232, 252
   arguments with respect to urban edge   199–201
   modern food system and urbanism and   12–15
   urban design and   236–238
   urban food growing and   164–167
sustainable urbanism *see* garden
Sweden   262
   café cultures and spaces in   113
   food markets in   85
   gardening in   52
Switzerland   204
symbolic capital   230–231, 257

table   19
   caring about   20–25
   domestic food space in review and   45–46
   experts on kitchen design and   39–41
   food at home and   41–45
   paradoxical kitchen and   25–39
tablescape   24, 45
Taco Bell food chains   224
tactical urbanism   108
Takashi Sugimoto   123
Tanzania
   urban agriculture in   163
   urban edge food cultivation in   202

Target   225
taste, justice, urbanism, and food   8–10
taste, manuals of   28–29
taverns   119–120
Taylorist principles   31, 32
terroir   256–259
Testaccio market   92, 93
Thailand, street food in   104, 105
third space, role of   114–115
T.K. Maxx stores   150
Torvehallerne   83
traditional agricultural landscapes   263
traditional design and modernism, and food   6–8
transect and urban edge   195–196
transect-based approaches   62, 63
Transition Cambridge   65–66
Transition Town Totnes   66
Transylvania
   peripheral food-growing space in   190
   street café in   112
   traditional productive urban space in   159
Trastevere market   91–92
Trussell Trust   235
Turkey
   gated communities in   220
   Slow Cities movement in   255
   urban food growing in ancient   158

United Kingdom   225
   business parks in   228
   café cultures and spaces in   112–115
   dark stores in   227
   decline and revival of allotments in   170–172
   EU quality to differentiate food products in   260
   farmer's markets in   88, 90
   food courts design at super-regional mall level in   153–154
   food festivals in   127, 128
   food markets in   77, 78, 81–83, 87, 92
   food shops and food streets in   98–101
   food strategy in   63
   Garden City movement in *see* Garden City
   gardening in   51–54, 61
   gated communities in   220
   gourmet food vans and pop up food spaces in   106–107, 123
   hybrid markets in   90
   inns, taverns, and modern restaurant spaces and   119–120
   kitchen design in   32, 36, 37
   London as example for diversity and adaptation in   86–88
   out-of-town shopping centres in   150
   population and economic activities in   216
   restaurant space in   123
   retrofitting design scenarios in   237
   roadside restaurants in   154–155
   shopping malls in   149
   Slow Cities movement in   255
   slow food spaces and courts in   110, 111
   snack bars in   117
   street food in   103, 104
   suburbia   131, 133–136, 138–139
   supermarkets in   140, 142
   super-regional mall café food space in   223
   urban agriculture in   159–160, 162, 165, 175, 177–178, 181–183
   public policy initiatives   184
   urban edge food cultivation in   191–193, 205
   works canteen in   125
   zero hunger strategy in   235
United States   258
   business parks in   228
   community gardens in   174, 176–177
   designed space of food streets in   108
   farmer's markets in   88–89
   fast food in   155, 156
   food deserts in   234, 235
   food festivals in   127
   food markets in   82, 85
   food shops and food streets in   102
   food trucks and pods in   107
   front yard farming in   64
   gardening in   51, 54, 63
   gated developments in   220, 221
   kitchen design in   32, 35
   malls in   225–226
   population and economic activities in   216, 218
   pop-up restaurants in   123

regional food policy in   245, 246, 266
regional food tourism and connections to
    spatiality and design in   264
right to farm' legislation in   204
roadside restaurants in   154–155
shopping malls in   144–146
street food in   104
suburbia   132
supermarkets in   139–144, 151
    redlining of   153
urban agriculture in   160–162, 165,
    169, 176, 179, 183
    public policy initiatives   184
urban edge food cultivation in   194,
    196, 203
Unwin, Raymond   34, 42, 133, 135, 137
urban edge   189–190
    contemporary
        exploring theoretically in food
            terms   194–197
        practice   197–199
    contemporary food
        transformations   202–204
    design responses to, as food
        space   211–213
    as gastronomic tourism
        landscape   206–209
    historic food practice at   190–192
    Italian case study of   209–211
    reconfiguration of food' relationship
        to   204–206
    space, for pleasure   193
    sustainability arguments with respect
        to   199–201
urban food growing   158–161
    convivial green space as public policy
        and   183–184
    convivial green spatiality and   167–170
    decline and revival of allotments
        and   170–173
    design requirements for   179–182

green space and food consumption
    and   182–183
as movement, and urban
    implications   173–175
projects   175–179
as resilient strategy   161–163
sustainable urbanism implications
    and   164–167
urban gardens   54–55
    see also garden
urban stores   151–152
Uruguay, street food in   104

vegetable gardening   50–51, 192
    see also under garden
Venice Charter   8
Vietnam
    kitchen design in   35
    street food in   105
    supermarkets in   141
    urban edge food cultivation in   191
villa garden   56
virtuous globalization   256
Volkskuchen (people's kitchens)   33
Voluntary Consortium of Typical Grana
    see Consorzio del Formaggio
    Parmigiano Reggiano

walled gardens   47–48, 59
Wal-Mart   149, 152, 221, 222, 236, 238
Wal-Mart effect   225
Williams-Ellis, Clough   136
Wirth, Louis
    'Urbanism as a Way of Life'   5
Wolfert, Paula
    Cooking of South West France, The   243
works canteens   125–126
*Works in Architecture of Robert and James
    Adam, The*   182
Wright, Henry   36, 136, 143
    Tomorrow's House   57